INTEGRATED MODELS IN PRODUCTION PLANNING, INVENTORY, QUALITY, AND MAINTENANCE

INTEGRATED MODELS IN PRODUCTION PLANNING, INVENTORY, QUALITY, AND MAINTENANCE

Edited by
M. A. RAHIM
University of New Brunswick

MOHAMED BEN-DAYA
King Fahd University of Petroleum & Minerals

Kluwer Academic Publishers
Boston/Dordrecht/London

Distributors for North, Central and South America:
Kluwer Academic Publishers
101 Philip Drive
Assinippi Park
Norwell, Massachusetts 02061 USA
Telephone (781) 871-6600
Fax (781) 871-6528
E-Mail <kluwer@wkap.com>

Distributors for all other countries:
Kluwer Academic Publishers Group
Distribution Centre
Post Office Box 322
3300 AH Dordrecht, THE NETHERLANDS
Telephone 31 78 6392 392
Fax 31 78 6546 474
E-Mail <orderdept@wkap.nl>

Electronic Services <http://www.wkap.nl>

Library of Congress Cataloging-in-Publication

Integrated models in production planning, inventory, quality, and maintenance / edited by M.A. Rahim, Mohamed Ben-Daya.
p. cm.
Includes bibliographical references and index.
ISBN 0-7923-7347-2
1. Production planning. 2. Inventory control. 3. Production scheduling. I. Rahim, M. A. II. Ben-Daya, M. (Mohamed)

TS176 .I5514 2001
658.5--dc21

2001029405

Printed on acid-free paper.

Printed in the United States of America

Contents

Part III PRODUCTION/INVENTORY AND QUALITY

List of Figures

List of Tables

Preface

Production planning, inventory management, quality control, and maintenance policy are important components of a manufacturing system. Effective integration of these four components will give an industry a competitive edge in the market place.

This book provides in one volume the latest developments in integrated production, quality, and maintenance models. Prominent researchers who are actively engaged in the area have contributed chapters covering a wide range of topics. We hope that this book will serve as a useful and informative introduction to these fields and foster further research and development in these fast growing areas.

This volume is divided into six parts and contains eighteen chapters. Each chapter has been subject to review by at least two experts.

In Part I, Ben-Daya and Rahim provide an overview of the literature dealing with integrated models for production, quality, and maintenance. Directions for future research are outlined.

Part II contains six chapters (chapters 2 to 6) dealing with integrated models for production and maintenance. In chapter 2, Okamura, Dohi, and Osaki develop computational algorithms to determine optimal policies that consider both Economic Manufacturing Quantity (EMQ) and Preventive Maintenance(PM) time for demand that occurs in accordance with a homogeneous Poisson process and for the general case with renewal process demand. Makis and Jiang present, in chapter 3, a general EMQ model with joint production /maintenance /quality control. They investigate the problem of determining jointly the optimal lot sizing, preventive maintenance, and inspection policy which minimizes the expected average cost per unit time. In chapter 4, Lee and Srinivasan consider the economic lot-sizing problem for an unreliable production facility. A simple approach for determining a practical operating policy regarding the scheduling of maintenance work and lot sizing of several products produced on a single machine is presented in chapter 5 by Goyal and Kahyaoglu. In chapter 6, Shemesh, Mehrez, and Rabinowitz solve a family of problems related to machine replacement and preventive maintenance that consider the possibility of skipping available technology in order to be ready for the next technological breakthrough.

Part III deals with integrated production/inventory and quality models in chapters 7-11. In chapter 7, Del Castillo and Salcedo present an optimization model for the design of a Shewhart X-bar chart, taking into account both inventory and statistical process control considerations. A generalized economic model for the joint determination of production quantity, an inspection schedule, and the economic design of the X-bar and R control charts are developed in chapter 8 by Ohta, Kimura, and Rahim. An integrated inventory quality problem is presented by Hariga and Azaiez in chapter 9. They investigate the single-period inventory problem when the lot received may contain defective items where good and poor-quality items are sold in primary and secondary markets. The analysis extends to investigate pricing policies in the different markets and their impact on the order size and the expected profit. In chapter 10, Boukas and Liu extend the continuous flow model of a stochastic manufacturing system to include deteriorating items, quality, and marketing control, where production quality varies with the change of the system mode. The last chapter of Part III, chapter 11 by Serel and Moskowitz, deals with an extension of the traditional newsboy problem that takes into account random demand surprises.

Part IV focuses on quality and maintenance integrated models and contains two chapters. In the first one, chapter 12, Collani and Kiesmüller deal with an integrated approach which enables an objective-oriented joint design of monitoring and maintenance policies. The aim is to minimize disturbances and inappropriate process adjustments which may lead to lower profitability of a production process. The second, chapter 13 by Tortorella, introduces a framework for studying the quantitative relationship between reliability and the quality of post-design product realization processes.

Part V deals with warranty, manufacturing, and quality and contains two chapters. Chapter 14 by Lyons and Murthy is a consideration of warranty and manufacturing. The authors develop a warranty management system that can assist in decision making at different stages and discusses the elements of the system and the management of information. In chapter 15, Murthy and Djamaludin deal with the link between warranty servicing cost, design quality, manufacturing quality, and service quality. The authors also review the literature linking warranty and quality

Part VI addresses issues related to quality and contains three chapters (chapters 16-18). In chapter 16, English, Alsein, and Cole develop a statistical process control approach to process diagnosis for scenarios in which multiple sub-processes can be concurrently dysfunctional. The Multiple Concurrent Dysfunctional Steps approach integrates existing SPC knowledge into the field of process diagnosis. In chapter 17, Liu develops models for the stepwise-programmed regulation of quality. In particular, a dispersal quality regulation model using the discrete approximation of Subbotin is developed. The last chapter of Part VI, chapter 18 by Pachano-Azuaje and Das, deals with simultaneous monitoring of mean and variance through optimally designed Sequential Probability Ratio Tests charts. This type of control charts possess four essential properties that a control chart must have namely, applicability, high sensitivity for detecting shifts, cost-optimality and the capability of jointly controlling all the important parameters that characterize a process. Although the chapters in Part VI do not deal with integrated models, they present novel ideas for addressing quality control problems.

The production of quality goods depends upon machine tools, and the performance of machine tools depends upon maintenance policy. Clearly, the reliability of operating equipment is an important issue. Warranties are particularly important to both buyers and sellers. Product quality and production quantity are equally important to them. A book which expresses these simple facts is overdue, and it is anticipated that the findings of the research that is outlined in this edited volume will prove very useful to end users and future researchers.

THE EDITORS

Acknowledgments

The Editors would like to acknowledge the contributing authors for their valuable submissions. This book would not have been possible without their participation and cooperation. Each chapter has gone through several stages of review. We would like to express our gratitude to all the reviewers whose comments and suggestions have led to substantial improvements in the quality of the text.

We are indebted to Gary Folven, Publishing Editor OR/MS areas at Kluwer, for his interest in this book and his full cooperation and assistance since the initiation of the project. During the review process of the proposal conducted by the publisher, four anonymous experts made many valuable and insightful comments. We would like to acknowledge their contributions.

The editorial assistance of Janet Gamache throughout the project has been greatly appreciated. It takes a lot of skill and patience to do the word processing, text conversion, proofreading, and typesetting necessary for the production of a text such as this. Special thanks go to Kian Keong Lam and Agnes Wong for retrieving the electronic versions of the manuscripts in various formats from the contributors, for converting the manuscripts to MS Word format for editing purposes, and for preparing the final version of the edited manuscripts to be converted into LaTex format. We would like to thank Shaikh Arifusalam for his support in converting Word manuscripts to LaTex and preparing the index. The assistance of Atique Siddiqui and Mohammed El-Hassen in proofreading the converted manuscripts and formatting references is highly appreciated.

This project has been funded by the Faculty of Administration, University of New Brunswick through the Faculty Development Fund, and by the King Fahd University of Petroleum and Minerals under project number SE/PIQM/217. These supports are greatly appreciated. The financial assistance of the Natural Science and Engineering Research Council (NSERC) of Canada in support of this collaborative research project is gratefully acknowledged.

Last but certainly not least, we wish to thank our families for their patience and encouragement. The preparation of this book has, on several occasions, required sacrifices on their part.

Contributing Authors

Mohammed Alsein
Transplace.com
Lowell AR, USA

M. N. Azaiez
Industrial Engineering Program,
King Saud University, Saudi Arabia

M. Ben-Daya
Dept. of Systems Engineering,
King Fahd University, Saudi Arabia

E. K. Boukas
Mechanical Engineering Department,
École Polytechnique de Montréal, Canada

E. Del Castillo
Dept. of Industrial & Manufacturing Eng.
The Pennsylvania State University, USA

M. H. Cole
Department of Industrial Engineering
University of Arkansas, USA

E. von Collani
Institut fur Angewandte Mathematik
und Statistik, Würzburg, Germany

T. K. Das
Dept. of Industrial & Managt. Sys. Eng.
University of South Florida, USA

I. Djamaludin
Dept. of Mechanical Engineering
The Univ. of Queensland, Australia

T. Dohi
Dept of Industrial & Systems Eng.
Hiroshima University, Japan

John R. English
Department of Industrial Engineering
University of Arkansas, USA

Suresh K. Goyal
Dept. of Decision Sciences and MIS
Concordia University, Canada

M. Hariga
Industrial Engineering Program,
King Saud University, Saudi Arabia

X. Jiang
Dept. of Mech. and Industrial Eng.
University of Toronto, Canada

Y. Kahyaoglu
Department of Management
Bogazici University, Turkey

Gudrun Kiesmüller
Eindhoven University of Technology,
Eindhoven, The Netherlands

Aritoshi Kimura
Dept. of Management Systems Eng.
Kinki University, Japan

Hyo-Seong Lee
Dept. of Industrial Engineering
Kyung-Hee University, Korea

John J. Liu
School of Business Administration
Univ. of Wisconsin-Milwaukee, USA

Z. K. Liu
Mechanical Engineering Department,
École Polytechnique de Montréal, Canada

K.F. Lyons
Dept. of Mechanical Engineering
The University of Queensland, Australia

V. Makis
Dept. of Mech. and Industrial Eng.
University of Toronto, Canada

A. Mehrez
Dept. of Industrial Eng. & Management,
Ben Gurion Univ. of The Negev, Isreal

H. Moskowitz
Krannert Graduate School of Management
Purdue University, USA

D.N.P. Murthy
Dept. of Mechanical Engineering
The University of Queensland, Australia

Hiroshi Ohta
Dept. of Industrial Engineering
Osaka Prefecture University, Japan

H. Okamura
Dept of Industrial & Systems Eng.
Hiroshima University, Japan

S. Osaki
Dept of Industrial & Systems Eng.
Hiroshima University, Japan

F. Pachano-Azuaje
Dept. of Industrial & Managt. Sys. Eng.
University of South Florida, USA

G. Rabinowitz
Dept. of Industrial Eng. & Management,
Ben Gurion Univ. of The Negev, Isreal

M. Abdur Rahim
Faculty of Administration
University of New Brunswick, Canada

I. T. Salcedo
Dept. of Industrial & Manufacturing Eng.
The Pennsylvania State University, USA

D. Serel
Krannert Graduate School of Management
Purdue University, USA

E. Shemesh
Dept. of Industrial Eng. & Management,
Ben Gurion Univ. of The Negev, Isreal

M. M. Srinivasan
College of Business Administration,
The University of Tennessee, USA

Michael Tortorella
Bell Laboratories
Lucent Techn., Holmdel, USA

I

INTRODUCTION

Chapter 1

INTEGRATED PRODUCTION, QUALITY & MAINTENANCE MODELS: AN OVERVIEW

M. Ben-Daya and
Dept. of Systems Engineering,
King Fahd Univerity of Petroleum
and Minerals, Saudi Arabia

M.A. Rahim
Faculty of Administration
University of New Brunswick
Fredericton, N.B., Canada E3B 5A3

Abstract Production, quality, and maintenance are three important aspects of any industrial process and are interrelated problems. Understanding and solving industrial operation-related problems requires adequate modeling of the dependencies between them. In the past, production, quality, and maintenance have been treated as separate issues. Recently, there has been a considerable interest in models attempting to integrate issues in these three areas. In this chapter, we review the literature on models which integrate production, quality, and maintenance, and we provide insights for future research.

Keywords: Production process, production system, deteriorating production process, maintenance, quality, integrated models, warranty, inventory.

1. INTRODUCTION

In today's technology driven and fast paced industrial environment, firms are facing difficult challenges. Mass production of a uniform product is being replaced by a system that favors a variety of products. Rapidly changing markets, an increase in product variety, and concerns for product quality have led to automation and sophisticated equipment, which in turn, have raised the cost of manufacturing processes. To meet quality demands and to ensure the success of new management system strategies such as the Just-In-Time (JIT) approach [34, 74], production processes have to be maintained in good operating condition, high quality standards maintained, and sound production planning adopted. Decision support systems that take all these aspects into consideration are required.

Production, quality, and maintenance are three important aspects in any industrial process and are interrelated problems. To understand and solve these problems, adequate modeling of the dependencies between production, quality, and maintenance is required. In the past, production, quality, and maintenance have been treated as three separate problems. Because of their interdependence, many attempts have been made to develop models that take into consideration more than one of the three aspects. This area of research has grown considerably in the last decade.

The purpose of this chapter is to review the literature which deals with the integration of more than one of these three main aspects of any modern production process. Our goal is to provide ideas about these integration models that foster further developments in the field. We also suggest useful directions for future research.

This chapter is organized as follows. In the next section, we discuss further the relationship between production, quality, and maintenance. Models dealing with the integration of production and quality are reviewed in Section 3. Sections 4 and 5 deal with production-maintenance and maintenance-quality literature, respectively. Models that attempt to integrate all three aspects are discussed in Section 6. Section 7 deals with integrated models in the context of multi-stage production systems. Other integrated models are reviewed in Section 8. The limitations of the existing models, future research directions, and concluding remarks are presented in Section 9.

2. PRELIMINARIES

Production, quality, and maintenance are three fundamental aspects in any modern production process. As such, they have received considerable attention from researchers. In fact, significant progress has been made in understanding and modeling problems in these areas. However, sound decision-making requires a good understanding of the dependencies between production, quality, and maintenance and the trade-offs involved. If production, quality, and maintenance problems are treated independently, sub-optimal solutions will result. This justifies the need to develop models that take into consideration more than one of these three aspects.

In production and inventory analysis literature, economic production quantity (EPQ) has been studied under various conditions. Many books have been written in the area of production planning [47, 52, 82, 104]. Most models dealing with EPQ, for instance, assume that the production process never fails and that it always produces items of acceptable quality. However, a more realistic situation is one in which the quality is not always acceptable because the condition of the production process may deteriorate with time. As discussed below, many authors [28, 92, 95, 96] have looked at production models taking into account the quality issue.

maintenance activities interfere with the production schedule and with the deterioration pattern of the production process. The absence of a unified production-maintenance model results in limited understanding of the existing relationships between production and maintenance parameters and the trade-offs between alternative maintenance and production policies. There is a growing interest in developing production models which considering the maintenance requirement, as can be seen below.

maintenance as a separate problem has received a lot of attention from researchers. Pierskalla and Voelker [91] have presented a survey of the literature on maintenance models up to 1976. Other surveys include Sherif and Smith [106], Valdez and Feldman [111], Cho and Parlar [32], Nakagawa [84] and Osaki and Nakagawa [86]. The reader is also referred to the classical book by Jardine [51]. Many reliability books contain chapters dealing with maintenance issues, see for example the book by Elsayed [40].

According to total quality management philosophy [34], quality can no longer be inspected into the product. Final control inspection is moved to the process level through adequate process control techniques. Consequently, variations are eliminated at their source. Control chart methods are generally used for the purpose of controlling the quality of the product and assessing the state of the production process [75]. The most commonly used control chart is the control chart. The design of a control chart has economic consequences since the costs involved are affected by the choice of the control chart parameters. The first model for the economic design of $\bar{x}$ control charts was introduced by Duncan [39]. Duncan's model stimulated most of the work in this area and has been extended by several authors [41, 48, 76, 112]. In most cases, statistical process control cannot be separated from preventive maintenance.

The maintenance and quality relationship is also well documented [7]. As stated by Tagaras [107], information obtained in the course of statistical process control determines possible restoration actions to be taken and thus affects the preventive maintenance schedule. Similarly, maintenance activities interfere with the deterioration pattern of the production process, thus changing the process control requirements. Many models addressing statistical process control and maintenance simultaneously have appeared in the literature [5, 11, 14].

Interest in integrated models addressing all three aspects at the same time is also evident in recent literature [11], [95]-[97]. This presents a natural development built on the research noted above.

Integrated models in the context of multi-stage production systems [6], [12]-[13] present additional modeling issues compared to single stage systems. Research in this area will also be reviewed. Other integrated models considering additional aspects, such as warranty and the integration of spare parts ordering policies with maintenance [79]-[81], will be discussed as well.

In the remainder of this chapter, we review the literature dealing with integrated models for production, quality, and maintenance.

3. PRODUCTION AND QUALITY

Many models integrating production and quality consider quality issues in the context of EPQ. Rosenblatt and Lee [100]-[101] have developed a model that addresses the problem of determining EPQ for imperfect production processes. In their model, the time of the process shift to the out-of-control state is assumed to follow an exponential distribution. Three different dynamic process deterioration schemes are investigated, namely, linear, exponential, and generalized multiple state deterioration. For both linear and exponential deterioration, they found that production run lengths are shorter than that of the classical EPQ model. For multiple state deterioration, they show that the production run length is even smaller than or equal to linear and exponential deterioration. Lin *et al.* [65] have studied Rosenblatt and Lees model for the case where deterioration of the process follows a general distribution. Hariga and Ben-Daya [46] have generalized Rosenblatt's model for the case where the time to shift to the out-of-control state follow a general distribution. Bounds on the optimal cost are also provided. Kim and Hong [56] have also presented an EMQ model which determines an optimal production run length in three deteriorating production process: constant, linearly increasing, and exponentially increasing.

Porteus [92]-[93] has also developed models which study the relationships between lot size and quality. He suggests three options for investment in quality improvement: (i) reducing the probability that the process moves to an out-of-control state, (ii) reducing setup costs and (iii) using the previous two options simultaneously. He has also proposed that lot sizes should be reduced to compensate for poor quality if no effective inspection is possible. Illustrative examples of his models using these three investment options demonstrate that lower costs will be yielded for all of the three options compared to the cost of the classical EPQ model. One of his most interesting findings is that the optimal lot size increases with the improvement in process quality. Vörös [113] has developed another model relating lot sizing, quality improvement, and setup time reduction under the assumptions that the production rate is finite and that the cost of attaining close to zero setup time is not infinite.

Lee [59] has proposed a model for finding the best lot size when the possibility exists that the inspection process will go out of control during the production of wafers in semiconductor manufacturing.

Liou *et al.* [66] have addressed the problem of determining the optimal inspection number and optimal EPQ of a single item. They have assumed that Type I and Type II inspection errors exist when inspecting the imperfect production system. Ohta and Ogawa [85] have also dealt with models involving inspection errors.

Cheng [26] has proposed a general equation to model the relationship between production set-up and process reliability and flexibility. Cheng [25] has also proposed an EPQ model with demand-dependent unit production cost and imperfect production processes. He formulated the EPQ problem as a geometric programming problem and derived closed loop solutions.

Khouja and Mehrez [55] have proposed an extension to the economic production lot size model to cases where the production rate is a decision variable. They assume that the quality of the production process deteriorates with an increased production rate. They show that, for cases where increases in the production rate lead to a significant deterioration in quality, the optimal production rate may be smaller than the rate that minimizes unit production cost. For cases where quality is largely independent of production rate, the optimal production rate may be larger than the rate that minimizes unit production cost. Mehrez *et al.* [72] also considered the effect of manufacturing technology decisions on lot size and quality in the context of the newsboy problem. The problem was solved using stochastic programming in two stages.

Keller and Noori [53] consider uniform and exponential lead time demand to determine when it is not economical to invest in a quality improvement program. They show that when the probability that the process will go out of control increases, the lot size and the shortage cost increase. Large shortages are avoided at the expense of quality.

Chand [23] has investigated the effect of learning in process quality on the optimal setup frequency. Hong [49] developed a model that considers the effect of lead time reduction and quality improvement on optimal lot size for an imperfect process.

Djamaludin *et al.* [37] consider product defects that are not readily detectable. Discovery occurs only through time testing or warranty claims. Optimal lot size for free replacement warranty and pro-rata warranty is selected to balance the trade-off of the assumption that decreasing the lot size decreases warranty costs and increases manufacturing costs because of the number of increased setups when small batches are used to control quality. This line of research integrates warranty with production and quality issues and will be discussed later. Yeh *et al.* [114] have also studied the optimal production run length for products sold with warranty for a deteriorating production system. The deterioration process of the system is characterized by a two-state continuous-time Markov chain.

The effect of imperfect production processes on the economic lot scheduling problem (ELSP) has been studied recently by Ben-Daya and Hariga [9].

Another approach to integrating production and quality is by developing joint production and quality control policies. Rahim [95] has considered the determination of optimal control chart design parameters and production quantity so as to minimize the total expected cost (the quality control cost and the inventory control cost) per unit time for a class of deteriorating processes where the in-control period follow a general probability distribution with increasing hazard rate. Both uniform and non-uniform schemes are considered. For non-uniform inspection schemes, the length of the inspection intervals is regulated to maintain a constant integrated hazard rate over each inspection interval. This model has been extended in Rahim and Ben-Daya [96] to address the more realistic case where production ceases in case of a false alarm.

Peters *et al.* [90] have developed a model for the joint determination of optimal inventory and quality control policy. They present a cost model that combines a fixed order quantity inventory control system with a Baysian quality control system for a lot-by-lot attribute acceptance sampling plan. Various simulation examples show that their model is superior to both inventory and quality control systems when they are considered separately. This superiority becomes even more significant when the difference between the cost of defective units exiting the quality control system and the cost of repairing defective units increases. Lee and Zipkin [64] considered issue related to production control in Kanban like system with defective ouputs.

Cheng [28] has considered the determination of EPQ for imperfect production processes. He developed a simple model which relates unit production cost with process capability and quality assurance, and he showed that unit production cost increases with increases in process capability and quality assurance expenses. Chen and Chung [24] have also considered the problem of jointly determining the optimal production run and quality selection. They have developed a model for the joint determination of the optimal process mean and the production run for a process having an exponential shift to failure distribution.

A closely related field, in which production and quality have been considered jointly, is the determination of optimal production runs for processes with random drift that can occur when a tool wears out or a major component deteriorates. The relationship between tool wear, production run, and quality has been considered by many authors. Rahim and Banerjee [98] have developed an optimal decision rule for tool resetting for processes with random linear drift to minimize the cost of rejected items and the loss of production time due to repetitive resetting. Schneider *et al.* [103] have also developed a model for optimal control of a production process subject to random deterioration caused by tool wear. They developed a decision criterion for selecting the starting level of the process mean and the level at which the process mean should be adjusted back to the starting level to minimize production cost, tool adjustment, and loss of defective items. Simple approximation solutions were developed to evaluate the optimal policy. A survey of this area is provided by Al-Sultan and Rahim [2].

Goyal *et al.* [42] have reviewed the literature available on integrating the various aspects of production, lot sizing, inspection, and rework under a suitable classification scheme. They also present a framework for developing an efficient production design system to encourage the development of new model for modern manufacturing. Gunasekaran *et al.* [45] also addressed isues related to optimal investment and lot sizing policies for improved productivity and quality. Wright and Mehrez [115] provided an overview of the relationship between quality and inventory. Salameh and Jaber [102] developed an economic production model for items with imperfect quality.

4. PRODUCTION AND MAINTENANCE

The problem of integrating production and maintenance has generally been approached in the literature in two different ways. Some authors have approached this problem by determining the optimal maintenance policy in the production system and others by taking maintenance as a constraint to the production schedule. Using the first approach, Lee and Rosenblatt [60] have developed a model to determine the EPQ of a single product and the schedule of inspection for process maintenance. An exponential probability distribution of failure and uniform sampling inspection intervals were assumed to compute the elapsed time for the process to be in the in-control state. The decision to perform maintenance or not, at a given inspection interval, was derived based on a comparison between the restoration cost and the cost of defective items that may occur during that inspection interval. Computation schemes for other decision variables such as production time, EPQ, and the number of inspection intervals were also derived based on the same comparison criterion. Lee and Rosenblatt [61] consider the costs of different policies for providing early detection of process shifts and restoration capabilities. They considered four different combinations of inspection and restoration policies: i) continuous inspection and restoration based on the process time to failure, ii) periodic inspection with restoration if needed, iii) continuous inspection and periodic restoration and iv) periodic inspection and restoration. The optimal cost models for all policies are derived for processes with in-control periods following exponential distributions. Lee and Rosenblatt [62] have also developed a joint model for production planning and maintenance schedule. They assume that the cost of process restoration is a function of the detection delay. They address the problem of simultaneous determination of the economic production quantity and the schedule of inspection of the process for maintenance. Process deterioration is assumed to follow an exponential distribution.

Groenevelt *et al.* [43] have also developed a model to study the problem of selecting the economic lot size for an unreliable manufacturing facility with a constant failure rate and general randomly distributed repair times. The quantity of the safety stock to be used over those time intervals when the machine is being repaired was derived based on the target service level (desired fraction of total demand satisfied) constraint. The optimal lot size which minimizes the average running cycle cost was also derived as a function of setup cost, demand rate, and the level of preventive maintenance which is carried out at the beginning of the production cycle. Three different functions of the mean time between failures (MTBF) were implemented to illustrate the relationships between the maintenance budget level and the expected total cost. Groenevelt *et al.* [44] have developed another model to study the effect of machine breakdowns and corrective maintenance on economic lot sizing decisions. Two production control policies are proposed to accommodate stochastic interruptions caused by unexpected machine breakdowns. The first policy assumes that production of the interrupted lots is not resumed after a breakdown and that the next cycle is initiated when the on hand inventory is depleted. In contrast, the second policy assumes that production is immediately resumed.

Zhang and Grechak [116] has provided joint lot sizing and inspection policy for an EOQ model with random yield. Makis [68] has studied the problem of the optimal lot sizing and scheduling of inspections in an economic manufacturing quantity (EMQ) model with a production process subject to random deterioration. The inspections are imperfect, and they do not necessarily reveal the true state of the process. The in-control periods of the process are assumed to follow a general distribution.

Makis and Fung [70] have studied the problem of the joint determination of the lot size, inspection interval, and preventive replacement time. They assume that the shift to failure time of the process follows a general distribution. They also have studied the relationship between the preventive replacement time and the mean time to failure of the process. Makis *et al.* [71] have also considered a production-inventory control problem for an EMQ model with machine breakdowns and planned preventive replacement of the production unit. Two types of failure were considered. When Type I failure occurs (major failure), the unit is replaced by a new one, and the interrupted lot is aborted. Starting a new lot after a replacement or after a complete production run incurs a setup cost. Type II failures (minor failures) are corrected by minimal repairs that take negligible time. After performing a minimal repair, production can be resumed immediately at a cost lower than the setup cost. Semi-Markov decision processes were used to analyze the problem.

Lee and Park [58] have also considered a production- maintenance policy for a deteriorating production system. The assumed that inspections are carried out at equally spaced intervals, and derived a cost model to determine the production cycle and inspection intervals jointly. This was done by considering the difference between the reworking cost before sale and the warranty cost after sale.

Remedies to machine breakdown include preventive maintenance and safety stocks. Cheung and Hausman [29] have formulated a model for the joint implementation of preventive maintenance and safety stocks in an unrelibale production environment.They have investigated the trade-offs between investing in the two options. Meller and Kim [73] studied the impact of preventive maintenance on system cost and buffer size.

Dohi *et al.* [38] have proposed an age dependent replacement policy for a stochastic EMQ model and have extended the model by Groenevelt *et al.* [44]. Later, Dohi *et al.* [36] considered two minimal repair models. In the first model, an infinite number of minimal repairs can be made until the predetermined inventory level is reached, when machine failures occur. In the second model, a limited number of minimal repairs are performed during the production period. The impact of machine breakdowns and repairs on production lot sizing has been illustrated.

Other EMQ models considering machine breakdown include those of Kuhn [57] and Abboud *et al.* [1].

A different line of research involving optimal control theory was used by Boukas *et al.* [16]-[21] for modeling production and maintenance planning for flexible manufacturing systems and other production systems. They used continuous-flow models and other control theory techniques to study various types of models that incorporate production, corrective maintenance, and preventive maintenance.

The optimal flow control for a manufacturing system subject to random failure and repairs has been studied by Kenne and Gharbi [54]. They extend the hedging point policy to non-exponential failure and repair time distributions and random demand rate models.

Ashayeri [3] has proposed a production and maintenance model for the process industry. He developed a mixed integer linear programming model to simultaneously plan preventive maintenance and production in a process industry environment. The model schedules production jobs and preventive maintenance jobs, while minimizing costs associated with production, backorders, corrective, and preventive maintenance.

Using the second approach, that is, taking maintenance as a constraint to the production schedule, Rishel and Christy [99] have studied the impact of incorporating alternative scheduled maintenance policies into the material requirements planning (MRP) system. They used four performance measures to evaluate the impact of incorporating maintenance into the production schedule, namely, the number of on-time orders, the number of scheduled maintenance actions, the number of equipment failures and the total maintenance cost. Also, they have studied the feasibility of incorporating predicted emergency maintenance into the production schedule. Six different material requirements plans were used in their study to determine if integrating scheduled maintenance with the production schedule improves the measures of performance, whether it is more appropriate to schedule maintenance as independent and/or dependent demand item, if it is beneficial to incorporate a predicted failure into the production schedule, and whether it is more appropriate to predict emergency maintenance as an independent or dependent demand item. Results of their simulation show that scheduled maintenance does make a significant impact on the number of failures and allows an increase in the production potential. Also, they show that including emergency maintenance in the MRP system and treating it as a dependent demand item will provide the ability to make more informed planning, scheduling, and control decisions in both production and maintenance functions. Brandolese *et al.* [22] have considered the problem of the planning and the management of a multi-product and one-stage production system made up of flexible machines operating in parallel. They developed a model to find the optimal schedule for both production and maintenance intervention, considered to be capacity-consuming jobs to be scheduled according to the production resources. Their model was designed to meet release and due dates, minimize the total cost (sum of the expected maintenance cost, the setup cost, and the production cost) and to minimize the total plant utilization time (sum of the total job processing time, the total machine idle time, and the total maintenance time). A Weibull reliability function was used to compute the expected preventive maintenance cost, and the system age was assumed to become as good as new after each preventive maintenance activity.

5. MAINTENANCE AND QUALITY

In their study, Ben-Daya and Duffuaa [7] discuss the relationship between maintenance and quality and proposed a broad framework for modeling the maintenance-quality relationship. They have proposed two approaches for linking and modeling this relationship. The first approach is based on the idea that maintenance affects the failure pattern of the equipment and that it should be modeled using the concept of imperfect maintenance. That is, the equipment failure rate is decreased by a certain amount after each preventive maintenance intervention, which in turn amounts to a reduction in the age of the equipment. The second approach is based on Taguchi's [108] approach to quality, i.e. using maintenance to reduce the deviation of product quality characteristics from their target value.

Statistical process control is one of the most common methods for controlling product quality. As mentioned earlier, for many processes, it is not possible to separate process control from preventive maintenance because of their interrelationships. Tagaras [107] has provided an integrated cost model for the joint analysis and optimization of process control and maintenance operations. The proposed model can be viewed as an improvement and extension of the models in Pate *et al.* [89] which addressed the inspection-maintenance problem with more emphasis on maintenance. Tagaras's model allows for an arbitrary number of out-of-control states and different sampling and maintenance intervals. Preventive maintenance is performed periodically on a subset of the quality control inspection epochs.

Banerjee and Rahim [4] and Moskowitz *et al.* [77] have investigated the effect of process failure mechanisms on economic control charts. Banerjee and Rahim [4] assume that assignable causes occur according to a Weibull distribution and the sampling interval is non-uniform and monotomically increasing. Using a continuous time model, Moskowitz *et al.* [77] show empirically that the smaller the shift, the more significant the effect on the design parameters.

Chiu and Huang [31] have developed a model that introduces preventive maintenance into the economic design of control charts. They assume a non-uniform distribution for the in-control period with an increasing hazard rate and fixed sampling intervals. Also, they assume that the system will become as good as new, i.e. the age of the system will drop to zero after each preventive maintenance schedule. Results of their simulation show that a small-scale restoration will significantly reduce the total expected cost and that relatively large-scale restoration will still yield lower costs than those of non-maintained systems. Chiu and Huang [30] have also been developed a model to design both $\bar{x}$ and S^2 control charts with preventive maintenance. Two different preventive maintenance policies with both uniform and non-uniform sampling schemes were compared to determine the advantages of maintained systems over non-maintained systems.

The effect of maintenance on the economic design of $\bar{x}$-control charts for general time to shift distribution with increasing hazard rate has also been studied by Ben-Daya and Rahim [14]. In their study, the age of the process is reduced due to preventive maintenance proportional to the PM cost. It is shown, through numerical examples, that preventive maintenance may lead to reductions in quality control costs.

Tapiero [109] has formulated the problem of continuous quality production and maintenance of a machine. He assumes that quality is a known function of the machine's degradation states. He uses different applications of a specific quality function to obtain analytical solutions to open loop and feedback stochastic control maintenance problems. Collani [35] stated that wear-out phenomena can be compensated for by means of continuous monitoring process.

6. JOINT MODELS FOR MAINTENANCE, PRODUCTION AND QUALITY

As a natural extension to the above models, papers have appeared in the literature addressing production, quality, and maintenance in a single model. Some of the models that integrate production and quality, reviewed above, consider corrective maintenance. In this section, we consider research dealing with preventive maintenance and modeling and its effect on production quantity and quality control design parameters.

Huang and Chiu [50] have attempted to integrate production planning, inspection, and preventive maintenance. They have developed two cost models for monitoring policies with and without preventive maintenance. They assume that preventive maintenance is performed at equal intervals and that the shift time of the production process follows a general distribution.

Ben-Daya [5] has extended Rahim's [95] model by introducing an imperfect preventive maintenance concept into the determination of EPQ and the economic design of an control chart for a class of deteriorating processes whcre the in-control period follows a general probability distribution with an increasing failure rate. Preventive maintenance activities are coordinated with quality control inspections and carried out as a subset of the inspection epochs. The effect of preventive maintenance is incorporated into Rahim's model by adapting the actual age of the machine at each inspection interval according to the preventive maintenance level. Linear and nonlinear relationships between the reduction in the actual machine age and the maintenance level are assumed. An integrated model which considers maintenance and quality has also been developed for the economic lot scheduling problem by Ben-Daya *et al.* [10].

Lin *et al.* [65] have generalized Rosenblatt and Lee [100, 101] and Lee and Rosenblatts models [60]-[62] by allowing the process shift distribution to be an increasing rate function. Tseng *et al.* [110] have investigated the effect of imperfect maintenance policies for deteriorating systems. Equally spaced and equally cumulative hazard policies are used.

7. INTEGRATED MODELS FOR MULTI-STAGE SYSTEMS

Integrating quality aspects in production planning, especially in lot sizing models, has not been adequately addressed in the context of multistage production systems. Imperfect multi-stage production processes present additional practical complications that are not present in perfect systems. The following important issues must be addressed when dealing with imperfect multi-stage production processes.

- Nonconforming items must be screened so that they are not passed to subsequent stages to avoid waste of resources and unnecessary processing.
- While screening nonconforming items, errors may be committed. Nonconforming items may be incorrectly accepted and good items incorrectly rejected.
- Processes shift to an out of control state where nonconforming items are produced must be prevented or at least delayed, and detected quickly if it occurs.

Ben-Daya and Rahim [12] have developed a lot-sizing model for a multi-stage system that addresses the first two issues. However, there is a need to build some quality improvement aspects into this model. In a subsequent paper [13], they have addressed the third issue and developed multi-stage models that take into account inspection errors and process inspection and restoration for imperfect processes. They show through numerical examples that inspection and restoration of the processes at various stages lead to a lower expected total cost due to reductions in quality related costs.

Lee and Rung [63] consider lot-sizing policies in a two-stage serial production system with machines prone to failure. They show that lot sizes for unreliable systems can be smaller or larger than the ones in the classical model. They also illustrate through numerical examples that lot sizing decisions in these systems can be counter-intuitive due to the interaction of machine breakdown and repair and inventory carrying effects.

8. OTHER INTEGRATED MODELS

8.1 Inventory and Maintenance. A critical factor in any maintenance system is the availability of spare parts and materials. Insufficient stock can lead to extended machine downtime. On the other hand, maintaining an excessive stock of spares can lead to large holding costs and can conceal basic problems that cause the need for spares in the first place. These factors clearly show the need for models that integrate inventory and maintenance decisions.

Generally, the literature in this area can be divided into two major categories:

1. Irreparable item inventory models.

2. Repairable (recoverable) items inventory models.

Two reviews have appeared in the literature in this area. The first review by Nahmis [83] covering most of the work that appeared prior to 1981. He organized his review in the following manner:

- The $(S-1, s)$ ordering policies and their importance in managing repairable items.
- The METRIC (multi-echelon technique for recoverable item control) model and its extensions.
- Continuous review models for deterministic and random demand.
- Periodic review models under random demand.

The second review by Cho and Parlor [32] focused on maintenance models for multi-unit systems. They included a section on inventory and maintenance models. Osaki and his group published a lot of work in this area (see for example Osaki and Nakagawa [87] and Osaki *et al.* [88], see also So and Tang [105] and Mabinin *et al.* [67].

Important models dealing with spare parts provisioning for maintenance have been reviewed by Ben-Daya and Duffuaa [8]. The reader is referred to this chapter for more details.

8.2 Production quality and warranty. Products that are sold with warranty present additional complications, especially when non-conforming items are shipped to customers. The performance of a non-conforming item is inferior to that of a conforming item which meets the design specification. When items are sold with warranty, the warranty cost to service non-conforming items is much bigger than that for conforming items. One way to control non-conforming items is through lot sizing. Smaller lot size results in a reduction in the warranty cost per item but this is achieved at the expense of increased manufacturing cost per unit. Djamaludin *et al.* [37] presented a model that determines the optimal lot sizing which achieves a proper trade-off between the two costs for repairable items sold with free replacement warranty (FRW) policy. A detailed discussion and review of various issues related to warranty policies can be found in [15], [78]-[81]. Part 5 of this book is also devoted to warranty, manufacturing, and quality issues.

8.3 inventory and detriorating items. Usually, analysis of inventory systems is performed without takong into account product deterioration. However, there are products that havea very high rate of deteriorartion such as volatile substances, radioactive materials, etc.

In this cases detrioration cannot be ignored. There is a vast literature in this area and we have no intention to cover it here. The reader is referred to a recent paper by Chung and Lin [33] or the complete survey presented by Raafat [94]

9. CONCLUSION AND DIRECTIONS FOR FUTURE RESEARCH

In this chapter, we reviewed the literature dealing with integrating production, maintenance, and quality. In most cases, integrated models lead to a better understanding of the relationships between these three aspects and the trade-offs between alternative policies for production, quality control, and maintenance policies.

It is clear from the review that this line of research is experiencing a considerable growth. Also, most of the research papers in these areas have appeared in print in recent years. This indicates a growing interest in integrated models for production, quality, and maintenance.

The issues of production, quality, and maintenance, when treated as separate problems, are well understood. In fact, many books and textbooks have been written on these subjects. However, integrated models have not matured to the same level. There is need for more work in this area, and many opportunities exist for developing better formulations and solution methods. The following suggestions for further research in various areas may be of interest.

Quality Control and Maintenance Policy

As stated earlier, maintenance and quality control are interrelated. Tagaras [107] has developed an economic statistical process control model considering maintenance policy under a Markovian shock model. Ben-Daya and Rahim [14] have revisited the problem under a non-Markovian shock model and an integrated hazard criterion for determining the maintenance schedule. Makis and Jardine [69] have achieved considerable success in scheduling proportional hazard models. For the sake of simplicity, inspection and preventive maintenance are performed concurrently in the work of Ben-Daya and Rahim [14]. Conducting preventive maintenance while monitoring the inspection policy may not be optimal; other policies are possible. Furthermore, it is assumed that time and cost parameters for inspection and maintenance are not time dependent. However in practice, they are time dependent and could be random. In future research, it would be interesting to study other policies by taking into account the age dependent salvage value of the equipment and the age dependent inspection time and inspection cost.

A comparative study of an integrated hazard model and a proportional hazard model would be an interesting extension.

Production / Inventory Control and Maintenance

In the literature, many production-inventory and maintenance models assume that process failure rate is constant or purely a function of time, irrespective of the use. The failure distribution is considered to be independent of how many units are produced or the number of settings. A suggestion for further work in this direction is the consideration of various probability distributions of the timing of major failure / maintenance and minor failure / maintenance. In Chapter 3 of this book, Makis and Jiang investigate the problem of determining jointly the optimal lot sizing, preventive maintenance, and inspection policy which will minimize the expected average cost per unit time. Their study may considered to be, for example, as a starting reference paper for further extension in the light of above suggestion.

Production / Inventory, Quality Control, and Maintenance

Integration of EPQ and quality control in the context of deteriorating inventory management and maintenance and inspection schemes is a very important issue. Considerable attention to the effects of a deteriorating process on EPQ has been devoted, but the effect of deteriorating products has not been considered. Recently Rahim and Ben-Daya [97] have developed an integrated model to determine economic production quantity, maintenance schedules, and quality control decision parameters for studying the simultaneous effects of both deteriorating production processes and deteriorating products. Deterioration times for both product and process are assumed to follow some known probability distribution. It is assumed that the process stops either at failure or at the mth inspection interval, whichever occurs first. The inventory is depleted to zero before starting a new cycle. In many manufacturing situations, manufacturers may wish to continue production of a predetermined fixed quantity and then stop the process to deplete the inventory to zero. This is an interesting and significant problem which may be considered for further study. A consideration of the situation in which demand rate, production rate, and maintenance and inspections are all probabilistic would be an interesting extension. The assumptions and basic structure of the model considered in the paper could easily be relaxed to include variable changes in both the process mean and process variance.

In addition, possible areas for future research include but are not limited to the following.

- alternative formulations of the effect of maintenance policies effect on production quantities and quality control parameters.

- modeling of the effect of maintenance policies on the deterioration pattern of production processes.
- simulation studies in these areas.
- expert systems may play a role in evaluating integrated policies
- integrated models for multi-stage production systems
- integrated optimization modeling by incorporating Taguchis quality loss function

It is hoped that this chapter and the book as a whole will be a useful reference for graduate students and researchers in the area.

References

[1] Abboud, N.E., Jaber, M.Y. and Noeihad, N.A. Economic Lot Sizing with the Consideration of Random Machine Unavailable Time. *Computers and Operations Research*, **27** (2000) 335-251.

[2] Al-Sultan, K.S. and Rahim, M.A. Economic selection of process parameters: A literature survey. *Working paper* Systems Engineering Department, KFUPM, Dhahran, Saudi Arabia (1996).

[3] Ashyeri, J., Teelen, A. and Selen, W. A production and maintenance planning model for the process industry. *International Journal of Production Research*, **34** No. 12 (1996) 3311-26.

[4] Banerjee, P.K. and Rahim, M.A. Economic design of $\bar{x}$-control chart under Weibull shock models. *Technometrics*, **30** (1988) 407-414.

[5] Ben-Daya, M. Integrated Production, Maintenance, and Quality Model for Imperfect Processes. *IIE Transactions*, **31** (1999) 491-501.

[6] Ben-Daya, M. (1999). Quality and reliability Improvement in Multistage Production Systems. *Proceedings of The 5th ISSAT International Conference on Reliability and Quality Design*, pp. 229-233.

[7] Ben-Daya, M. and Duffuaa, S.O. Maintenance and Quality: The Missing Link. *Journal of Quality in Maintenance Engineering*, **1** No.1 (1995) 20-26.

[8] Ben-Daya, M. and Duffuaa, S.O. Overview of Maintenance Modeling Areas. In *Maintenance Modeling and Optimization*, M. Ben-Daya, S. Duffuaa, and A. Raouf (eds). Kluwer Academic Publishers, 2000.

[9] Ben-Daya, M. and Hariga, M. Economic Lot Scheduling problem with Imperfect Production Processes. *Journal of the Operational Research Society*, **51** (2000) 875-881.

[10] Ben-Daya, M., Hariga, M., and El-Hassan, M. An Integrated Maintenance and Quality Model for the Economic Lot Scheduling prob-

lem. Working Paper, Systems Engineering, King Fahd University of Petroleum & Minerals.

[11] Ben-Daya, M. and Makhdoom, M. Integrated Production and Quality Model Under Various Preventive Maintenance Policies. *Journal of the Operational Research Society*, **49** (1998) 840-853.

[12] Ben-Daya, M. and Rahim, A. Multi-Stage Lot sizing Models with Imperfect Processes and Inspection Errors. *Production Planning and Control*, **10** (1999) 118-126.

[13] Ben-Daya, M. and Rahim, A. (1998) Optimal Lot Sizing, Quaity Improvement, and Inspection Errors for Multi-Stage Production Systems. Systems Engineering, King Fahd University of Petroleum & Minerals.

[14] Ben-Daya, M. and Rahim, M.A. Effect of Maintenance on the Economic Design of $\bar{x}-$Chart. *The European Journal of Operational Research.*, **120** (2000) 131-143.

[15] Blischke, W.R. and Murthy, D.N.P. Product warranty man- agement - I: A taxonomy for warranty policies. *European Journal of Operational Research*, **62** (1992) 127-148.

[16] Boukas, E. K. and Haurie, A. Manufacturing Flow Control and Preventive Maintenance : a stochastic control approach, *IEEE Transaction on Automatic Control*, **35-9** (1990) 1024-1031.

[17] Boukas, E. K. and Kenne, J. P. Maintenance and production control of manufacturing system with setups, *Lectures in Applied Mathematics*, **33** (1997) 55-70.

[18] Boukas E. K., and Yang, H. Manufacturing Flow Control and Preventive Maintenance : A Stochastic Control Approach, *IEEE Transaction on Automatic Control*, **41-6** (1996) 881-885.

[19] Boukas, E. K., Zhang, Q. and Yin, G. On Robust Design for a Class of Failure Prone Manufacturing System , in *Recent advances in control and optimization of Manufacturing Systems*, Yin and Zhang ed. Lecture Notes in control and information sciences 214, Springer-Verlag, London, 1996.

[20] Boukas, E. K., Zhu, Q. and Zhang, Q. Piecewise Deterministic Markov Process Model for Flexible Manufacturing Systems with Preventive Maintenance . *Journal of Optimization Theory and Application* , **81-2** (1994) 258-275.

[21] Boukas, E.K., Zhang, Q. Stochastic Manufacturing Systems: Production and Maintenance Control. In *Maintenance Modeling and Optimization*, M. Ben-Daya, S. Duffuaa, and A. Raouf (eds). Kluwer Academic Publishers, 2000.

[22] Brandolese, M., M. Francei and A. Pozzetti Production and Maintenance Integrated Planning. *International Journal of Production Research*, **34** No. 7 (1996) 2059-75.

[23] Chand, S. Lot Sizes and Setup Frequency with Learning in Setups and Process Quality. *European Journal of Operations Research*, **42** (1989) 190-202.

[24] Chen, S.L. and Chung, K.J. Determination of the optimal production run and the most profitable process mean for a production process. *International Journal of Production Research*, **34** No.7 (1996) 2051-58.

[25] Cheng, T.C.E. An Economic Order Quantity Model with Demand Dependent Unit Production Cost and Imperfect Production Processes *IIE Transactions*, **23** No. 1 (1991) 23-28.

[26] Cheng, T.C.E. An Economic Production Quantity Model with Flexibility and Reliability Considerations. *European Journal of Operational Research*, **39** (1989) 174-179.

[27] Cheng, T.C.E. An Economic Production Quantity Model with Flexibility and Reliability considerations. *European Journal of Operational Research*, **39** (1989) 174-79.

[28] Cheng, T.C.E. EPQ with Process Capability and Quality Assurance Considerations. *Journal of the Operational Research Society*, **42** No.8 (1991) 713-20.

[29] Cheung, K.L. and Hausaman, W.H. Joint Determination of Preventive Maintenance and Safety Stocks in an Unreliable Production Environment. *Navel Research Logistics*, **44** (1997) 257-272.

[30] Chiu, H.N. and Huang, B.S. The Economic Design of $\bar{x}$ and S^2 Control Charts with Preventive Maintenance and Increasing Hazard Rate. *Journal of Quality in Maintenance Engineering*, **1** No.4 (1995) 17-40.

[31] Chiu, H.N. and Huang, B.S. The Economic Design of $\bar{x}$-Control Charts under Preventive Maintenance Policy. *Journal of Quality in Maintenance Engineering*, **13** No.1 (1996) 61-71.

[32] Cho, D. and M. Parlar. A Survey of Maintenance Models for Multi-Unit systems. *European Journal of Operational research* **51** (1991) 1-23.

[33] Chung, K.J. and Lin, C.N. Optimal Inventeory Replenishment Models for Deteriorating Items Taking account of Time Discounting. *Computers and Operations Research*, **28** (2001) 67-83.

[34] Crosby, L.B. The Just-in-Time manufacturing Processes: Control of Quality and Quantity. *Production and Inventory Management*, **4th Qtr** (1984) 21-33.

[35] Collani, E.v. Control of Production Process Subject to Random Shocks. *Anals of Operation Research*, **91** (1999) 289-304.

[36] Dohi, T., Kaio, N. and Osaki, S. Minimal Repair Policies for an Economic Manufacturing Porcess *Journal of Quality in Maintenance Engineering*, **4**(4) (1998) 248-262.

[37] Djamaludin, I., Murthy, D.N. P. and Wilson, R.J. Quality Control through Lot Sizing for Items Sold with Warranty. *International Journal of Production Economics*, **33** (1994) 137-144.

[38] Dohi, T., Yamada, Y., Kaio, N. and Osaki, S. The optimal lot sizing for unreliable economic manufacturing models. *International Journal of reliability Quality and Safety Engineering*, **4** (1997) 413-426.

[39] Duncan, A.J. The Economic design of $\bar{x}$-charts used to maintain current control of a process. *Journal of the American Statistical Association*, **51** (1956) 228-42.

[40] Elsayed, E. A. *Reliability Engineering.* New York, Addison Wesley Longman, 1996.

[41] Gibra, I.N. Recent Development in control chart techniques. *Journal of Qualtiy Technology*, **7** (1975) 192-193.

[42] Goyal, S.K., Gunasekaran A., Martikainen T. and Yli-Olli, P. Integrating production and quality control policies: A survey. *European Journal of Operational Research*, **69** (1993) 1-13.

[43] Groenevelt, H.A., Pintelon L. and Seidmann, A. Production Batching with Machine Breakdowns and Safety Stocks. *Operations Research*, **40** No.5 (1992) 959-71.

[44] Groenevelt, H.A., Pintelon L. and Seidmann A. Production Lot Sizing with Machine Breakdowns. *Management Science*, **38** No.1 (1992) 104-23.

[45] Gunasekaran, A., Korukonda, A.R., Virtanen, I. and Yli-Olli, P. Optimal Investment and Lot Sizing Policies for Improved Productivity and Quality *International Journal of Produciton Research*, **33** No.1 (1995) 261-278.

[46] Hariga, M. and Ben-Daya, M. The Economic Manufacturing Lot-Sizing Problem with Imperfect Manufacturing Processes: Bounds and Optimal Solutions. *Naval Research Logistics*, **45** (1998) 423-433.

[47] Hax, A.C. and Candea, D. *Production and Inventory Management*, a-Hall, Englewood Cliffs, New Jersey, (1984).

[48] Ho, C. and Case, K.E. Economic Design of Control Charts: A Literature Review for 1981-1991. *Journal of Qualtiy Technology*, **26** (1994) 39-53.

[49] Hong, J. The Effects of Manufacturing Lead Time and Process Quality in a Multi-Process Produciton System *Production Planning and Control*, **6** No.4 (1995) 311-319.

[50] Huang, B. and Chiu, H. The Quality Management of the Imperfect Production Process under Two Monitoring Policies. *International Journal of Quality & Reliability Management*, **12** No.3 (1995) 19-31.

[51] Jardine, A.K.S. *Maintenance, replacement and Reliability*, Pitman, Canada, (1973).

[52] Johnson, L.A. and Montgomery, D.C. *Operations Research in Production Planning, Scheduling, and Inventory control*, Wiley and Sons, New York, (1974).

[53] Keller, G. and Noori, H. Impact of Investing in the Quality Improvement on the Lot size Model. *OMEGA International Journal of Management Scince*, **16** No.6 (1988) 595-601.

[54] Kenne, J.P. and Gharbi, A. Production Planning Problem in Manufactuing Systems with General Failure and Repair Time Distributions. *Production Planning and Control*, **11** No.6 (2000) 581-588.

[55] Khouja, M. and Mehrez, A. Economic Production Lot Size Model with Variable Production Rate and Imperfect Quality. *Journal of Operational Research Society*, **45** No.12 (1994) 1405-17.

[56] Kim, C. H. and Hong, Y. An Optimal Production Run Length in Deteriorating Production Processes. *International Journal of Production Economics*, **58** No.2 (1999) 183-189.

[57] Kuhn, H A dynamic Lot Sizing Model with Exponential Machine Breakdowns. *European Journal of Operations Research*, **100** No.3 (1997) 514-536.

[58] Lee, J.S. and Park, K. Joint Determination of Production Cycle and Inspection Intervals in a Deteriorating Production. *Journal of Operational Research Society*, **42** No.9 (1991) 775-83.

[59] Lee, H.L. Lot Sizing to Reduce Capacity Lot Utilization in a Production Process with Defective Items, Process correction and Rework. *Management Science*, (1992) 1314-1328.

[60] Lee, H.L., and Rosenblatt M.J. A Production and Maintenance Planning Model with Restoration Cost Dependent on Detection Delay. *IIE Transactions*, **21** No.4 (1989) 368-75.

[61] Lee, H.L., and Rosenblatt M.J. Economic Design and Control of Monitoring Mechanisms in Automated Production Systems. *IIE Transactions*, **20** No.2 (1988) 201-9.

[62] Lee, H.L., and Rosenblatt M.J. Simultaneous Determination of Production Cycles and Inspection Schedules in a Production System. *Management Science*, **33** (1987) 1125-36.

[63] Lee, S.D. and Rung, J.M. Production Lot Sizing in Failure Prone two Stages Serial Systems *European Journal of Operations Research*, **123** (2000) 42-60.

[64] Lee Y. J. and Zipkin, P. H. Production Control in a Kanban-Like System with Defective Outputs. *International Journal of Production Economics*, **28** (1992) 143-155.

[65] Lin, T.M., Tseng, S.T. and Liou, M.J. Optimal Inspection Schedule in the Imperfect Production System Under General Shift Distribution. *Journal of the Chinese Institute of Industrial Engineers*, **8** No.2 (1991) 73-81.

[66] Liou, M.J., Tseng, S.T. and Lin, T.M. The Effects of Inspection Errors to the Imperfect EMQ Model. *IIE Transactions*, **26** No.2 (1994) 42-51.

[67] Mabinin, M.C. Pintelon, L.M. and Gelders, L.F. EOQ Type Formulations for Controlling Repairable Inventories. *International Journal of Production Economics*, **28** (1992) 21-33.

[68] Makis, V. Optimal Lot Sizing and Inspection Policy for an EMQ Model with Imperfect Inspections *Naval Research Logistics*, **45** (1998) 165-186.

[69] Makis, V. Jardine, A.K.S. Optimal Replacement in Proportional Hazards Model. *INFOR*, **30** (1992) 172-183.

[70] Makis, V. and J. Fung Optimal Preventive Maintenance Replacement, Lot Sizing and Inspection Policy for a Deteriorating Production System. *Journal of Quality in Maintenance Engineering*, **1** No.4 (1995) 41-55.

[71] Makis, V., X. Jiang and Tse, E. A General EMQ Model with Machine Breakdowns. In *Maintenance Modeling and Optimization*, M. Ben-Daya, S. Duffuaa, and A. Raouf (eds). Kluwer Academic Publishers, 2000.

[72] Mehraz, A., Myers, B. and Khouja, M.J. Quality and Inventory Issues within the Newsboy Problem. *Computers Operations Research*, **5** (1991) 397-410.

[73] Meller, R.D. and Kim, D.S. The Impact of Preventive Maintenance on System Cost and Buffer Size. *European Journal of Operations Research*, **95** (1996) 577-591.

[74] Miltenburg, J. A Theoretical Framework for Understanding why JIT Reduces Cost and Cycle Time and Improves Quality. *International Journal of Production Economics* , **30-31** (1993) 195-204.

[75] Montgomery, D.C. *Introduction to Statistical Quality Control*, 2nd Edition, Wiley and Sons, New York, (1991).

[76] Montgomery, D.C. The Economic Design of Control Charts: A Review and Literature Survey. *Journal of Quality Technology*, **12**, (1980) 75-87.

[77] Moskowitz, H., Plante, R. and Chun, Y. H. Effect of Process Failure Mechanisms on Economic $\bar{x}$-Control Charts. *IIE Transactions*, **26** No.6 (1994) 12-21.

[78] Murthy, D.N.P. Optimal reliability choice in product design. *Engineering Optimization*, **15** (1990) 281-294.

[79] Murthy, D.N.P. and Blischke, W.R. Product warranty man- agement - II: An integrated framework for study. *European Journal of Operational Research*, **62** (1992) 261-281.

[80] Murthy, D.N.P. and Blischke, W.R. Product warranty man- agement - III: A review of mathematical models. *European Journal of Operational Research*, **63** (1992) 1-34.

[81] Murthy, D.N.P., Wilson, R.J. and Djamaludin, I. Product warranty and quality control, *Quality and Reliability Engineering*, **9** (1993) 431-443.

[82] Nahmias, S. *Production and Operations Analysis*, Irwin Inc., New York, (1989).

[83] Nahmias, S. Managing Repairable Item Inventory Systems: a Review. *TIMS Studies in the Management Sciences*, **16** (1981) 253-277.

[84] Nakagawa, T. A Summary of Imperfect Preventive Maintenance Policies with Minimal Repair. *Operations Research*, **14** No.3 (1980) 249-55.

[85] Ohta, H. and Ogawa, S. (1991). Joint Determination of Production Cycle and Inspection Schedules Under Inspection Errors. In: Pridhan M. and Brien C.O. (eds) *Production Research: Approaching the 21st Century.* Taylor and Francis, pp. 179-188.

[86] Osaki, S. and Nakagawa, T. Bibliography of Reliability and availability of Stochastic Systems. *IEEE Transaction on Reliability*, **R-25** (1976) 284-287.

[87] Osaki, S., and Nakagawa, T. Two Models for Ordering Policies. *Notes in Decision Theory*, No.34, University of Manchester, Manchester, 1977.

[88] Osaki, S., Kaio, N., and Yamada, S. A summary of optimal ordering policies. *IEEE Transactions on Reliability*, **30** (1981) 272-277.

[89] Pate'-Cornell, Lee, H.L. and Tagaras, G. Warnings of Malfunction: The Decision to Inspect and Maintain Production Processes on Schedule or on Demand. *Management Science*, **33** (1987) 1277-90.

[90] Peters, M.H., Schneider H. and Tang, K. Joint Determination of Optimal Inventory and Quality Control Policy. *Management Science*, **34** No.8 (1988) 991-1004.

[91] Pierskalla, W.P. and Voelker, J.A. A Survey of Maintenance Models: The Control and Surveillance of Detriorating Systems. *Naval Research Logistics Quarterly*, **23** (1976) 353-388.

[92] Porteus, E.L. Optimal Lot Sizing, Process Quality Improvement and Setup Cost Reduction. *Operations Research*, **34** (1986) 137-144.

[93] Porteus, E.L. The Impact of Inspection Delay on Process and Inspection Lot Sizing. *Management Science* **8** (1990) 999-1007.

[94] Raafat, F. Survey of Literature of Continuously Deteriorating Inventory Models. *Journal of the Operational research Society*, **42** (1991) 27-37.

[95] Rahim, M.A. Joint Determination of Production Quantity, Inspection Schedule and Control Chart Design. *IIE Transactions*, **26** No.6 (1994) 2-11.

[96] Rahim, M.A. and Ben-Daya, M. A Generalized Economic Model for Joint Determination of Production Run, Inspection Schedules and Control Chart Design. *International Journal of Production Research*, **36** (1998) 277-289.

[97] Rahim, M.A. and Ben-Daya, M. A Joint Optimization of Production Quantity, Inspection Schedule and Quality Control for an Imperfect Process with deterioration Inventory Items. *Journal of Operations Research Society.* In review process (2000).

[98] Rahim, M.A and Banerjee, P.K. Optimal Production Run for a Process with Random Linear Drift. *Omega: International Journal of Management Science*, **16** No.4 (1988) 347-51.

[99] Rishel, T.D. and Christy, D.P. Incorporating Maintenance Activities into Production Planning; Integration at the Master Schedule Versus Material Requirements Level. *International Journal of Production Research*, **34** No.2 (1996) 421-446.

[100] Rosenblatt, M.J., and Lee, H.L. A comparative Study of Continuous and Periodic Inspection Policies in Deteriorating Production Systems. *IIE Transactions*, **18** (1986) 2-9.

[101] Rosenblatt, M.J., and Lee, H.L. Economic Production Cycles with Imperfect Production Process. *IIE Transactions*, **18** (1986) 48-55.

[102] Salameh, K. M. and Jaber, M. Y. Economic Production Quantity Model for Items with Imperfect Quality. *International Journal of Production Economics*, **64** (2000) 59-64.

[103] Schneider, H., Tang, K. and C. O'Cinneide Optimal Control of a Production Process Subject to Random Deterioration. *Operations Research*, **38** No.6 (1990) 1116-22.

[104] Silver, E.A., Pyke, D.F. and Peterson, R. *Inventory Management and Production Planning & Scheduling*, Wiley and Sons, New York, 1998.

[105] So K.C. Tang, C.S. Optimal Batch Sizing and Repair Strategies for Operations with Repairable Jobs. *Management Scineces*, **41** No.5 (1995) 894-908.

[106] Sherif, V.S., and M.L. Smith Optimum Maintenance Models for Systems Subject to Failure: A Review. *Naval Research Logistics Quarterly*, **28** (1981) 47-74.

[107] Tagaras, G. An Integrated Cost Model for the Joint Optimization of Process Control and Maintenance. *J. Opl Res. Soc.* **39** No.8 (1988) 757-766.

[108] Taguchi, G. Introduction to Quality Engineering. *Asian Productivity Organization* American Supplier Institute (1986).

[109] Tapiero, C.S. Continuous Quality Production and Machine Maintenance. *Naval Research Logistics Quarterly*, **33** (1986) 489-99.

[110] Tseng, S., Yeh, R.H. and Ho, W. Imperfect Maintenance Policies for Deteriorating Production Systems. *International Journal of Production Economics*, **55** (1998) 191-201.

[111] Valdez-Flores, C. and Feldman, R.M. A Survey of Preventive Maintenance Models for Stochastically Deteriorating Single Unit Systems. *Naval Research Logistics Quarterly.* **36** (1989) 419-446.

[112] Vance, L.C. A biblography of Statistical Quality Control Chart Techniques 1970-1980. *Journal of Qualtiy Technology*, **36** (1989) 419-446.

[113] Vörös, J. Lot Sizing with Quality Improvement and Setup Time Reduction. *European Journal of Operations Research*, **113** (1999) 568-574.

[114] Yeh, R.H., Ho, W. and Tseng, S. Optimal Production Run Length for Products Sold with Warrenty. *European Journal of Operations Research*, **120** (2000) 575-582.

[115] Wright, C.M. and Mehrez A. An Overview of Representative Research of the Relationships Between Quality and Inventory. *Omega: International Journal of Management Science*, **26** (1998) 29-47.

[116] Zhang, X. and Grechak, Y. Joint Lot Sizing and Inspection Policy in an EOQ Model with Random Yeild. *IIE Transactions*, **22** (1990) 41-47.

II

PRODUCTION & MAINTENANCE

Chapter 2

COMPUTATION ALGORITHMS OF COST-EFFECTIVE EMQ POLICIES WITH PM

H. Okamura, T. Dohi, and S. Osaki

Department of Industrial and Systems Engineering,

Hiroshima University

Higashi-Hiroshima 739-8527, Japan

Abstract This chapter concerns EMQ (Economic Manufacturing Quantity) policies with PM (Preventive Maintenance). Specifically, we develop computational algorithms to calculate the optimal policies which consider both EMQ and PM time using a cost effectiveness criterion. In the case where the demand occurs in accordance with a homogeneous Poisson process, cost effectiveness can be derived in the closed form. For the general case with renewal process demand, two approximation algorithms, diffusion approximation and phase approximation, are proposed to express cost effectiveness analytically. In the numerical examples, we compare quantitatively the approximation performance of the two algorithms and investigate their applicability to Production/inventory management with unreliable machines.

Keywords: EMQ, PM, cost effectiveness, computation algorithms, $G/G/1$ queue, diffusion approximation, phase approximation, Production/inventory management.

1. INTRODUCTION

When dealing with stocked items, it is well known that the question of how much to order is answered by reference to economic order quantity (EOQ) (see [38]). In the basic EOQ model, the resulting order quantity assures the most economic inventory policy in an ideal situation without stock out and lead time. Similarly, the economic manufacturing quantity (EMQ) model [38], which is a natural extension of the EOQ, provides the proper production lot size. Since these simple inventory models are formulated under very restrictive assumptions, it is difficult to apply the

EOQ or EMQ model directly to actual inventory management. Hence, several authors extended the basic EOQ and/or EMQ models from various points of view. Axäter and Elmaghraby [2] and Muth and Spremann [30] took account of the effects of learning and forgetting in EMQ management. Kuzdrall and Britney [26] and Sethi [42] considered lot sizing models with quantity discounts. Murdeshwar and Sathe [29] proposed an inventory model with limited storage capacity. Jesse *et al.* [20] questioned the basic EOQ, minimizing the cost ratio over an infinite time horizon and reformulating the EOQ under inflation conditions. Also, inventory models with deteriorating items [43], nonlinear holding costs [6], purchase price fluctuations [15], and joint purchaser/vendor replenishment decisions [3] were developed in the EOQ and/or EMQ modeling framework.

An alternative direction for extending the basic EMQ model is needed to challenge the assumption that the production facility is completely reliable. In fact, the most important issue in inventory management is the establishment of a stable production circumstance without production machine breakdown and accurate prediction of the demand under uncertainty. In that sense, the application of preventive maintenance (PM) in production/inventory management plays a significant role in the effective operation of manufacturing systems subject to failures. Comprehensive surveys on PM theory are found in Sherif and Smith [44] and Valdez-Flores and Feldman [49]. It should be noted, however, that the optimal inspection and maintenance policies presented in the literature [44, 49] can not be applied directly to EMQ without modification since the traditional maintenance models do not take account of the tradeoff between the inventory holding cost and the shortage loss. In other words, the manufacturer should design the production lot from the standpoint of the desire to increase both holding cost and shortage loss caused by machine breakdown. Thus, the effect of both machine breakdown and preventive maintenance in economic production lot sizing decisions should be examined jointly in an uncertain environment with unreliable manufacturing facilities.

Several papers are concerned with the EMQ models with PM policies. Lee and Rosenblatt [27] addressed the problem of joint control of production cycle and maintenance by periodic inspection using an EMQ model under the assumption that some items produced were defective. They derived the optimal lot size and the optimal inspection schedule and showed that the resulting optimal lot size was smaller than the basic EMQ. Groenevelt *et al.* [16] focused on the effects of stochastic machine breakdowns and corrective maintenance on economic lot sizing decisions. They proposed two production control policies. In the first policy, the production of interrupted lots is not resumed after a breakdown, and the on-hand inventory is depleted before a new cycle is initiated. In the second policy, production is immediately resumed after a breakdown if the current on-hand inventory is below a certain threshold level. Under these control policies, they showed that the optimal lot size was larger than the basic EMQ, and that the optimal lot size increased in the failure rate. In a subsequent paper [17], Groenevelt *et al.* also considered the safety stocks to meet the managerially prescribed service level, and derived the exact and approximate expressions for safety stock holding costs based on the workload process of a special single server queueing system. The stochastic EMQ models of Groenevelt *et al.* [16, 17] were extended in the literature [7, 8, 9, 10, 12]. Furthermore, Dohi *et al.* [11] introduced a new control method, called an *order limit policy* in an (r, Q) inventory system, based on the idea of repair limit replacement [44, 49].

In the EMQ models mentioned above, constant demand and production rates are assumed. Of course, it is questionable whether these simplified assumptions represent a real production/inventory system. Berg *et al.* [5] considered an $M/M/1$ queue-like production system with Poisson process demand and exponential processing times. In their model, a number of production machines are susceptible to machine failure, and the failed machines have to be repaired so as to be in the operative state. Furthermore, if machine failure occurs, production has to be stopped deliberately due to stocking capacity limitations. Berg *et al.* [5] analyzed the inventory system with multiple machines by employing level-crossing techniques and computed performance measures that characterized the operation of the production/inventory system with respect to its service-level to customers, the expected inventory stocked, the machines' utilization, repairmen utilization, *etc.* Srinivasan and Lee [46] developed an EMQ model with PM to guarantee higher system maintainability. In their model, two inventory threshold levels, *PM level* and *production restarting levels are assumed* in order to investigate the combined effects of lot sizing and PM decisions on the cost of operation. These threshold levels correspond to the control limits for starting PM and restarting production, respectively.

In this chapter, we further extend the Srinivasan and Lee model [46]. Note in [46] that the demand is assumed to follow a homogeneous Poisson process. This assumption is, of course, too restrictive. Actually, it is known that in real production/inventory management the demand rate is not always constant in time. Thus, it is necessary to develop a production/inventory process with a general demand pattern. Using these generalized assumptions, we consider the combined effects of PM and lot sizing decisions on the cost of operation of a facility.

Our production/inventory model is similar to that in Srinivasan and Lee [46] but assumes that both demand and production processes follow a continuous-time renewal counting processes. In addition, we suppose that machine breakdown occurs according to a non-homogeneous Poisson process. This generalized framework for modeling involves the results of Srinivasan and Lee [46] as a special case. In this chapter, we employ *cost effectiveness* criterion as a criterion of optimality, though Srinivasan and Lee [46] minimize the expected cost per unit time in steady-state. We define cost effectiveness as the mean length of operative time of the system per unit cost, where the operative time is when production items are not out of stock. This criterion of optimality is rather unique in the inventory literature. Since cost effectiveness is defined as the ratio of steady-state availability to the expected cost per unit time in the steady-state, the EMQ policy which maximizes cost effectiveness takes into account not only the availability of the production system but also the economic justification. Hence, in Production/inventory management subject to failures, cost effectiveness is useful to guarantee higher system reliability.

This chapter is organized as follows. In the following section, we describe the EMQ model with general demand and production processes. The underlying stochastic process is essentially similar to the $G/G/1$ queue with service breakdown and vacation [47]. As a special case, we formulate cost effectiveness in the case of Poisson demands in Section 3. This is an analytically solvable case and is compared with the result of Srinivasan and Lee [46]. In general, it is difficult to formulate the cost effectiveness explicitly when demand follows renewal stochastic counting processes. Thus, an approximation method to express cost effectiveness has to be developed to calculate the optimal EMQ and PM policies. We apply two approximation schemes: diffusion approximation [35] and phase approximation [37]. In diffusion approximation, the underlying stochastic process can be approximated by continuous-state stochastic processes driven by Brownian motion processes. This technique is useful to approximate complex discrete-event systems [18, 25]. On the other hand, the phase method approximates the renewal process by a phase-type renewal process. This method, called matrix-geometric analysis [1, 21, 31, 32, 33], plays an important role in the analysis of $G/G/1$ queue-like stochastic processes. In the numerical examples, we compare quantitatively the approximation performance for the two algorithms and investigate their applicability in Production/inventory management with unreliable machines. Specifically, we develop computational algorithms for cost-effective EMQ and PM policies based on the respective approximation methods. Finally, the chapter is concluded with some remarks and future work.

2. MODEL DESCRIPTION

Consider the EMQ model with PM. Let $\{D(t);\ t > 0|D(0) = 0\}$ denote the cumulative quantity of demands at time t. Demand occurs at any time, but demands intervals are distributed randomly. We assume that $D(t)$ is an arbitrary ergodic stochastic process in continuous time to simplify the analysis. The demand rate and its variance in the steady-state are defined by $\lambda = \lim_{t\to\infty} D(t)/t$ and $\sigma_D^2 = \lim_{t\to\infty}(D(t) - \lambda t)^2/t$, respectively. The production time for the n-th item is a random variable θ_n, having the probability distribution function $H(t)$, the mean time length $\mathrm{E}[\theta_n] = 1/\mu$, and the variance $\mathrm{Var}[\theta_n] = \sigma_\theta^2$. Production machinery sequentially produces the items one by one, and the produced items are stored in the buffer as inventory so that the inventory process forms the $G/G/1$ queueing process. Define the utilization factor as the fraction of demand rate to production rate, *i.e.* $\rho = \lambda/\mu$. Generally speaking, it is well known that the stationary queue length is infinite if the utilization factor is larger than or equal to unity. Therefore, it is assumed without any loss of generality that $\rho < 1$. Suppose that the production machine failure occurs according to a non-homogeneous Poisson process with mean value function $m(t)$. More specifically, assume that the intensity function of the non-homogeneous Poisson process is the Weibull type: $r(t) = dm(t)/dt = \kappa\omega t^{\omega-1}$. Since Weibull-type failure rate has both scale and shape parameters, and the scale parameter can characterize the monotone failure property called aging (increasing failure rate or decreasing failure rate), it will be useful to represent production machinery failure. Now, suppose that the failure rate has an IFR (increasing failure rate) property, *i.e.* $r(t)$ is increasing in t for an arbitrary $\omega \geq 1$.

When production machinery failure occur, a minimal repair is started immediately. By minimal repair, the failed machines can be returned to the state prior to failure, where the time for minimal repair can be negligible. The assumption of minimal repair seems to be plausible since production machines, in practice, never become as good as new by repair. Also, if the repair time is much shorter than the production time, the assumption of no repair time will be validated [44, 49]. In this manufacturing circumstance, the production facility degenerates gradually as time elapses. Thus, the manufacturer should perform the PM at any time to return state of the machine to as good as new. Srinivasan and Lee [46] proposed the following PM policy.

1. When the stock level is at PM level S (> 0) (pre-determined constant), the PM is started after suspending production.
2. The rule for restarting production is as follows.
   (a) If the stock level is less than production restarting level s ($0 < s < S$) (pre-determined constant), then production is restarted just after the completion of PM.
   (b) If the stock level is higher than the production restarting level s, then production is restarted after the stock level decreases to the level s.
3. After restarting production, production machines continues to produce the items until the stock level is at level S again.

Although this policy is not necessarily optimal for all inventory management, in terms of practical use, the policy proposed by Srinivasan and Lee [46] is useful because it is an admissible control in practice. Furthermore, the inventory process under consideration behaves in a manner similar to the traditional (s, S) inventory process. Hence, the model in Srinivasan and Lee [46] can be interpreted as an extension of the (s, S) inventory model. The time length of the PM is a random variable, having a probability distribution function $\xi(\cdot)$, a finite mean $1/\eta$, a variance σ_P, and a hazard rate function (failure rate function) $\phi(\cdot)$. To simplify the model description, the PM period and the manufacturing period are called *Phase I* and *Phase II*, with time lengths denoted by τ_1 and τ_2, respectively (see Fig. 18.1).

We define cost component as follows:

c_k (> 0) : fixed set-up cost/lot

c_f (> 0) : fixed repair cost/machine failure

c_m (> 0) : PM cost/time

c_b (> 0) : shortage (penalty) cost/time/item

c_h (> 0) : inventory holding cost/time/item.

As a mathematical preliminary, we introduce the concept of ***virtual stock level***. The virtual stock level is defined by changing the indices of stock level as follows:

$$
\begin{array}{lcccccccc}
\text{actual stock level:} & S & S-1 & \cdots & s & \cdots & 0 & \cdots \\
 & \downarrow & \downarrow & & \downarrow & & \downarrow & \\
\text{virtual stock level:} & 0 & 1 & \cdots & r & \cdots & S & \cdots,
\end{array}
$$

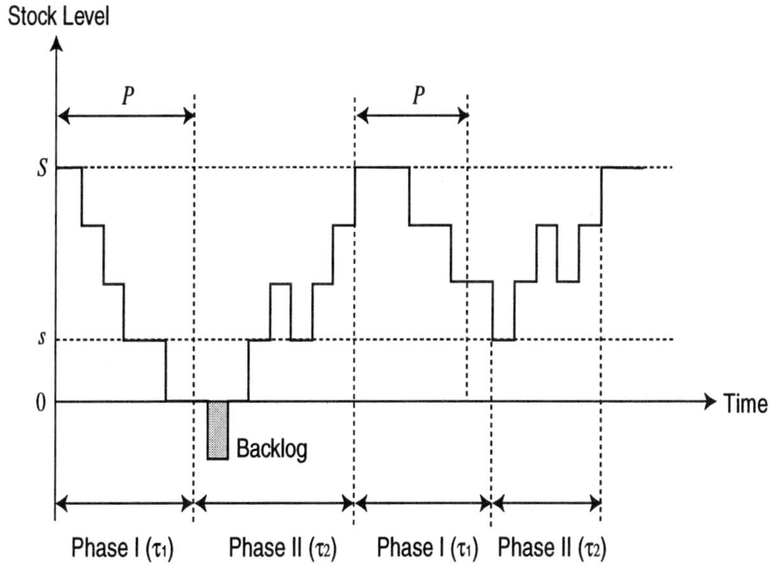

Figure 2.1 Configuration of the EMQ model with PM.

where $r = S - s$ (see Fig. 18.2).

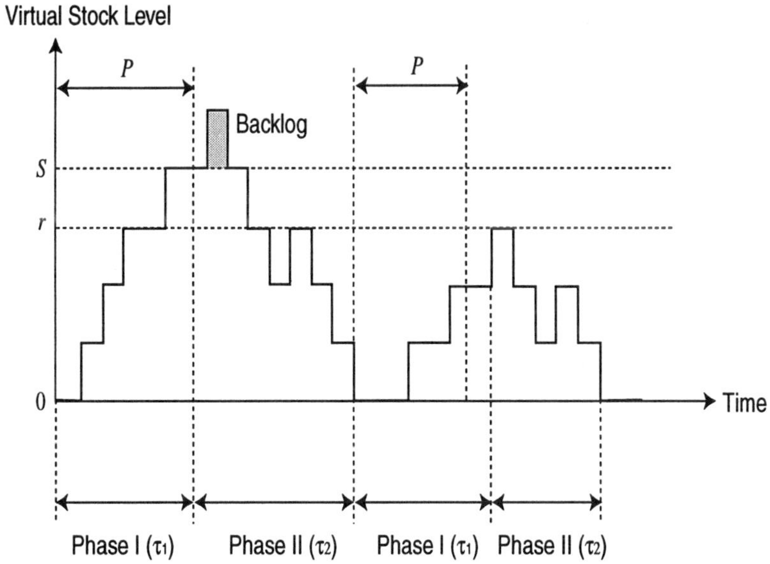

Figure 2.2 Configuration of the EMQ model with PM using the virtual stock level.

Thus, the inventory process under consideration can be regarded as a *G/G/1* queueing process with single vacation [4, 19, 23, 45, 47, 51]. The problem is to find the optimal pair (r^*, S^*) which maximize the cost effectiveness criterion $C_e(r, S)$.

3. POISSON PROCESS DEMAND

In this section, we re-consider the Srinivasan and Lee model [46] according to the cost effectiveness criterion. Let $\{X_n;\ n \geq 1\}$ denote a sequence of i.i.d. random variables which represents an inter-arrival time between the $(n-1)$-st and n-th demands, having the probability distribution function $F(t) = 1 - \exp\{-\lambda t\}$. The expected number of minimal repairs for one cycle, the expected inventory holding cost per unit time in the steady-state, the mean time length of one cycle, and the mean operative time length (the time length of holding items) for one cycle are defined by $M_{r,S}$, $I_{r,S}$, $L_{r,S}$ and $A_{r,S}$, respectively. For the inventory model, Srinivasan and Lee [46] minimize the expected cost per unit time in steady-state. The expected cost per unit time in steady-state is one of the most familiar criterion. However, in practical Production/inventory management, it is important to provide items without a shortage from the standpoint of reliability. Therefore, we employ a *cost effectiveness* criterion as a criterion of optimality, defined as the mean time length of holding inventory per unit cost. If cost effectiveness is used, safety stock can be larger than that in the expected cost per unit time. Using the renewal reward theorem [40], we define cost effectiveness as follows:

$$\begin{aligned} C_e(r,S) &= \lim_{t\to\infty} \frac{\text{E[cumulative time of holding items in } [0,t)]}{\text{E[total cost incurred in } [0,t)]} \\ &= \frac{\text{E[time length of holding items for one cycle]}}{\text{E[total cost incurred for one cycle]}} \\ &= \frac{A_{r,S}}{c_m/\eta + c_K + c_f M_{r,S} + I_{r,S} L_{r,S}}. \end{aligned} \tag{2.1}$$

(i) Mean time length of one cycle

Let γ_n denote the probability that n demands occur while the production system undergoes a PM. That is,

$$\gamma_n = \int_0^\infty \frac{(\lambda t)^n}{n!} e^{-\lambda t} d\xi(t), \quad n \geq 0. \tag{2.2}$$

Therefore, the Laplace Stieltjes transform (LST) of the random variable τ_2 is given by

$$\mathrm{E}[e^{-\beta\tau_2}] = \sum_{n=0}^{r} \gamma_n\{\theta^*(\beta)\}^r + \sum_{n=r+1}^{\infty} \gamma_n\{\theta^*(\beta)\}^n, \qquad (2.3)$$

where $\theta^*(\beta)$ is the LST of a *busy period* in the *M/G/1* queueing system. From the well-known result of queueing theory, it is seen that

$$\begin{aligned} \theta^*(\beta) &= \int_0^\infty e^{-\beta t} \sum_{n=0}^{\infty} (\theta^*(\beta))^n \frac{(\lambda t)^n}{n!} e^{-\lambda t} dH(t) \\ &= H^*(\beta + \lambda\{1 - \theta^*(\beta)\}), \end{aligned} \qquad (2.4)$$

where $H^*(\beta)$ is the LST of $H(t)$. Since the mean time length of one cycle is the sum of the mean time lengths of Phase I and Phase II, we obtain

$$\begin{aligned} L_{r,S} &= \mathrm{E}[\tau_1] + \mathrm{E}[\tau_2] \\ &= \frac{1}{1-\rho}\left\{\frac{1}{\eta} + \sum_{n=0}^{r-1} \frac{\gamma_n(r-n)}{\lambda}\right\}. \end{aligned} \qquad (2.5)$$

(ii) Probability for the stock level in the steady-state

Let N be the virtual stock level in the steady-state, where $\Pr\{N = n\} = p_m(n)$. Consider the embedded Markov chain at the completion of production; the transition matrix of virtual stock level is given by

$$P = \begin{bmatrix} b_0 & b_1 & b_2 & \cdots \\ a_0 & a_1 & a_2 & \cdots \\ 0 & a_0 & a_1 & \cdots \\ 0 & 0 & a_0 & \cdots \\ \cdot & \cdot & \cdot & \\ \cdot & \cdot & \cdot & \end{bmatrix}, \qquad (2.6)$$

where a_n and b_n are the probabilities that n demands occur during a production time period and during a maintenance time period plus a production time period, respectively. That is,

$$a_n = \int_0^\infty \frac{(\lambda t)^n}{n!} e^{-\lambda t} dH(t) \qquad (2.7)$$

and

$$b_n = \begin{cases} 0, & n = 0, \cdots, r-2, \\ \sum_{k=0}^{r-1} \gamma_k a_{n-r+1} + \sum_{k=r}^{n+1} \gamma_k a_{n+1-k}, & n = r-1, \cdots. \end{cases} \tag{2.8}$$

Since the stationary probability row vector $\boldsymbol{\nu} = (p_m(0), p_m(1), \cdots)$ is the corresponding eigenvector to the matrix P, the following relations have to be satisfied:

$$\begin{aligned} p_m(0) &= b_0 p_m(0) + a_0 p_m(1), \\ p_m(1) &= b_1 p_m(0) + a_1 p_m(1) + a_0 p_m(2), \\ &\vdots \\ p_m(n) &= b_n p_m(0) + \sum_{k=1}^{n+1} a_{n+1-k} p_m(k), \\ &\vdots \end{aligned} \tag{2.9}$$

where $p_m(0)$ is represented as the fraction of $1/\lambda$ to the mean time length of one cycle $L_{r,S}$, *i.e.*

$$p_m(0) = \frac{1-\rho}{\lambda/\eta + \sum_{n=0}^{r-1} a_n (r-n)}. \tag{2.10}$$

Therefore, using Equations (2.9) and (2.10), we derive the following recursive equations:

$$\begin{aligned} p_m(1) &= \frac{1}{a_0}(1-b_0)p_m(0), \\ p_m(n) &= \frac{1}{a_0}\left(p_m(n-1) - b_{n-1}p_m(0) - \sum_{k=1}^{n-1} a_{n-k} p_m(k)\right), \\ &\quad \text{for } n = 1, 2, \cdots. \end{aligned} \tag{2.11}$$

(iii) The mean time of holding items for one cycle

The mean time of holding items for one cycle is equal to the mean time period when the virtual stock level is less than S during one cycle. Using the stationary probability $p_m(n)$, the mean time of holding items for one cycle is

$$A_{r,S} = L_{r,S} \sum_{n=0}^{S-1} p_m(n). \tag{2.12}$$

(iv) The expected number of minimal repairs for one cycle

Let $m(t)$ denote the expected number of minimal repairs up to time t. Using an intuitive argument, the expected number of minimal repairs for one cycle is given by

$$M_{r,S} = \int_0^\infty m(t)\,\Pr\{\tau_2 \in dt\}. \tag{2.13}$$

If the failure rate function is the Weibull-type, $r(t) = dm(t)/dt = \kappa\omega t^{\omega-1}$, then the expected number of minimal repairs becomes

$$M_{r,S} = \kappa \mathrm{E}[\tau_2^{\omega}]. \tag{2.14}$$

Calculating the ω-th order derivative of $\mathrm{E}[\exp\{-\beta\tau_2\}]$ with respect to β and letting $\beta \to 0$ yield the explicit form of $M_{r,S}$, for instance, if $\omega = 2$, we obtain

$$\begin{aligned} M_{r,S} &= \kappa\left\{\frac{\lambda}{\eta}\mathrm{E}[B^2] + \lambda^2\left(\sigma_P^2 + \frac{1}{\eta^2}\right)\mathrm{E}[B]^2\right. \\ &\quad \left. + \sum_{n=0}^{r-1} a_n\left\{(r^2-n^2)\mathrm{E}[B]^2 + (\mathrm{E}[B^2]-\mathrm{E}[B]^2)(r-n)\right\}\right\}, \end{aligned} \tag{2.15}$$

where $\mathrm{E}[B]$ and $\mathrm{E}[B^2]$ are the first and second moments of the busy period in the *M/G/1* queueing system, *i.e.*

$$\mathrm{E}[B] = \frac{1}{\mu(1-\rho)} \tag{2.16}$$

and

$$\mathrm{E}[B^2] = \frac{1}{1-\rho}\left(\frac{1+\lambda\mathrm{E}[B]}{\mu}\right)^2, \tag{2.17}$$

respectively.

(v) The expected holding and backlog costs in the steady-state

In regard to the *M/G/1* queueing system, we can adopt the well-known *PASTA* (Poisson Arrival See Time Average) theorem [50], so that the stationary queue length distribution at arbitrary time is equal to that at the completion of a service. Thus, the expected holding and backlog costs can be represented by using the stationary probability $p_m(n)$. The sum of holding cost and backlog cost in the steady-state is

$$\begin{aligned} I_{r,S} &= c_h \sum_{n=0}^{S} (S-n) p_m(n) + c_b \sum_{n=S+1}^{\infty} (n-S) p_m(n) \\ &= c_b(\mathrm{E}[N] - S) + (c_h + c_b) \sum_{n=0}^{S} (S-n) p_m(n). \end{aligned} \tag{2.18}$$

To derive E[N] in the above equation, consider the probability generating function of $p_m(n)$, *i.e.* $p_m^*(z) = \sum_{n=0}^{\infty} z^n p_m(n)$. From Equation (2.9), it is not difficult to obtain

$$p_m^*(z) = b_z^* p_m(0) + z^{-1} \left\{ a_z^* p_m^*(z) - a_z^* p_m(0) \right\}, \tag{2.19}$$

where a_z^* and b_z^* are the probability generating functions of a_n and b_n, namely,

$$a_z^* = \sum_{n=0}^{\infty} z^n a_n = H^*(\lambda(1-z)), \tag{2.20}$$

$$\begin{aligned} b_z^* &= \sum_{n=0}^{\infty} z^n b_n \\ &= z^{-1} \left\{ \xi^*(\lambda(1-z)) + \sum_{n=0}^{r-1} \gamma_n (z^r - z^n) \right\} H^*(\lambda(1-z)), \end{aligned} \tag{2.21}$$

respectively. In the above expression, $\xi^*(\beta)$ is the LST of PM time. Differentiating Equation (2.19) with respect to z and letting $z = 1$, it is straightforward to obtain the expected stock level in the steady-state.

The computation algorithm to find the optimal (r^*, S^*) which maximizes the cost effectiveness $C_e(r, S)$ is as follows.

Algorithm Poisson:

(Step 1) $r := 1$ (initial value)

(Step 2) Calculate the optimal PM level S_1 for $r = 1$

(Step 3) $C_{max} := C_e(1, S_1)$

(Step 4) $r := r + 1$

(Step 5) Calculate the optimal PM S_r for renewed r.

(Step 6) If $C_{max} < C_e(r, S_r)$ then $C_{max} := C_e(r, S_r)$ and go to Step 4.

(Step 7) $(r^*, S^*) := (r - 1, S_{r-1})$.

Note that the optimal policy (r^*, S^*) based on the above algorithm is the global optimum if $C_e(r, S)$ is concave in r and S. Thus, one has to check carefully the shape of the function $C_e(r, S)$.

In this section, we have formulated the cost effectiveness of the Srinivasan and Lee model [46]. In the following sections, we consider the general case where the demand follows the renewal counting process with i.i.d. inter-arrival times and develop approximation methods to represent cost effectiveness.

4. DIFFUSION APPROXIMATION

In this section, we propose a diffusion approximation method to represent cost effectiveness. The presented approximation method regards the inventory level process with renewal demand as usual diffusion processes. In fact, Kimura [24] and Okamura *et al.* [34, 36] applied a similar method to approximate vacation queueing systems. Generally, it is known that the accuracy of diffusion approximation strongly depends on the utilization factor. That is, if the utilization factor ρ is close to unity, then the approximation performance can be expected to be good (see [13, 14]). In addition, note that, in the diffusion approximation, no complex calculation is needed to develop the approximation formula. This is a powerful advantage of applying the present method to the problem.

Let $\{X_n;\ n \geq 1\}$ denote a sequence of i.i.d. random variables which represent the inter-arrival times between the $(n-1)$-st and n-th demands, where X_n obeys the absolutely continuous probability distribution function $F(t)$ with mean $\mathrm{E}[X_n] = 1/\lambda$ and variance $\mathrm{Var}[X_n] = \sigma_D^2$. Let $\{Z_1(t);\ t \geq 0 | Z_1(0) = 0\}$ and $\{Z_2(t);\ t \geq 0 | Z_2(0) = Z_1(\tau_1)\}$ denote the virtual stock levels at time t in Phase I and II, respectively. We approximate the stochastic counting processes $Z_1(t)$ and $Z_2(t)$ using the diffusion processes, $W_1(t)$ and $W_2(t)$, with absolutely continuous sample paths, that is,

$$\begin{aligned} Z_1(t) &\approx W_1(t), \qquad W_1(0) = 0, \\ Z_2(t) &\approx W_2(t), \qquad W_2(0) = W_1(\tau_1). \end{aligned}$$

Denoting the standard Brownian motion process (standard Wiener process) by $\{B(t), t \geq 0\}$, the diffusion processes $W_1(t)$ and $W_2(t)$ can be defined by the solutions of the following stochastic differential equations:

$$dW_1(t) = \mu_1 dt + \sigma_1 dB(t), \qquad W_1(0) = 0, \tag{2.22}$$

$$dW_2(t) \quad = \quad \mu_2 dt + \sigma_2 dB(t), \qquad W_2(0) = W_1(\tau_1), \tag{2.23}$$

where the drift parameters μ_i and the diffusion parameters σ_i $(i = 1, 2)$ are given by

$$\mu_1 = \lambda \ (> 0), \qquad \sigma_1^2 = \lambda^3 \sigma_D^2 \ (> 0),$$

$$\mu_2 = \lambda - \mu \ (< 0), \quad \sigma_2^2 = \lambda^3 \sigma_D^2 + \mu^3 \sigma_\theta^2 \ (> 0).$$

Define the sum of both time lengths of Phase I and Phase II as one cycle. Furthermore, define $L_r(x)$ as the mean time length when the virtual stock level is over the level x for one cycle. Thus, we can formulate

$$\begin{aligned} L_r(x) \ = \ & \mathrm{E}\left[\int_0^{\tau_1} \chi(W_1(t) > x)dt \middle| W_1(0) = 0\right] \\ & + \mathrm{E}\left[\mathrm{E}\left[\int_0^{\tau_2} \chi(W_2(t) > x)dt \middle| W_2(0) = W_1(\tau_1)\right] \middle| W_1(0) = 0\right], \end{aligned} \tag{2.24}$$

where

$$\chi(\mathcal{A}) = \begin{cases} 1 & \text{if event } \mathcal{A} \text{ occurs} \\ 0 & \text{otherwise} \end{cases} \tag{2.25}$$

is the indicator function. To derive $L_r(x)$ explicitly, define the function

$$f_i(x, z; \alpha) = \int_0^\infty e^{-\alpha t} \mathrm{E}[\chi(W_i(t) > x) | W_i(0) = z] dt, \tag{2.26}$$

for $i = 1, 2$. Applying the change of measure theorem [18], these can be derived as follow:

$$\begin{aligned} f_i(x, z; \alpha) \ &= \ \int_0^\infty e^{-\alpha t} \mathrm{E}[\chi(W_i(t) > x - z) | W_i(0) = 0] dt \\ &= \int_0^\infty e^{-\alpha t} \mathrm{E}^* \left[\exp\left(\frac{\mu_i}{\sigma_i^2} W_i(t) - \frac{1}{2} \frac{\mu_i^2}{\sigma_i^2} t \right) \chi(W_i(t) > x - z) \right] dt \\ &= \int_0^\infty \int_{x-z}^\infty e^{-\alpha} \exp\left(\frac{\mu_i}{\sigma_i^2} u - \frac{1}{2} \frac{\mu_i^2}{\sigma_i^2} t \right) g_i(t, u) du dt \\ &= \int_{x-z}^\infty \exp\left(\frac{\mu_i}{\sigma_i^2} u \right) \int_0^\infty \exp\left\{ -\left(\alpha + \frac{1}{2} \frac{\mu_i^2}{\sigma_i^2} \right) t \right\} g_i(t, u) dt du \\ &= \frac{1}{\psi_i(\alpha)} \int_{x-z}^\infty \exp\left(\frac{\mu_i}{\sigma_i^2} u - \frac{\psi_i(\alpha)}{\sigma_i^2} |u| \right) du, \end{aligned} \tag{2.27}$$

where E* denotes the expectation operator associated with the measure where the drift of the original diffusion process is zero, and where

$$g_i(t,x) = \frac{1}{\sqrt{2\pi\sigma_i^2 t}} \exp\left(-\frac{x^2}{2\sigma_i^2 t}\right), \tag{2.28}$$

and

$$\psi_i(\alpha) = \sqrt{\mu_i^2 + 2\alpha\sigma_i^2}. \tag{2.29}$$

From Equation (2.27), the function $f_i(x, z; \alpha)$ has two forms.

In the case of $x - z \geq 0$,

$$f_i(x,z;\alpha) = \frac{\sigma_i^2}{\psi_i(\alpha)\{\psi_i(\alpha)-\mu_i\}} \exp\left\{-\frac{1}{\sigma_i^2}(\psi_i(\alpha)-\mu_i)(x-z)\right\}. \tag{2.30}$$

In the case of $x - z < 0$,

$$\begin{aligned} f_i(x,z;\alpha) &= \frac{\sigma_i^2}{\psi_i(\alpha)\{\psi_i(\alpha)-\mu_i\}} + \frac{\sigma_i^2}{\psi_i(\alpha)\{\psi_i(\alpha)+\mu_i\}} \\ &\quad -\frac{\sigma_i^2}{\psi_i(\alpha)\{\psi_i(\alpha)+\mu_i\}} \exp\left\{\frac{1}{\sigma_i^2}(\psi_i(\alpha)+\mu_i)(x-z)\right\}. \end{aligned} \tag{2.31}$$

Using Dynkin's formula [18], we obtain the following relations on two functions f_1 and f_2:

$$\begin{aligned} &\mathrm{E}\left[\int_0^{\tau_1} \chi(W_1(t) > x)dt \,\middle|\, W_1(0) = 0\right] \\ &= \lim_{\alpha\to 0}\{f_1(x,0;\alpha) - \mathrm{E}[e^{-\alpha\tau_1} f_1(x, W_1(\tau_1);\alpha)]\} \\ &= \frac{1}{\mu_1}\mathrm{E}[W_1(\tau_1)\chi(W_1(\tau_1) > x)] - \left(\frac{x}{\mu_1} - \frac{\sigma_1^2}{2\mu_1^2}\right)\Pr\{W_1(\tau_1) > x\} \\ &\quad -\frac{\sigma_1^2}{2\mu_1^2}\mathrm{E}\left[\exp\left\{-\frac{2\mu_1}{\sigma_1^2}(W_1(\tau_1)-x)\right\}\chi(W_1(\tau_1) > x)\right] \end{aligned} \tag{2.32}$$

and

$$\mathrm{E}\left[\mathrm{E}\left[\int_0^{\tau_2} \chi(W_2(t) > x)dt \,\middle|\, W_2(0) = W_1(\tau_1)\right]\right]$$

$$
\begin{aligned}
&= \lim_{\alpha\to 0} \mathrm{E}[f_2(x, W_1(\tau_1);\alpha) - \mathrm{E}[e^{-\alpha\tau_2} f_2(x,0;\alpha)]]. \\
&= \frac{1}{|\mu_2|}\mathrm{E}[W_1(\tau_1)\chi(W_1(\tau_1) > x)] - \left(\frac{x}{|\mu_2|} - \frac{\sigma_2^2}{2\mu_2^2}\right)\Pr\{W_1(\tau_1) > x\} \\
&\quad -\frac{\sigma_2^2}{2\mu_2^2}\exp\left\{\frac{2\mu_2}{\sigma_2^2}x\right\}\left\{1 - \mathrm{E}\left[\exp\left(-\frac{2\mu_2}{\sigma_2^2}W_1(\tau_1)\right)\chi(W_1(\tau_1) \le x)\right]\right\}.
\end{aligned}
\tag{2.33}
$$

Thus, the mean time length when the virtual stock level is over the level x for one cycle can be obtained as follows.

In the case of $x < r$,

$$
\begin{aligned}
L_r(x) &= \left(\frac{1}{\mu_1} + \frac{1}{|\mu_2|}\right)\left\{r\Phi(r) + \int_r^\infty u d\Phi(u) - x\right\} + \left(\frac{\sigma_1^2}{2\mu_1^2} + \frac{\sigma_2^2}{2\mu_2^2}\right) \\
&\quad -\frac{\sigma_1^2}{2\mu_1^2}\exp\left\{-\frac{2\mu_1}{\sigma_1^2}(r-x)\right\}\Phi(r) - \frac{\sigma_2^2}{2\mu_2^2}\exp\left\{\frac{2\mu_2}{\sigma_2^2}x\right\} \\
&\quad -\frac{\sigma_1^2}{2\mu_1^2}\exp\left\{\frac{2\mu_1}{\sigma_1^2}x\right\}\int_r^\infty \exp\left\{-\frac{2\mu_1}{\sigma_1^2}u\right\}d\Phi(u).
\end{aligned}
\tag{2.34}
$$

In the case of $x \ge r$,

$$
\begin{aligned}
L_r(x) &= \left(\frac{1}{\mu_1} + \frac{1}{|\mu_2|}\right)\int_x^\infty (u-x)d\Phi(u) + \left(\frac{\sigma_1^2}{2\mu_1^2} + \frac{\sigma_2^2}{2\mu_2^2}\right)\{1 - \Phi(x)\} \\
&\quad -\frac{\sigma_1^2}{2\mu_1^2}\exp\left\{\frac{2\mu_1}{\sigma_1^2}x\right\}\int_x^\infty \exp\left\{-\frac{2\mu_1}{\sigma_1^2}u\right\}d\Phi(u) \\
&\quad -\frac{\sigma_2^2}{2\mu_2^2}\exp\left\{\frac{2\mu_2}{\sigma_2^2}x\right\}\left(1 - \exp\left\{-\frac{2\mu_2}{\sigma_2^2}r\right\}\Phi(r)\right) \\
&\quad +\frac{\sigma_2^2}{2\mu_2^2}\int_r^x \exp\left\{\frac{2\mu_2}{\sigma_2^2}(x-u)\right\}d\Phi(u),
\end{aligned}
\tag{2.35}
$$

where $\Phi(\cdot)$ is the virtual stock level distribution at the completion time of PM. Applying the change of measure theorem [18], the function $\Phi(\cdot)$ is also given by

$$
\begin{aligned}
\Phi(u) &= \int_0^\infty \Pr\{W_1(t) \le u\}d\xi(t) \\
&= \int_0^\infty \mathrm{E}[\chi(W_1(t) \le u)]d\xi(t) \\
&= \int_0^\infty \mathrm{E}^*\left[\exp\left(\frac{\mu_1}{\sigma_1^2}W_1(t) - \frac{1}{2}\frac{\mu_1^2}{\sigma_1^2}t\right)\chi(W_1(t) \le u)\right]d\xi(t)
\end{aligned}
$$

$$= \int_0^\infty \int_{-\infty}^u \exp\left(\frac{\mu_1}{\sigma_1^2}x - \frac{1}{2}\frac{\mu_1^2}{\sigma_1^2}t\right) g(t,x)dxd\xi(t)$$
$$= \int_0^\infty \int_{-\infty}^u \phi(t) \exp\left(\frac{\mu_1}{\sigma_1^2}x - \frac{1}{2}\frac{\mu_1^2}{\sigma_1^2}t - \int_0^t \phi(y)dy\right) g(t,x)dxdt$$
$$= \int_{-\infty}^u \exp\left(\frac{\mu_1}{\sigma_1^2}x\right) \int_0^\infty \phi(t) \exp\left(-\frac{1}{2}\frac{\mu_1^2}{\sigma_1^2}t\right.$$
$$\left. - \int_0^t \phi(y)dy\right) g(t,x)dtdx. \tag{2.36}$$

Using the function $L_r(x)$, the mean time length when the system holds inventory for one cycle is given by $L_r(0) - L_r(S)$. Also, the sum of the expected holding cost and the expected shortage cost for one cycle is given by

$$I(r,S) = c_h\left(SL_r(0) - \int_0^S L_r(x)dx\right) + c_b \int_S^\infty L_r(x)dx. \tag{2.37}$$

Next, consider the cost incurred by failures and repairs for one cycle. The failure rate function $r(t) = dm(t)/dt = \kappa\omega t^{\omega-1}$ depends on the elapsed time in Phase II. Thus, it is seen that the expected number of failures for one cycle is

$$\mathrm{E}\left[\int_0^{\tau_2} r(t)dt\right] = \kappa \mathrm{E}[\tau_2^\omega]. \tag{2.38}$$

The ω-th order moment of τ_2 can be derived using the function $f_2(\cdot)$ in the following equation:

$$\mathrm{E}[\tau_2^\omega] = \lim_{\alpha\to 0} \frac{d^\omega}{d\alpha^\omega} \frac{1 - \alpha \mathrm{E}[f_2(-\infty, W_1(\tau_1);\alpha)]}{1 - \alpha f_2(-\infty, 0;\alpha)}. \tag{2.39}$$

However, because the procedure for deriving the higher order moments is quite complicated, we calculate the lower bound instead of deriving the exact moment. Applying Jensen's inequality and the balance equation in the $G/G/1$ queue $\rho\mathrm{E}[\tau_1] = (1-\rho)\mathrm{E}[\tau_2]$ (see [47]), we can derive the lower bound of the expected failure-repair cost as follows:

$$c_f \kappa \rho^\omega L_r(0)^\omega. \tag{2.40}$$

Consequently, from the well-known renewal reward theorem [40], the upper bound of cost effectiveness based on the diffusion approximation is given by

$$C_e(r,S) \le \frac{L_r(0) - L_r(S)}{c_K + c_m(1/\eta) + c_f\kappa\rho^\omega L_r(0)^\omega + I(r,S)}. \tag{2.41}$$

The optimal EMQ and PM policies can be approximated by (r^*, S^*), maximizing the lower bound of cost effectiveness. In fact, this approach can be used to approximate the optimal policies in some cases.

As a special case, consider the optimal PM level S^* for a fixed r. To calculate the optimal EMQ policy numerically, it is necessary to search the optimal solution S^* which maximizes the cost effectiveness in the range $0 < S < \infty$. More precisely, the Newton-Raphson method may be useful to find S^* for a fixed r. However, since the Newton-Raphson method strongly depends on the initial value, a globally optimum solution can not be necessarily guaranteed. Actually, the initial value always has to be adjusted carefully, and much computation effort will be required to maximize the cost effectiveness. Hence, it is important to effectively reduce the search space in order to find the optimal solution S^*. The following result provides a useful basis for improving the computation procedure of S^*.

Theorem 1 The lower bound of S^* is given by $\tilde{S}$ which satisfies

$$L_r(\tilde{S}) = \frac{c_h}{c_h + c_b} L_r(0). \tag{2.42}$$

Proof: Partially differentiating $I(r, S)$ with respect to S, we obtain

$$\frac{\partial}{\partial S} I(r, S) = c_h(L_r(0) - L_r(S)) - c_b L_r(S). \tag{2.43}$$

Since $L_r(S)$ is the strictly decreasing function of S, the stock level which minimizes $I(r, S)$ is defined by

$$\tilde{S} = \inf\{S > 0;\ c_h(L_r(0) - L_r(S)) - c_b L_r(S) = 0\}. \tag{2.44}$$

For any $\delta > 0$, let S_δ^- and S_δ^+ ($S_\delta^- < S_\delta^+$) be the solutions of

$$S_\delta^- = \inf\{S > 0; I(r, S) = I(r, \tilde{S}) + \delta\} \tag{2.45}$$

and

$$S_\delta^+ = \sup\{S > 0; I(r, S) = I(r, \tilde{S}) + \delta\}, \tag{2.46}$$

respectively. From the values of $C_e(r, S_\delta^-)$ and $C_e(r, S_\delta^+)$, it is found that $C_e(r, S_\delta^-) < C_e(r, S_\delta^+)$ since $L_r(0) - L_r(S)$ is the strictly increasing function of S. Thus, the lower bound is given by $C_e(r, \tilde{S}) < C_e(r, S_\delta^+)$, and the proof is completed. ■

Finally, we develop a computation algorithm to find the optimal policy (r^*, S^*) as follows.

Algorithm Approximation:

(Step 1) $r := 1$

(Step 2) Calculate the lower bound $\tilde{S}_1$ for $r = 1$

(Step 3) Find the optimal PM level S_1 for $S_1 > \tilde{S}_1$

(Step 4) $C_{max} := C_e(1, S_1)$

(Step 5) $r := r + 1$

(Step 6) Calculate the lower bound $\tilde{S}_r$ for renewed r

(Step 7) Find the optimal PM level S_r for $S_r > \tilde{S}_r$

(Step 8) If $C_{max} < C_e(r, S_r)$, then $C_{max} := C_e(r, S_r)$ and go to Step 5.

(Step 9) $(r^*, S^*) := (r - 1, S_{r-1})$.

Like the Algorithm Poisson in the previous section, note that the above computational algorithm does not always guarantee the global optimum. However, this algorithm is quite simple and can be used to calculate the global optimal policy if cost effectiveness is a unimodal function of r and S.

5. PHASE APPROXIMATION

As mentioned in the previous section, the diffusion approximation does not function well if the utilization factor ρ is far from unity. Also, since the diffusion approximation is based on only the first two moments of the underlying stochastic counting process, estimation errors may increase as the tail of the probability distribution function for the renewal process spreads. To the best of our knowledge, phase approximation is the most applicable approximation method for discrete event systems. In fact, it is known that the renewal counting process can be approximated at arbitrary accuracy by the phase-type renewal process [1, 21, 31, 32, 33]. The problem of this approximation method is that it requires too much computation time if the number of phases for the phase-type renewal process increases.

Let $\{X_n;\ n \geq 1\}$ denote a sequence of i.i.d. random variables which represents an inter-arrival time between the $(n-1)$-st and n-th demands, where X_n has the arbitrary probability distribution function $F(t)$, the mean $\mathrm{E}[X_n] = 1/\lambda$, and the variance $\mathrm{Var}[X_n] = \sigma_D^2$. First, we describe the phase type renewal process. Consider a Markov process for the state space $\{1, 2, \cdots, m+1\}$, where $\{1, 2, \cdots, m\}$ denotes the transient states (phases), and $\{m+1\}$ is the absorbing one. The initial probability vector for the Markov process is given by $(\boldsymbol{\alpha}, 0)$. Until absorption in the state $m+1$, the process behaves similar to a Markov process with an infinitesimal generator T, where T is a matrix with components λ_{ij} (> 0), $1 \leq i, j \leq m$, $j \neq i$ and $-\lambda_{ii}$ (< 0). In our model, absorption implies the occurrence of a demand. After entering the absorption, the process is restarted with an initial state having the probability vector $\boldsymbol{\alpha}$. The inter-arrival time distribution is called the *phase-type distribution* with parameter $(\boldsymbol{\alpha}, T)$, so that

$$F_{PH}(t) = 1 - \boldsymbol{\alpha} \exp(Tt)\boldsymbol{e}, \tag{2.47}$$

where $\boldsymbol{e}$ is a column vector of 1s.

Let us now return to the phase approximation. N_t and J_t are the cumulative number of demands and the phase in demand at time t, respectively, where the phase can be interpreted as the state of various factors which cause the occurrence of demands. Define the transition probability

$$P_{ij}(n,t) = \Pr\{N_t = n, J_t = j | N_0 = 0, J_0 = i\} \tag{2.48}$$

and the matrix $P(n,t)$ with components $P_{ij}(n,t)$. Since both the number of demands N_t and the phase process J_t construct an embedded Markov chain at the points of the completion of production, it can be seen that the underlying virtual stock level process is regarded as a *PH/G/1* queueing process with N-policy [22].

The expected number of minimal repairs for one cycle, the expected inventory holding and shortage costs incurred for one cycle, and the mean time length of one cycle are defined as $M_{r,S}$, $I_{r,S}$ and $L_{r,S}$, respectively. For the virtual stock level, the transition probability matrix is given by

$$P = \begin{bmatrix} B_0 & B_1 & B_2 & \cdots \\ A_0 & A_1 & A_2 & \cdots \\ O & A_0 & A_1 & \cdots \\ O & O & A_0 & \cdots \\ \cdot & \cdot & \cdot & \\ \cdot & \cdot & \cdot & \end{bmatrix}, \tag{2.49}$$

where A_n and B_n are the $m \times m$ matrices with components $[A_n]_{ij}$ and $[B_n]_{ij}$, respectively, for $n \geq 0$, and where O is a zero matrix. The component $[A_n]_{ij}$ is the probability that the phase moves from i to j and that n demands occur during a single production. Thus, it is straightforward to obtain

$$A_n = \int_0^\infty P(n,t) dH(t). \tag{2.50}$$

The component $[B_n]_{ij}$ is also the probability of a transition from the phase i to j and that n items remain in the buffer. Hence, the matrix B_n is given by

$$B_n = \begin{cases} O, & 0 \leq n < r-1, \\ \sum_{k=0}^{r-1} \xi_k A^0 A_{n-(r-1)} + \sum_{k=r}^{\infty} \xi_k A_{n-k+1}, & n \leq r-1, \end{cases} \tag{2.51}$$

where $A^0 = \boldsymbol{e\alpha}$, and

$$\xi_k = \int_0^\infty P(k,x)d\xi(x). \tag{2.52}$$

In the above equations, ξ_k is the probability that k demands occur during the PM period.

(i) Phase probability in the steady-state

From the property of phase-type distribution, we can derive the probability vector of phases when the virtual stock level is zero. More precisely, it is calculated as the eigenvector of the matrix $\sum_{n=0}^{\infty} B_n G^n$, namely,

$$\begin{aligned} \boldsymbol{\pi} &= \boldsymbol{\pi} \sum_{n=0}^{\infty} B_n G^n \\ &= \boldsymbol{\pi} \left(\sum_{k=0}^{r-1} \xi_k A^0 G^r + \sum_{k=r}^{\infty} \xi_k G^k \right), \end{aligned} \tag{2.53}$$

where the matrix G is given by $G = \sum_{n=0}^{\infty} A_n G^n$. In the queueing literature, the matrix G is known as the *fundamental matrix*, indicating that the phase moves from i to j with the probability $[G]_{ij}$, while the stock level decreases one item at a time. The computation algorithm for the fundamental matrix was developed by Lucantoni and Ramaswami [28], Ramaswami [39] and Takine *el al.* [48]. Below, we give a summary of the algorithm. Let $G_0 = O$. For $n = 0, 1, \cdots$, compute

$$G_{n+1} = \int_0^\infty \exp\{(C + DG_n)x\} dH(x). \tag{2.54}$$

It can be proven that G_n converges to G ultimately. Thus, the matrix G can be computed by applying Equation (2.54) repeatedly.

(ii) Mean time length of one cycle

From the balance equation arising in the queueing theory, we obtain

$$\mathrm{E}[\tau_2]/(\mathrm{E}[\tau_1] + \mathrm{E}[\tau_2]) = \rho. \tag{2.55}$$

Hence, the mean time length of one cycle is as follows:

$$\begin{aligned} L_{r,S} &= \frac{1}{1-\rho}\mathrm{E}[\tau_1] \\ &= \frac{1}{1-\rho}\left\{\frac{1}{\eta} - \sum_{k=0}^{r-1}\pi\xi_k\left[I + (r-k-1)A^0\right]T^{-1}e\right\}. \end{aligned} \quad (2.56)$$

(iii) Lower bound of the expected number of minimal repairs for one cycle

Let $m(t)$ denote the expected number of minimal repairs up to time t. Thus, the expected number of minimal repairs for one cycle is given by

$$M_{r,S} = \int_0^\infty m(t)\Pr\{\tau_2 \in dt\}. \quad (2.57)$$

Since the failure rate function is the Weibull-type, $r(t) = \kappa\omega t^{\omega-1}$, the expected number of minimal repairs becomes

$$M_{r,S} = \int_0^\infty m(t)\Pr\{\tau_2 \in dt\} = \kappa\mathrm{E}[\tau_2^\omega]. \quad (2.58)$$

In a way similar to the diffusion approximation, the higher order moments of τ_2 can be derived in principle, but the forms are not explicit. Since the computation procedure is also quite complicated, the lower bound will be useful to approximate the expected number of minimal repairs. Thus, we derive the lower bound from Jensen's inequality as

$$\mathrm{E}[\tau_2^\omega] \geq \mathrm{E}[\tau_2]^\omega. \quad (2.59)$$

From Equation (2.59), the lower bound of $M_{r,S}$ is given by

$$M_{r,S} \geq \kappa\left(\rho L_{r,S}\right)^\omega. \quad (2.60)$$

(iv) Upper bound of the inventory holding cost and shortage cost in the steady-state

Let N denote the virtual stock level at the completion point of a production in the steady-state. Although, in the case of Poisson demands, it is found that the stock level distribution at the completion of prooduction is the same as that at an arbitrary time because of the *PASTA* property, this property does not always hold in the case of i.i.d. demands. However, in the queueing literature, the *PASTA* property can be assumed to hold approximately for a queueing system with non-homogeneous arrivals (see [41]), and is used to develop the approximation formulae for several queueing systems. Accordingly, we approximate the virtual stock level at arbitrary time points by using N. The inventory holding cost in the steady-state is derived approximately as follows:

$$\begin{aligned} I_{r,S} &\approx c_h \sum_{n=0}^{S} (S-n)\Pr\{N=n\} + c_b \sum_{n=S+1}^{\infty} (n-S)\Pr\{N=n\} \\ &= c_h(S-\mathrm{E}[N]) + (c_h+c_b)\sum_{n=S}^{\infty} \Pr\{N>n\}, \end{aligned} \tag{2.61}$$

where $\mathrm{E}[N]$ is the expected stock level in the steady-state. It is known that the probability $\Pr\{N = n\}$ can be calculated by the following algorithm [39]. Let $\boldsymbol{x}_n = (\Pr\{N=n,\ J=1\},\cdots,\Pr\{N=n,\ J=m\})$ denote the probability row vector in the steady-state. Thus, we have the following relationship:

$$\boldsymbol{x}_i = \left(\boldsymbol{x}_0 \overline{B}_i + \sum_{j=1}^{i-1} \boldsymbol{x}_j \overline{A}_{i-j+1} \right) (I - \overline{A}_1)^{-1}, \tag{2.62}$$

where

$$\overline{A}_i = \sum_{k=i}^{\infty} A_k G^{k-i}, \tag{2.63}$$

and

$$\overline{B}_i = \sum_{k=i}^{\infty} B_k G^{k-i}. \tag{2.64}$$

In what follows, we derive the approximation form of the function $I_{r,S}$. From the Markov inequality, we obtain

$$\Pr\Big\{N > \mathrm{E}[N] + c\sqrt{\mathrm{Var}[N]}\Big\} \le \frac{1}{1+c^2} \tag{2.65}$$

for arbitrary $c > 0$. Hence, the upper bound of the inventory holding cost in the steady-state, under the assumption $S > \mathrm{E}[N]$, is given by

$$\begin{aligned} I_{r,S} &\le c_h(S-\mathrm{E}[N]) \\ &\quad + (c_h+c_b)\sum_{n=S}^{\infty} \frac{\mathrm{Var}[N]}{\mathrm{Var}[N] + (n-\mathrm{E}[N])^2}. \end{aligned} \tag{2.66}$$

Letting $\boldsymbol{X}(z)$ be the vector generating function of N, it can be seen that [22]

$$\boldsymbol{X}(z) = \boldsymbol{\pi} B(z) + [\boldsymbol{X}(z) - \boldsymbol{\pi}]\frac{A(z)}{z}, \tag{2.67}$$

where $B(z)$ and $A(z)$ are the matrix generating functions of B_n and A_n, respectively. Thus, E$[N]$ and Var$[N]$ become

$$\mathrm{E}[N] = \frac{d}{dz}\boldsymbol{X}(z)\bigg|_{z\to 1} \boldsymbol{e}, \tag{2.68}$$

$$\mathrm{Var}[N] = \frac{d^2}{dz^2}\boldsymbol{X}(z)\bigg|_{z\to 1} \boldsymbol{e} + \mathrm{E}[N] - \mathrm{E}[N]^2. \tag{2.69}$$

Finally, the approximation form of cost effectiveness is given by

$$\tilde{C}_e(r,S) \approx \frac{(S - \mathrm{E}[N])^2 L_{r,S}}{\{\mathrm{Var}[N] + (S - \mathrm{E}[N])^2\}\tilde{C}(r,S)}, \tag{2.70}$$

where

$$\begin{aligned}\tilde{C}(r,S) &= \frac{c_m/\eta + c_K}{L_{r,S}} + c_f \kappa \rho^{\omega} L_{r,S}^{\omega-1} + c_h(S - \mathrm{E}[N]) \\ &\quad + (c_h + c_b)\sum_{n=S}^{\infty} \frac{\mathrm{Var}[N]}{\mathrm{Var}[N] + (n - \mathrm{E}[N])^2}.\end{aligned} \tag{2.71}$$

Thus, the approximated optimal policy becomes

$$(r^*, S^*) \approx \mathrm{argmax}\, \tilde{C}_e(r,S). \tag{2.72}$$

In a fashion similar to that in the previous section, we develop the lower bound of the optimal PM level for a fixed r. Since S is assumed to be distinctly larger than E$[N]$, the search space of S^* is limited to the region E$[N] < S < \infty$. The following result gives the lower bound of the optimal PM level under the phase approximation.

Theorem 2 The lower bound of S^* is given by

$$S^* > \mathrm{E}[N] + \sqrt{\frac{c_b}{c_h}\mathrm{Var}[N]}. \tag{2.73}$$

Proof: Define

$$q(S) = \tilde{C}(r, S+1) - \tilde{C}(r,S)$$

$$= \quad c_h - (c_h + c_b)\frac{\text{Var}[N]}{\text{Var}[N] + (S - \text{E}[N])^2}. \qquad (2.74)$$

By differentiating Equation (2.74) with respect to S, we obtain the level which minimizes $\tilde{C}(r, S)$ as

$$S = \text{E}[N] + \sqrt{(c_b/c_h)\text{Var}[N]}. \qquad (2.75)$$

If there exists an optimal PM level $\hat{S}$ for

$$\text{E}[N] < \hat{S} < \text{E}[N] + \sqrt{(c_b/c_h)\text{Var}[N]}, \qquad (2.76)$$

then it can be seen that there exists $\hat{S}'$ satisfying $\tilde{C}(r, \hat{S}') = \tilde{C}(r, \hat{S})$ for $\text{E}[N] + \sqrt{(c_b/c_h)\text{Var}[N]} < \hat{S}'$. Since $(S - \text{E}[N])^2/\{\text{Var}[N] + (S - \text{E}[N])^2\}$ is strictly increasing in S, it is clear that $\tilde{C}_e(r, \hat{S}') > \tilde{C}_e(r, \hat{S})$. Thus, the optimal PM level S^* satisfies Equation (2.73), and the proof is completed. ■

6. NUMERICAL ILLUSTRATIONS

6.1 Case of Poisson Process Demand.

First, we consider cost-effective EMQ and PM policies in accordance with Poisson demands and compare then with the policy in Srinivasan and Lee [46]. The production time and PM time distributions are

$$H(t) = \int_0^t \frac{\alpha^\alpha x^{\alpha-1}}{\Gamma(\alpha)} e^{-\alpha x} dx, \qquad \alpha = 0.5,\ 2.0 \qquad (2.77)$$

and

$$\xi(t) = 1 - e^{-0.5t}, \qquad (2.78)$$

respectively, where $\Gamma(\cdot)$ is the standard gamma function. These imply that production time obeys a gamma distribution with the scale parameter α having a fixed mean 1.0, and that the PM time obeys an exponential distribution with mean 2.0. The cost parameters are as follows:

$$\begin{array}{lll} c_K = 300.0, & c_f = 50.0, & c_m = 100.0, \\ c_b = 20.0, & c_h = 10.0, & \kappa = 0.05, \\ \omega = 2.0. & & \end{array} \qquad (2.79)$$

In Tables 2.1-2.3, the optimal PM policies for $\lambda = 0.1, 0.9$ and $\alpha = 0.5, 2.0$ are presented. Also, the optimal EMQ and PM policies are calculated in Table 2.4. From Table 2.4, it is observed that the optimal EMQ policies are very sensitive to the traffic intensity ρ. In contrast, the shape parameters of service time have little effect on the optimal

EMQ policies. Specifically, in the traffic light case ($\rho = 0.1$), r^* is almost equal to S^*. This is because the arrival rate is relatively small; that is, many demands do not arrive during the PM period. Therefore, in the case of heavy traffic ($\rho = 0.9$), the difference between r^* and S^* is remarkable. In other words, the optimal EMQ policies should be designed such that the stock level is close to zero at the completion of PM. It can be concluded that, according to the cost effectiveness criterion, the stock level at the completion of PM strongly depends on the EMQ decision with PM and is not to be ignored. Table 2.5 shows the comparison with the policies of Srinivasan and Lee [46]. These results indicate that the optimal PM policies based on cost effectiveness tend to be sensitive to a shortage in all cases. In other words, the optimal production restarting levels, $s = S - r$, for cost effectiveness are higher than those for expected per unit cost in the steady-state.

Table 2.1 The PM policy under IFR failure distribution ($\rho = 0.1, \alpha = 2.0$).

r	S^*	$C_e(r, S^*)$
1	1	0.0159261
2	2	0.0238555
3	3	0.0252541
4	4	0.0240324

Table 2.2 The PM policy under IFR failure distribution ($\rho = 0.9, \alpha = 2.0$).

r	S^*	$C_e(r, S^*)$
1	27	0.0005079
2	27	0.0004918

Table 2.3 The PM policy under DFR failure distribution ($\rho = 0.1, \alpha = 0.5$).

r	S^*	$C_e(r, S^*)$
1	2	0.0102524
2	2	0.0109557
3	3	0.0095634

Table 2.4 The optimal EMQ and PM policies and their associated cost effectiveness.

ρ	α	r^*	S^*	cost effectiveness
0.1	2.0	3	3	0.0252541
0.9	2.0	1	27	0.0005079
0.1	0.5	2	2	0.0109557

Table 2.5 Comparison between the optimal PM policies based on cost effectiveness (CE) and expected cost per unit time in the steady-state (CU).

		CE		CU	
ρ	α	r^*	S^*	r^*	S^*
0.1	2.0	3	3	4	2
0.9	2.0	1	27	1	10
0.1	0.5	2	2	4	2

6.2 Approximation Performance. Next, we investigate the accuracy of two approximation methods. The inter-arrival time of demands, production time, and PM time are distributed as

$$F(t) = 1 - e^{-(\beta t)^m}, \qquad \beta = \rho\Gamma\left(1 + \frac{1}{2.0}\right), \tag{2.80}$$

$$H(t) = \int_0^t \frac{\alpha^\alpha x^{\alpha-1}}{\Gamma(\alpha)} e^{-\alpha x} dx, \qquad \alpha = 2.0 \tag{2.81}$$

and

$$\xi(t) = 1 - e^{-0.5t}, \tag{2.82}$$

respectively. That is, we assume that the inter-arrival of demands obeys a Weibull distribution having a mean ρ and scale parameter 2.0, production time obeying an Erlang-2 distribution with a mean 1.0, and the PM time length obeying an exponential distribution with mean 2.0. The cost parameters are as follows:

$$\begin{aligned} &c_K = 300.0, \quad c_f = 50.0, \quad c_m = 100.0, \\ &c_b = 20.0, \qquad c_h = 10.0, \quad \kappa = 0.05, \\ &\omega = 2.0. \end{aligned} \tag{2.83}$$

Consider the diffusion approximation. Tables 2.6-2.8 present the optimal PM policies for $\rho = 0.1, 0.5, 0.9$. From these results, the approximated optimal EMQ policy with PM for $\rho = 0.1, 0.5, 0.9$, based on the diffusion approximation, are $(r^*, S^*) = (3,3), (4,4), (1,11)$, respectively.

Table 2.6 The PM policy based on diffusion approximation ($\rho = 0.1$).

r	S^*	lower bound	$\tilde{C}_e(r, S^*)$
1	1	0.96	0.208851
2	2	1.90	0.303746
3	3	2.83	0.320416
4	4	3.76	0.302598

Table 2.7 The PM policy based on diffusion approximation ($\rho = 0.5$).

r	S^*	lower bound	$\tilde{C}_e(r, S^*)$
1	3	1.89	0.071590
2	3	2.35	0.106854
3	4	3.09	0.122341
4	4	3.96	0.125030
5	5	4.89	0.122609

Table 2.8 The PM policy based on diffusion approximation ($\rho = 0.9$).

r	S^*	lower bound	$\tilde{C}_e(r, S^*)$
1	11	9.53	0.049964
2	11	9.74	0.044568

Similarly, consider the phase approximation. In Tables 2.9-2.11, we calculate the optimal PM policy for $\rho = 0.1, 0.5, 0.9$. From these results, the approximated optimal EMQ with PM for $\rho = 0.1, 0.5, 0.9$, based on the phase approximation, are $(r^*, S^*) = (3,1), (4,6), (5,8)$, respectively.

We compare the two approximated PM policies with the results based on the simulation. For $r = 1, \cdots, 20$ and $S = 1, \cdots, 20$, the Monte Carlo simulation was performed to calculate the optimal EMQ policy with PM.

Table 2.9 The PM policy based on phase approximation ($\rho = 0.1$).

r	S^*	lower bound	$\tilde{C}_e(r, S^*)$
1	1	0.2	0.01976258
2	1	0.4	0.02928690
3	1	0.7	0.03173251
4	2	1.0	0.03036237

Table 2.10 The PM policy based on phase approximation ($\rho = 0.5$).

r	S^*	lower bound	$\tilde{C}_e(r, S^*)$
1	5	2.6	0.007405181
2	5	2.6	0.009238761
3	5	3.0	0.010356031
4	6	3.5	0.010692222
5	7	4.1	0.010638699

Table 2.11 The PM policy based on phase approximation ($\rho = 0.9$).

r	S^*	lower bound	$\tilde{C}_e(r, S^*)$
1	6	3.2	0.00645242
2	6	3.2	0.00792299
3	6	3.5	0.00896993
4	7	4.1	0.00939465
5	8	4.7	0.00948017
6	9	5.4	0.00929416

In Table 2.12, the optimal EMQ policies with PM based on the simulation are presented for $\rho = 0.1, 0.5, 0.9$, where the optimal solution was calculated heuristically by simulating all sample paths for an inventory level process with breakdown. The value of maximum cost effectiveness resulting from the simulation is presented in Table 2.12. The optimal pair (r^*, S^*) in this table are calculated by applying the Algorithm Approximation. In these results, it is found that the approximated optimal EMQ policy based on the diffusion approximation is not extremely good, compared to that based on the phase approximation, especially in the case of $\rho = 0.9$. In other words, the optimal EMQ policies based on the diffusion approximation are similar to the results in the Poisson demand case (see Table 2.4). This is the reason why the diffusion coefficients are time homogeneous. In fact, it is known that the accuracy of the diffusion approximation is superior if one approximates the transient probability in the $M/M/1$ queue. It should be noted that the case in this chapter corresponds to non-homogeneous arrival and service processes. For this reason, it can be concluded that a diffusion approximation is not effective for approximating the optimal EMQ policies. That is, further approximation schemes using the diffusion process with non-homogeneous drift parameters should be developed. On the other hand, from the results in Table 2.12, it is seen that the errors in the cost effectiveness with phase approximation are 0.017553, 0.004448, and 0.000482, respectively, for $\rho = 0.1, 0.5, 0.9$. Therefore, we can conclude that phase approximation may be useful to calculate the optimal EMQ policies with PM in the case where the utilization factor is relatively large.

7. CONCLUSION

In this paper, we have considered optimal EMQ policies with PM according to the cost effectiveness criterion. In the case of Poisson demand, we have reformulated the Srinivasan and Lee model [46] based on cost effectiveness. For the general case with a renewal demand pattern, we have developed two approximation methods: diffusion approximation and phase approximation. In the numerical examples, we have calculated the optimal policy numerically and have shown that phase approximation could provide good performance in the case with a relatively larger utilization factor. Note that the method developed in this article is based on an approximation with an upper bound. Actually, since production machine failure can not be expected in advance, it will be necessary to determine the optimal EMQ and PM policies under an adoptive circumstance in actual Production/inventory management.

Table 2.12 The comparison between optimal EMQ policies with PM based on the simulation.

ρ	method	r^*	S^*	cost effectiveness
0.1	simulation	3	3	0.027256
	diffusion	3	3	0.027256
	phase	3	1	0.009703
0.5	simulation	6	6	0.018137
	diffusion	4	4	0.015607
	phase	4	6	0.013689
0.9	simulation	5	7	0.006283
	diffusion	1	11	0.004303
	phase	5	8	0.005801

Thus, the computational algorithm for the optimal policy should be improved in terms of computation time as well as accuracy.

Acknowledgments

This work was partially supported by a Grant-in-Aid for Scientific Research from the Ministry of Education, Sports, Science and Culture of Japan under Grant No. 10558059 and No. 11780332. Also, the second author was supported by the Telecommunication Advancement Foundation, Tokyo, Japan. The authors thank both editors, Dr. M. A. Rahim and Dr. M. Ben-Daya, for their kind invitation to contribute to this volume and the three anonymous referees for their valuable comments.

References

[1] Altiok, T. On the phase-type approximations of general distributions. *IIE Transactions*, **17** (1985) 110-116.

[2] Axäter, S. and Elmaghraby, S. E. A note on EMQ under learning and forgetting. *IIE Transactions*, **13** (1981) 86-90.

[3] Banerjee, A. A joint economic-lot-size model for purchaser and vendor. *Decision Science*, **17** (1986) 292-311.

[4] Bell, C. E. Characterization and computation of optimal policies for operating an *M/G/1* queueing system with removable server. *Operations Research*, **19** (1971) 208-218.

[5] Berg, M., Posner, M. J. M. and Zhao, H. Production-inventory systems with unreliable machines. *Operations Research*, **42** (1994) 111-118.

[6] Brown, R. M., Conine, T. E. and Tamarkin, M. A note on holding costs and lot-size errors. *Decision Science*, **17** (1986) 24-42.

[7] Dohi, T. and Osaki, S. The optimal policy for an economic manufacturing model with stochastic machine breakdown and repair. *Proceedings of the 6th IFIP TC5/WG5.7 International Conference on Advances in Production Management Systems*, November, Kyoto, Japan, (1996) 225-230.

[8] Dohi, T., Yamada, Y., Kaio, N. and Osaki, S. The optimal lot sizing for unreliable economic manufacturing models. *International Journal of Reliability, Quality and Safety Engineering*, **4** (1997) 413-426.

[9] Dohi, T., Yun, W. Y., Kaio, N. and Osaki, S. Optimal design of economic manufacturing process with machine failure rate depending on production speed. *Proceedings of the International Symposium on Manufacturing Strategy'98*, November, Tokyo, Japan, (1998) 404-409.

[10] Dohi, T., Kaio, N. and Osaki, S. Minimal repair policies for an economic manufacturing process. *Journal of Quality in Maintenance Engineering*, **4** (1998) 248-262.

[11] Dohi, T., Okamura, H. and Osaki, S. Optimal order limit policies for an (r, Q) inventory system. *IMA Journal of Mathematics Applied in Business and Industry*, **10** (1999) 127-145.

[12] Dohi, T., Okamura, H. and Osaki, S. Optimal control of preventive maintenance schedule and safety stocks in an unreliable manufacturing environment. to appear in *International Journal of Production Economics*, (2000).

[13] Giorno, V., Nobile, A. G. and Ricciardi, L. M. On some diffusion approximations to queueing systems. *Advances in Applied Probability*, **18** (1986) 991-1014.

[14] Giorno, V., Nobile, A. G. and Ricciardi, L. M. On some time-non-homogeneous diffusion approximations to queueing systems. *Advanced in Applied Probability*, **19** (1987) 974-994.

[15] Goyal, S. K. An inventory model for a product for which purchase price fluctuates. *New Zealand Operations Research*, **30** (1979) 43-53.

[16] Groenevelt, H., Pintelon, L. and Seidmann, A. Production lot sizing with machine breakdowns. *Management Science*, **38** (1992) 104-123.

[17] Groenevelt, H., Pintelon, L. and Seidmann, A. Production batching with machine breakdowns and safety stocks. *Operations Research*, **40** (1992) 959-971.

[18] Harrison, M. J. *Brownian Motion and Stochastic Flow Systems*, John Wiley & Sons, New York 1985.

[19] Heyman, D. P. Optimal operating policies for *M/G/1* queueing systems. *Operations Research*, **16** (1968) 362-383.

[20] Jesse, R. R., Mitra, A. and Cos, J. F. EOQ formula: is it valid under inflationary conditions?. *Decision Science*, **14** (1983) 370-374.

[21] Kao, E. P. C. Computing the phase-type renewal and related function. *Technometrics*, **30** (1988) 87-93.

[22] Kasahara, S., Takine, T., Takahashi, Y. and Hasegawa, T. *MAP/G/1* queues under *N*-policy with and without vacations. *Journal of the Operations Research Society of Japan*, **39** (1996) 188-212.

[23] Kella, O. The threshold policy in the *M/G/1* queue with server vacations. *Naval Research Logistics Quarterly*, **36** (1989) 111-123.

[24] Kimura, T. Optimal control of an *M/G/1* queueing system with removable server via diffusion approximation. *European Journal of Operational Research*, **8** (1981) 390-398.

[25] Kobayashi, H. Application of the diffusion approximation to queueing networks I: equilibrium queue distributions. *Journal of the ACM*, **21** (1974) 316-328.

[26] Kuzdrall, P. and Britney, R. Total setup lot sizing with quantity discounts. *Decision Science*, **13** (1982) 101-112.

[27] Lee, H. L. and Rosenblatt, M. J. Simultaneous determination of production cycle and inspection schedules in a production system. *Management Science*, **33** (1987) 1125-1136.

[28] Lucantoni, D. M. and Ramaswami, V. Efficient algorithms for solving the nonlinear matrix equations arising in phase type queues. *Stochastic Models*, **1** (1985) 29-52.

[29] Murdeshwar, T. M. and Sathe, Y. S. Some aspects of lot size models with two levels of storage. *Opsearch*, **22** (1985) 255-262.

[30] Muth, E. J. and Spremann, K. Learning effects in economic lotsizing. *Management Science*, **29** (1983) 264-269.

[31] Neuts, M. F. Renewal processes of phase type. *Naval Research Logistics Quarterly*, **25** (1978) 445-454.

[32] Neuts, M. F. *Matrix-Geometric Solutions in Stochastic Models: An Algorithmic Approach*, Johns Hopkins University Press, Baltimore, 1981.

[33] Neuts, M. F. (1989). *Structured Stochastic Matrices M/G/1 Type and Their Applications*, Marcel Dekker, New York.

[34] Okamura, H., Dohi, T. and Osaki, S. Optimal timing strategies for controlled *M/G/1* queueing system via diffusion approximation approach. *Proceedings of 28th International Symposium on Stochastic Systems Theory and Its Applications*, November, Kyoto, Japan, (1996) 143-148.

[35] Okamura, H., Hamamoto, K., Dohi, T. and Osaki, S. The diffusion approximation method for the $G/G/1$ type of economic manufacturing quantity process with preventive maintenance. *Proceedings of The 2nd Asia-Pacific Conference on Industrial Engineering and Management Systems*, October, Kanazawa, Japan, (1999) 141-144.

[36] Okamura, H., Dohi, T. and Osaki, S. Optimal policies for a controlled queueing system with removable server under a random vacation circumstance. *Computers & Mathematics with Applications*, **39** (2000) 215-227.

[37] Okamura, H., Hamamoto, K., Dohi, T. and Osaki, S. The optimal preventive maintenance schedule for an economic manufacturing quantity process: phase type approximation method. *Proceedings of 2nd International Conference on Mathematical Methods in Reliability*, July, Bordeaux, France, (2000) 832-835.

[38] Peterson, R. and Silver, E. A. *Decision System for Inventory Management and Production Planning*, John Wiley & Sons, New York, 1979.

[39] Ramaswami, V. Stable recursion for the steady state vector in Markov chains of *M/G/1* type. *Stochastic Models*, **4** (1988) 183-188.

[40] Ross, S. M. *Applied Probability Models with Optimization Applications*, Holden-Day, San Francisco, 1970.

[41] Sakasegawa, H., Miyazawa, M. and Yamazaki, G. Evaluating the overflow probability using the infinite queue. *Management Science*, **39** (1984) 1238-1245.

[42] Sethi, S. P. A quantity discount model with disposals. *IIE Transactions*, **22** (1984) 31-39.

[43] Shah, Y. K. An order-level lot-size model for deteriorating items. *IIE Transactions*, **9** (1977) 108-112.

[44] Sherif, Y. and Smith, M. L. Optimal maintenance models for systems subject to failure — review. *Naval Research Logistics Quarterly*, **28** (1981) 47-74.

[45] Sobel, M. Optimal average cost policy for a queue with start-up and shut-down costs. *Operations Research*, **17** (1969) 145-162.

[46] Srinivasan, M. M. and Lee, H.-S. Production-inventory systems with preventive maintenance. *IIE Transactions*, **28** (1996) 879-890.

[47] Takagi, H. (1991). *Queueing Analysis, Vol. 1*, Elsevier, Amsterdam.

[48] Takine, T., Matsumoto, Y., Suda, T. and Hasegawa, T. Mean waiting times in nonpreemptive priority queues with Markovian arrival and i.i.d service processes. *Performance Evaluation*, **20** (1994) 131-149.

[49] Valdez-Flores, C. and Feldman, R. M. A survey of preventive maintenance models for stochastically deteriorating single-unit systems. *Naval Research Logistics*, **36** (1989) 419-446.

[50] Wolff, R. W. Poisson arrivals see time averages. *Operations Research*, **30** (1982) 223-231.

[51] Yadin, M. and Naor, P. Queueing systems with a removable service station. *Operational Research Quarterly*, **14** (1963) 393-405.

Chapter 3

OPTIMAL CONTROL POLICY FOR A GENERAL EMQ MODEL WITH RANDOM MACHINE FAILURE

V. Makis and X. Jiang

Department of Mechanical and Industrial Engineering
University of Toronto
5 King's College Road,
Toronto, Canada M5S 3G8

Abstract In this paper, we present a general EMQ model with joint production /maintenance /quality control. We assume that the production facility is subject to failure and that the production process may shift to an out-of-control state randomly, resulting in a higher proportion of defective items produced.

The model includes, as special cases, many of the models considered previously in the literature that focus on a particular class of policies, derive the cost function, and find the best policy in that class without investigating the structure of the optimal policy generally.

Our objective is to investigate the problem of determining jointly the optimal lot sizing, preventive maintenance, and inspection policy minimizing the expected average cost per unit time. The problem is formulated and analyzed in the framework of semi-Markov decision processes and renewal reward processes. An algorithm is developed for computing the optimal policy and the optimal expected average cost, and a numerical example is presented to illustrate the computational algorithm.

Keywords: EMQ model, production-inventory control, quality control, preventive replacement.

1. INTRODUCTION

Economic manufacturing quantity models (EMQs) have been investigated in the literature for several decades. The classical EMQ model assumes constant production and demand rate, infinite horizon, linear inventory holding cost, a failure-free production facility, and a production process that is always in control, producing good quality items. The model is widely used in industry although some of the assumptions are not very realistic and rarely occur in practice.

Recently, considerable attention has been paid to EMQ models with an unreliable production facility (joint production and maintenance control) and to the models with production processes subject to random deterioration (joint production and quality control).

Joint production and maintenance control problems have been studied recently by Groenevelt *et al.* (1992a, 1992b), Tse and Makis (1994) and Makis (1998a), among others. Groenevelt *et al.* studied two extensions of the classical EMQ model. In their first paper (1992a), they assumed that the production unit is maintained at failure time or after completing a production run, whichever occurs first. It was assumed that the maintenance operation brings the machine to the as-good-as new state. In their second paper (1992b), they considered an EMQ model with exponentially distributed failure times, generally distributed repair times, and safety stocks used to meet a prescribed service level. In both papers, the objective was to find the optimum lot size.

Tse and Makis (1994) studied an EMQ model with preventive replacement and two types of machine failure, major and minor. When a major failure occurs, the failed unit is replaced by a new one, and the interrupted lot is aborted. Minor failure is corrected with minimal repair which takes negligible time, and production is resumed after the repair.

Makis (1998a) investigated an EMQ model with lost sales, machine breakdowns, and minimal repairs. He proved that the optimal preventive replacement is an age replacement and that the optimal lot size is generally a function of the operating age of the machine.

He showed that for an EMQ model with lost sales and a constant lot size, a good approximation to the optimal policy can be found by solving two nonlinear equations.

The effect of process deterioration on the optimal lot size was first investigated by Rosenblatt and Lee (1986) and Porteus (1986). In both papers, authors assume that the production process is subject to random deterioration, and no process monitoring was considered. They found that the optimal lot size in their models is smaller than that in the classical EMQ model. Lee and Rosenblatt (1987) and Lee and Park (1991) considered the situation where the state of the process can be monitored through inspections. They assumed that in-control periods were distributed exponentially and that inspections were perfect. Lee and Rosenblatt (1987) considered the cost of defectives to be the same regardless of whether the nonconforming item was reworked or sold, and Lee and Park (1991) considered a warranty cost incurred by selling defective items. The objective in both papers was the joint determination of the optimal lot size and the number of periodic inspections in a production cycle.

Makis (1998b) studied the problem of the joint determination of the production quantity and inspection schedule for an EMQ model with imperfect inspections. He investigated both discrete and continuous cases, and in the exponential case, he found the structure of the optimal inspection policy using Lagrange's method and showed that the optimal inspection times can be found by solving a nonlinear equation. Makis and Fung (1995 and 1998) studied the effect of machine failures and preventive maintenance on optimal lot size and an inspection schedule. They found that preventive maintenance considerably reduces the dependency of the lot size on the frequency of machine failures and also stabilizes the inspection schedule.

Joint control chart design for process monitoring and the preventive replacement problem was considered by Rahim and Banerjee (1993) and Rahim(1994). Rahim and Banerjee (1993) investigated the problem of determining jointly the optimal design parameters of an $\overline{X}$ control chart and a preventive replacement time for a production system with an increasing failure rate. Rahim (1994) considered the joint determination of production quantity, inspection schedule, and control chart design. He chose lengths of the sampling intervals that would make the cumulative hazards over different intervals equal.

In this paper, we consider a general lost-sales EMQ model with inspections, machine breakdowns, two types of failure, and planned replacement. The description of the model and the notation are given in Section 2 where we show that the original problem can be decomposed into a two-level optimization problem without loss of optimality. In Section 3, we solve the problem of finding the optimal inspection schedule for a given production cycle, and in Section 4, we show how to solve the optimal lot sizing and preventive replacement problem. A computational procedure for finding the optimal policy and the corresponding expected average cost is presented in Section 5.

2. PROBLEM FORMULATION

We study the problem of joint optimization of lot sizing, scheduling of inspections, and preventive replacement in an EMQ model with a production process subject to random quality deterioration and machine breakdowns. We assume that inspections reveal the true state of the process: "in control" or "out of control". Through the setup procedure, the process is always in control at the beginning of a production period, and the distribution of an in-control period is assumed to be exponential.

Planned preventive replacement and two types of machine breakdowns are considered. When a failure occurs, it is either type I (major failure) with probability p or type II (minor failure) with probability q. When type I failure occurs, the machine is replaced by a new one (or overhauled) at a cost $C_f + C_p$. Production resumes when all available inventory is depleted, i.e., we assume that the unfinished production run is aborted. Starting a new lot after a replacement or after a complete production run incurs a positive setup cost K. On the other hand, when type II failure occurs and machine age is t, minimal repair is performed at a cost $C(t)$. Production can be resumed immediately at a cost R_1, which is lower than the setup cost K. The planned preventive replacement (overhaul) can be performed only in an idle period after completing a production run at a cost C_p. It is assumed that the machine cannot fail in an idle period. Other basic assumptions include a deterministic demand at a constant rate D, production rate P, and inventory holding cost proportional to the on-hand inventory. We assume that the inspections, restorations, repair, and replacement take negligible time.

The following notation will be used.

P : the production rate

D : the demand rate

Y : the time between successive failures if no planned replacement is considered

$F(y)$: the distribution function of Y

$r(y)$: the failure rate of Y

X : the length of an in-control period

$G(x)$: $G(x) = 1 - exp(-\beta x)$ is the distribution of X

p : the probability of type I failure

q : $q = 1 - p$ is the probability of type II failure

$C(t)$: the cost of minimal repair after operating for t hours

C_p : the cost of preventive replacement

$C_f + C_p$: the cost of unplanned failure replacement

v : the operating cost of the system per unit time

K : the production setup cost

R_1 : the production resumption cost after minimal repair

C_I : the cost of inspection

R_2 : the cost of restoring the production process if the process is identified to be out of control.

$h(x)$: the holding cost per unit time when x units are present

s : the cost incurred by producing a defective item

γ_1 : the proportion of defective units produced when the process is in control

γ_2 : the proportion of defective units produced when the process is out of control

To simplify the analysis of the general setting, we reduce the original problem in the following way.

First, observing that a new renewal cycle starts after a replacement (or an overhaul) of the machine, we can apply the renewal reward process approach to show that one can deal with only one machine life cycle, instead of considering the whole time horizon to find the optimal policy. The optimal policy obtained for the one life cycle problem applies to each cycle and forms a stationary optimal policy for the original problem.

Second, it is easy to see that for a given production cycle, its inspection schedule is independent of those for the other cycles. Therefore, we can divide the optimization procedure into two stages.

1). Determine the optimal inspection schedule for a given production run starting at age T_1 and ending at $T_2 > T_1$.

2). Determine the optimal lot sizing and preventive replacement policy.

In the next section, we solve the optimal inspection problem. The optimal lot-sizing and preventive replacement problem will be solved in Section 4. A dynamic programming approach plays an important role in solving both problems.

3. OPTIMAL INSPECTION SCHEDULE FOR ONE PRODUCTION CYCLE

Consider a production cycle (T_1, T_2) starting at age T_1 where the planned production time is equal to $T_2 - T_1$ (we assume that during the idle time the machine does not age and cannot fail). The expected cycle length, denoted by $L(T_1, T_2)$, is given by the following formula:

$$L(T_1, T_2) = \frac{P}{D} \int_{T_1}^{T_2} \overline{F}_p(t)dt / \overline{F}_p(T_1). \tag{3.1}$$

Note that the cycle includes both the production time, also referred to as the production run, and the idle time. The expected holding cost is

$$\frac{P}{D} \int_{T_1}^{T_2} h((P-D)(t-T_1))\overline{F}_p(t)dt/\overline{F}_p(T_1),$$

and the expected cost of minimal repair, production resumption, operating cost, and setup cost is

$$\int_{T_1}^{T_2} [((q(c(t)+R_1)+pC_f)r(t)+v]\overline{F}_p(t)dt/\overline{F}_p(T_1)+K,$$

where $\overline{F}_p(t) = (\overline{F}(t))^p$, is the survival function of the time to the type I failure.

Denote $W^\pi(T_1, T_2)$ as the expected cost of inspections, restorations of the process, and the cost of defectives in a production cycle under policy π. We obtain the following formula for the total expected cost, denoted by $C(T_1, T_2)$:

$$\begin{aligned} C(T_1,T_2) &= \int_{T_1}^{T_2} [\frac{P}{D}h((P-D)(t-T_1)) + (q(c(t)+R_1)+pC_f)r(t) \\ &\quad +v]\overline{F}_p(t)dt/\overline{F}_p(T_1) + K + W^\pi(T_1,T_2). \end{aligned} \tag{3.2}$$

Therefore, for given T_1 and T_2, the minimization of the total expected cost in a production cycle (T_1, T_2) is equivalent to the minimization of $W^\pi(T_1, T_2)$.

Define

$$W(T_1,T_2) = min_{\pi \in \mathcal{P}} W^\pi(T_1,T_2), \tag{3.3}$$

and let $V^\pi(T_1,T_2) = \overline{F}_p(T_1)W^\pi(T_1,T_2)$, where $\mathcal{P}$ denotes the set of all inspection policies for this production cycle. Obviously, it is equivalent to minimize $V^\pi(T_1,T_2)$ instead of $W^\pi(T_1,T_2)$.

Lemma 3.1. We obtain the following result.

$$V(T_1,T_2) = min\{v_0(T_1,T_2), min_{\{T_1<x<T_2\}}\{v(T_1,x)+V(x,T_2)\}\}, \tag{3.4}$$

where

$$v_0(T_1,T_2) = \int_{T_1}^{T_2} Ps(\gamma_1 + (\gamma_2-\gamma_1)(1-exp(-\beta(t-T_1))))\overline{F}_p(t)dt \tag{3.5}$$

is the cost of defectives incurred in (T_1, T_2), and

$$v(T_1, T_2) = v_0(T_1, T_2) + R_2 \int_{T_1}^{T_2} (1 - exp(-\beta(t - T_1)))pr(t)\overline{F}_p(t)dt + \overline{F}_p(T_2)C_I \quad (3.6)$$

is the total cost of inspection, restoration and defectives incurred in (T_1, T_2) if the first inspection after time T_1 is planned at time T_2.

We note that, for this model, the periodic inspection policy is not optimal in general, even though the inspections are perfect and the "in-control" period is exponentially distributed.

If the time to machine failure is also exponentially distributed and the repair cost is constant, then the periodic inspection policy is optimal. In this case, no preventive replacement is planned, and the optimum lot size is constant for each production run.

Let $W_n(T_1, T_2)$ be the minimum cost when the number of scheduled inspections is less than or equal to n, and let $V_n(T_1, T_2) = \overline{F}_p(T_1)W_n(T_1, T_2)$.

Lemma 3.2. For any $0 < T_1 < T_2$, there exists an $n < \infty$ such that for $N \geq n$, $W_n(T_1, T_2) - W(T_1, T_2), V_n(T_1, T_2) = V(T_1, T_2)$,

$$V_m(T_1, T_2) = min\{V_0(T_1, T_2), min_{\{T_1 < x < T_2\}}(v(T_1, x) + V_{m-1}(x, T_2))\} \quad (3.7)$$

for $m > 0$, and $V_0(T_1, T_2) = v_0(T_1, T_2)$.

In principle, we can obtain $V(T_1, T_2)$ and $W(T_1, T_2)$ by solving recursively the dynamic equation in Lemma 3.2 starting with $V_0(T_1, T_2)$.

After obtaining $V(T_1, T_2)$ for the production cycle (T_1, T_2), we can consider the problem of finding the optimal lot-sizing and preventive replacement policy (i.e., the optimal number and the lengths of production cycles before preventive replacement of the machine).

4. OPTIMAL LOT SIZING AND REPLACEMENT POLICY

We use the $\lambda-$maximization technique (see e.g. Aven & Bergman[1]) to equivalently represent the optimization problem.

Suppose we plan n production cycles for a machine and that the planned production times of these cycles are $S_1, S_2 - S_1, ..., S_n - S_{n-1}$, respectively. Thus the expected total cost incurred during the machine's life cycle is

$$A(S_1, ..., S_n) = C_p + \sum_{i=1}^{n} C(S_{i-1}, S_i)\overline{F}_p(S_{i-1})$$

$$
\begin{aligned}
= & \sum_{i=1}^{n}\{\int_{S_{i-1}}^{S_i}\left[\frac{P}{D}h((P-D)(S-S_i))+(q(c(s)+R_1)\right. \\
& +pC_f)r(s)+v]\overline{F}_p(s)ds \\
& +K\overline{F}_p(S_{i-1})+V(S_{i-1},S_i)\}+C_p, \qquad (3.8)
\end{aligned}
$$

and the expected length of the life cycle is

$$
B(S_1,...,S_n)=\sum_{i=1}^{n}L(S_{i-1},S_i)\overline{F}_p(S_{i-1})=\frac{P}{D}\int_0^{S_n}\overline{F}_p(s)ds. \qquad (3.9)
$$

Consequently, the expected average cost is given by

$$
\lambda=\frac{A(S_1,...,S_n)}{B(S_1,...,S_n)}. \qquad (3.10)
$$

The optimization problem can then be described as follows: find n^* and $S_1^* \leq S_2^* \leq ... \leq S_n^*$, minimizing the right-hand side of (3.10), and let λ^* be the corresponding infimum.

For given $\lambda > 0$,

$$
W(\lambda)=\sup_{n,(S_1,S_2,..,S_n)}(\lambda B(S_1,...,S_n)-A(S_1,...,S_n)). \qquad (3.11)
$$

It follows from the boundedness of B, that for $\lambda=\lambda^*$, the optimization problem in (3.11) is equivalent to the problem of minimizing the expected average cost in (3.5) (the so-called $\lambda-$maximization technique). In fact, one can prove that there exists a unique λ_* such that

$$
W(\lambda_*)=0, \qquad (3.12)
$$

and $\lambda_*=\lambda^*$. Thus, the optimal set $(S_1,...,S_n)$ can be found by solving iteratively the problem in (3.11) for several values of λ until equation (3.12) is satisfied.

The advantage of the $\lambda-$maximization technique is that for each λ, the objective function in (3.11) is additive, and a dynamic equation for finding the optimal policy can be readily derived.

Define

$$
W^\lambda(t)=C_p+\sup_{n;\ (S_1,S_2,..,S_n)}\{\lambda B(t;S_1,...,S_n)-A(t;S_1,...,S_n)\}, \quad (3.13)
$$

where $B(t;S_1,...,S_n)$ and $A(t;S_1,...,S_n)$ are the expected length of time and the corresponding cost for a machine starting at age t, respectively, with planned preventive replacement at age S_n. The production cycles start when the age of the machine is $S_1, S_2, ..., S_{n-1}$. $W^\lambda(t)$ can be interpreted as the residual value of the machine at age t. In particular, $W^{\lambda^*}(0) = C_p$. $W^\lambda(t)$ satisfies the following dynamic equation:

$$W^\lambda(t) = 0 \vee \max_{t_1>t}\{G'(t,t_1) + W^\lambda(t_1)\overline{F}_p(t_1)/\overline{F}_p(t)\}, \tag{3.14}$$

where

$$\begin{aligned} G'(t,t_1) &= \int_t^{t_1} [\frac{P}{D}\lambda - \frac{P}{D}h((P-D)(s-t)) - (C(s) + R_1) \\ &\quad + C_f p)r(s) + v)]\overline{F}_p(s)/\overline{F}_p(t)ds - K - W(t,t_1). \end{aligned} \tag{3.15}$$

We obtain for $V^\lambda(t) = \overline{F}_p(t)W^\lambda(t)$,

$$V^\lambda(t) = 0 \vee \max_{t_1>t}\{G(t,t_1) + V^\lambda(t_1)\}, \tag{3.16}$$

where

$$\begin{aligned} G(t,t_1) &= \int_t^{t_1} \overline{F}_p(s)[\frac{P}{D}\lambda - \frac{P}{D}h((P-D)(s-t)) - (C(s) + R_1 \\ &\quad + C_f p)r(s) + v)]ds - K\overline{F}_p(t) - V(t,t_1). \end{aligned} \tag{3.17}$$

Define $V_0(t) = 0$, and for $n \geq 1$, let

$$V_n(t) = \text{the optimum value in the class of policies with at most } n \text{ production cycles} \tag{3.18}$$

We obtain the following results.
Lemma 4.1. For $n \geq 1$,

$$V_n(t) = \max\{V_{n-1}(t), \max_{t_1>t}\{G(t,t_1) + V_{n-1}(t_1)\}\}. \tag{3.19}$$

Furthermore, there exists $n < \infty$, such that for all $N \geq n, t \geq 0, V_N(t) = V(t)$.

Using Lemma 4.1 and Lemma 4.2, we develop an algorithm to solve this optimization problem.

The results are summarized in the following theorem.
Theorem 4.2. An optimal policy exists for this problem, and it has the following form. It is optimal to run n^* production cycles starting when the machine age is $0 = S_0 < S_1 < ... < S_{n^*-1}$ and to replace the machine at time S_{n^*}, after completing the n^*th production run.

In each production run, the optimal inspection schedule is age dependent, i.e., there exists an $m_i \geq 0$ and $S_{i-1} < t_1 < ... < t_{m_i} < S_i$, such that the inspections are performed at times t_j, $j \leq m_i$.

5. COMPUTATIONAL ALGORITHM AND NUMERICAL EXAMPLE

The computational procedure is based on the dynamic equation and developed by applying a discretization scheme. An accuracy bound of the discretization is provided to guarantee the precision of the solution.

The Computational Procedure

Step 1. Choose upper bound $\overline{\lambda}$ and lower bound $\underline{\lambda}$ of λ, and set $\lambda_i = (\overline{\lambda} + \underline{\lambda})/2$ for $i = 1$.

We can start with the following values:

$$\overline{\lambda} = \min_t \{ \frac{\bar{A}}{\frac{P}{D}\int_0^t \overline{F}_p(s)ds} \},$$

where

$$\begin{aligned} \bar{A} &= C_p + S + \int_0^t [\frac{P}{D} h((P-D)(s)) + (q(c(s) + R_1) + pC_f)r(s) \\ &\quad + v]\overline{F}_p(s)ds + v_0(0,t) \end{aligned}$$

and

$$\underline{\lambda} = \min_t \{ \frac{C_p + S + \int_0^t [(q(c(s) + R_1) + pC_f)r(s) + v + Ps\gamma_1]\overline{F}_p(s)ds}{\frac{P}{D}\int_0^t \overline{F}_p(s)ds} \}.$$

$\overline{\lambda}$ is the average cost of the optimal one-cycle policy for the case in which there are no inspections in a production cycle. $\underline{\lambda}$ is the value of the optimal policy for the case in which the holding cost is zero, and the out-of-control period has the same rate of defective products as the in-control period.

Step 2. Choose the maximum length of the horizon. Thus,

$$T = \min_t \{ r(t) > (\overline{\lambda}\frac{P}{D} - v - Ps\gamma_1)/((C(t) + R_1)q + C_f p) \}.$$

Obviously, for the optimal policy, the machine never runs after age T because the rate of the expected gain is lower than the expected loss.

Step 3. For the discretization of $[0, T]$, choose M, such that $S_i's$ are in the set $\{0, T/M, 2T/M, ..., T\}$.

Step 4. For all $T_1 < T_2$ and $T_1, T_2 \in \{0, T/M, 2T/M, ..., T\}$, find an approximation to the optimal inspection schedule in (T_1, T_2), i.e., $V_0(T_1, T_2) = v_0(T_1, T_2)$. Thus,

$$V_m(T_1, T_2) = min\{V_0(T_1, T_2), min_{\{T_1 < x < T_2\}}(v(T_1, x) + V_{m-1}(x, T_2))\}, \quad (3.20)$$

where v_0, and v are given by (3.5) and (3.5), respectively.

Stop calculating $V_m(T_1, T_2)$ when $V_m(T_1, T_2) = V_{m-1}(T_1, T_2)$, and set $V(T_1, T_2) = V_m(T_1, T_2)$.

Step 5. Calculate $V_n^\lambda(t)$ for given λ, i.e., $V_0(t) = 0$. Thus,

$$V_n^\lambda(t) = \max_{t_1 > t}\{V_{n-1}^\lambda(t), G(t, t_1) + V_{n-1}^\lambda(t_1)\}, \tag{3.21}$$

where $t, t_1 \in \{0, T/M, ..., T\}$, and G is given by (3.17).

Stop when $V_n^\lambda(t) = V_{n-1}^\lambda(t)$, and set $V^\lambda(t) = V_n^\lambda(t)$.

Step 6. The iteration procedure for λ is as follows.

If $V^{\lambda_i}(0) > C_p + \epsilon/2$, set $\overline{\lambda} = \lambda_i$.

If $V^{\lambda_i}(0) < C_p - \epsilon/2$, set $\underline{\lambda} = \lambda_i$.

Set $i = i + 1$ and $\lambda_i = (\overline{\lambda} + \underline{\lambda})/2$.

Step 7. The stopping rule for λ is as follows.

If $|V^{\lambda_i}(0) - C_p| < \epsilon/2$, and $|\lambda_i - \lambda_{i-1}| < \epsilon/2$, stop, and set $\lambda^* = \lambda_i$. Otherwise, go to Step 5.

The following lemma guarantees that the computational procedure summarized in Steps 1-7 above gives us an $\epsilon-$optimal policy, providing the length of the discretization interval is small enough.

Lemma 5.1. For any $\epsilon > 0$, there exists a $\delta > 0$, such that for any $\Delta < \delta$, the optimal policy using discretization length Δ is an $\epsilon/2-$optimal policy.

Example.

For $C_p = 3000$, $C_f = 2000$, $K = 500$, $C_I = 120$, $h = 0.5, S = 5$, $P = 1000$, $D = 600$, $p = 0.1$, $q = 0.9$, $v = 50$, $R_1 = 200, R_2 = 100$, $\beta = 0.5$, $\gamma_1 = 0.1$, $\gamma_2 = 0.6$, $r = 0.2$, $r(t) = 2rt, C(t) = 100t + 50$, we obtained an optimal average cost $\lambda^* = 1332.4$.

The optimal number of planned production cycles is 3, i.e., preventive replacement is carried out after finishing the third production run if the machine has not failed before.

The optimal production run lengths are $S_1 = 1.3$, $S_2 - S_1 = 1.4$, and $S_3 - S_2 = 1.4$, i.e., the optimal lot sizes of production runs are 1,300, 1,400 and 1,400, respectively.

The optimal inspection times are 0.6 for the first production run, 2.0 for the second run, and 3.4 for the third production run. Note that these times represent the machine age.

References

[1] Aven, T. and Bergman B. Optimal Replacement Times - A General Set-Up. *Journal of Applied Probability*, **23** (1986) 432-442.

[2] Groenevelt, H., L. Pintelon, and Seidmann A. Production Lot Sizing with Machine Breakdowns. *Management Science*, **38** (1992a) 104-

123.

[3] Groenevelt, H., A. Pintelon, and Seidmann A. Production Batching with Machine Breakdowns and Safety Stocks. *Operations Research*, **40** (1992b) 959-971.

[4] Lee, H.L. and Rosenblatt, M. J. Simultaneous Determination of Production Cycle and Inspection Schedules in a Production System. *Management Science*, **33**, (1987) 1125-1136.

[5] Lee, J.S. and Park, K.S. Joint Determination of Production Cycle and Inspection Intervals in a Deteriorating Production System. *Journal of the Operational Research Society*, **42**, (1991) 775-783.

[6] Makis, V. Optimal Lot-Sizing/ Preventive Replacement Policy for an EMQ model with Minimal Repair. *International Journal of Logistics: Research and Applications*, **1** (1998a) 173-180.

[7] Makis, V. Optimal Lot Sizing and Inspection Policy for an EMQ Model with Imperfect Inspections. *Naval Research Logistics*, **45** (1998b) 165-186.

[8] Makis, V. and Fung J. An EMQ Model with Inspections and Random Machine Failures. *Journal of the Operational Research Society*, **49** (1998) 66-76.

[9] Makis, V. and Fung, J. Optimal Preventive Replacement, Lot Sizing and Inspection Policy for a Deteriorating Production System. *Journal of Quality in Maintenance Engineering*, **1** No. 4 (1995) 41-55.

[10] Porteus, E.L. Optimal lot Sizing, Process Quality Improvement and Setup Cost Reduction. *Operations Research*, **34**, (1986) 137-144.

[11] Rahim, M.A. Joint Determination of Production Quantity, Inspection Schedule and Control Chart Design. *IIE Transactions*, **26** No. 6 (1994) 2-11.

[12] Rahim, M.A. and Banerjee P.K. A Generalized Model for the Economic Design of $\overline{X}$ Control Charts for Production Systems with Increasing Failure Rate and Early Replacement. *Naval Research Logistics*, **40** (1993) 787-809.

[13] Rosenblatt, M.J. and Lee H. J. Economic Production Cycles with Imperfect Production Processes. *IIE Transactions*, **18** (1986) 48-55.

[14] Tse, E. and Makis V. Optimization of the Lot Size and the Time to Replacement in a Production System Subject to Random Failure. *Wang M.T. (ed.): Proceedings of the Third International Conference on Automation Technology* 4: 163-169, College of Engineering, National Taiwan University, Taipei, Taiwan, 1994.

Chapter 4

A PRODUCTION/INVENTORY POLICY FOR AN UNRELIABLE MACHINE

Hyo-Seong Lee and

Department of Industrial Engineering

Kyung-Hee University,

1 Seocheon, Kiheung, Yongin, Kyunggi-do, 449-701

S. Korea

Mandyam M. Srinivasan

Management Science Program

College of Business Administration,

The University of Tennessee

Knoxville, TN, 37996-0562.

Abstract

In this paper, the economic lot-sizing problem for an unreliable production facility is considered. As in the standard Economic Manufacturing Quantity (*EMQ*) model, the demand and production rates are assumed to be constant. Unlike the *EMQ* model, the facility is assumed to deteriorate while it is in operation, with an increasing failure rate. However, a preventive maintenance overhaul of the facility is assumed to restore the facility to its original condition.

A production run is initiated as soon as the inventory level drops down to zero, and it continues until the inventory reaches a predetermined level S. If the facility fails during the production run, it is minimally repaired and put back into commission. The facility is set aside for preventive maintenance every N production runs. The preventive maintenance activity takes a random amount of time; if the inventory level drops to zero during this activity, then production resumes as soon as the activity is completed. Any demand that arrives during a stock-out period is assumed to be lost.

The control policy is thus the pair (S, N). Under a cost structure that includes a preventive maintenance cost, a minimal repair cost, a setup cost, a holding cost, and a penalty cost due to lost sales, an expression for the expected cost per unit time is obtained for a given policy. Following that, some properties of the cost function are developed to characterize the optimal policy. Based on these properties, an efficient search procedure to find the optimal policy is presented.

Keywords: Threshold policies, pull systems, preventive maintenance, renewal reward theorem, renewal theory.

1. INTRODUCTION

Increasingly, pull production systems are gaining acceptance and being used in the manufacturing industry. Typically, these systems work on a triggering mechanism; the trigger is activated when the inventory level drops below a threshold limit and initiates the start of a production cycle that continues until a predetermined inventory level is reached. Such a triggering mechanism has been the focus of research for a number of years, in the form of threshold policies (such as (s, S)-type policies) for controlling production/inventory systems. The widespread acceptance and use of pull production systems has thus provided a fresh impetus to research on such control policies.

The success of pull systems is affected, to a considerable extent, by the presence (or absence) of effective preventive maintenance programs to support production facilities. Researchers of threshold policy models of production/inventory systems should, therefore, consider including preventive maintenance in the model, if they expect to find applications for modeling pull production systems. There are, however, very few papers that consider preventive maintenance policies within the context of production/inventory systems. Most of the existing research on production/inventory systems assumes that the facility is completely reliable. At the same time, there is a large body of work that deals with machine inspection, reliability, and maintenance. These models typically do not incorporate production/inventory decisions (see Sherif and Smith [15] and Valdez-Flores and Feldman [18] for comprehensive surveys on the literature of inspection and maintenance policies).

In this paper, we consider the combined effects of preventive maintenance and production/inventory decisions on the cost of operation of the facility. We are thus interested in integrating the reliability aspects of a production facility with the production/inventory decisions made on it. The integration of these two aspects presents a challenging opportunity for research (for example, see Tseng *et al.* [17], and Rahim and Ben-Daya [12] and reference therein). Planned downtime, viz., preventive maintenance, decreases productive time, but it also reduces uncertainty concerning failures. In the following discussion, we discuss the existing research on production models with unreliable facilities.

Hsu and Tapeiro [5] consider the maintenance of an unreliable facility and model it as a $M/G/1$ queue. In their model, a preventive maintenance activity is scheduled as soon as n jobs are processed or if the machine breaks down, whichever comes first. Sharafali [14] considers a model of a facility that is subject to failure that includes the threshold inventory levels between which the facility is allowed to operate. Demand is assumed to arrive according to a Poisson process, and the processing times are assumed to be exponentially distributed. Failures arrive according to a Poisson process, and the repair time follows an arbitrary distribution. By identifying a semi-regenerative process, expressions for the probability mass function of the inventory process is obtained. In a related paper, Parthasarathy and Sharafali [9] consider a model of an unreliable facility that operates at two different production rates. In this model, the facility operates at a rate greater than the demand rate as long as the inventory level is below a threshold level; beyond this level, the facility produces at the same rate as the demand rate. They determine the optimal threshold level under a given cost structure that includes the inventory holding cost per unit time, repair cost, and the cost of lost sales.

Lee and Rosenblatt [7] analyze a production system where the production process can move into an "out-of-control state," in which some of the items produced are defective. Periodic inspections are assumed to determine the state of the process. If the process is found to be in an out-of-control state, then corrective maintenance is performed to restore it to an "in-control state." They determine the optimal lot size, as well as the optimal inspection schedule, and show that the optimal lot size is smaller than the lot size obtained by the classical economical manufacturing quantity (*EMQ*) model [10].

Groenevelt, Pintelon and Seidmann [3] analyze the effects of machine breakdowns and corrective maintenance on economic lot sizing decisions. In the policy they are considering, corrective maintenance is carried out as soon as the machine breaks down. If the on-hand inventory at this time is at or above a certain threshold value, then production is delayed until all available inventory is depleted. On the other hand, if the on-hand inventory is below a threshold value, then production is immediately resumed. Assuming a constant failure rate and instantaneous repair times, they show that the optimal lot sizes are always greater than the *EMQ*. In a subsequent paper [4], they investigate the issue of safety stocks required to meet the managerially prescribed service level in the presence of machine breakdowns. An easily implementable production control policy is proposed, and a closed form expression is obtained for the optimal lot size under this policy.

Berg, Posner and Zhao [2] consider a production system with Poisson demands and exponential processing times. They analyze a system with multiple identical machines devoted to producing a single part type, thereby generalizing work done in an earlier paper [11] in which a single machine producing a single part type is consiered. Their analysis assumes that each machine operates for an exponential amount of time before it fails and that repair times are also exponentially distributed. The analysis also accommodates phase-type distributions. The objective of the paper is to determine the stationary distribution of the inventory levels. It is assumed that inventory is built up to a level M, that all operative machines are simultaneously producing whenever the inventory level is below M, and that production on all machines is halted as soon as the inventory reaches level M. Any demand which cannot be met is assumed to be lost; although, it is subsequently indicated how the analysis extends to allow demands to be backordered.

These papers consider machine failures within the context of production/inventory systems, but they do not consider preventive maintenance strategies to reduce the failure rate. Meller and Kim [8] consider an operating policy for an unreliable production system which incorporates preventive maintenance. Their operating policy specifies that the system is repaired whenever it breaks down and that production resumes immediately following the completion of repair, continuing until the inventory level reaches a certain threshold value b^*. At this time, a preventive maintenance activity is initiated. The next production cycle begins as soon as preventive maintenance is completed. The time to failure is assumed to follow an exponential distribution. Although one could question the effectiveness of preventive maintenance in the presence of constant failure rates (see, for instance, Jorgenson, McCall and Radner [6]), Meller and Kim justify its usefulness by assuming that the failure rate is a non-decreasing function of the preventive maintenance frequency. Assuming exponential repair times and deterministic maintenance times, they use an embedded Markov chain model to determine the performance measures of interest for a given inventory threshold value, b^*. These measures are used to determine the average cost of operating under the given threshold value.

Srinivasan and Lee [16] present a model in which the facility is allowed to fail at an increasing failure rate while it is in operation. A preventive maintenance overhaul of the facility is assumed to restore the facility to its original condition. The demand process is assumed to be Poisson, and processing times follow an arbitrary distribution. The facility is set aside for preventive maintenance as soon as the inventory level is built up to a prespecified level, S. Production is re-initiated as soon as the inventory level drops below a level, s, unless the facility is still undergoing a preventive maintenance overhaul, in which case production resumes immediately following the completion of this activity. An algorithm is developed to find the optimal inventory threshold values, (s, S), within which the facility should operate, with the objective of minimizing a given cost function.

Similar to that in [16], the model considered in this paper assumes that the production facility deteriorates while it is in operation, with an increasing failure rate. Again, as in [16], a preventive maintenance overhaul is performed on the facility as soon as the inventory level is raised to a level S, and this overhaul is assumed to restore the facility to its original condition. Unlike [16], however, the demand and production rates are assumed to be constant, and a production run begins as soon as the inventory drops to zero. Furthermore, the facility is not set aside for preventive maintenance every time the inventory level is raised to S. Indeed, one of the control variables we consider here is the number, N, which specifies the number of production cycles the machine should go through before it is set aside for preventive maintenance.

The control policy considered in this paper is thus the pair, (S, N). Under a very general cost structure that includes a setup cost, a holding cost, a cost of lost sales, and a preventive maintenance cost, we first develop an expression for the expected cost per unit time for a given policy. Some properties of the cost functions are then developed to characterize the optimal policy. Based on these properties, an efficient algorithm to find the optimal policy is presented.

2. THE MODEL

As in the standard *EMQ* model, the demand rate, r, and the production rate, p, are both assumed to be constants, with $p > r$. The control policy we consider initiates production as soon as the inventory level drops down to zero. The production run continues until the inventory reaches a predetermined level, S. There is a setup cost of K_1 incurred for each production run, and a holding cost of C_h per unit held in inventory per unit time. The setup time is assumed to be negligible. Unlike the *EMQ* model, the production facility is assumed to deteriorate, while it is in operation, with an increasing failure rate, $h(t)$. If the facility fails during operation, it is assumed to be repaired at a cost, C_m. This is a minimal repair, assumed to take a negligible amount of time, and it only restores the facility to the condition it was in immediately before the failure. A preventive maintenance overhaul of the facility, however, is assumed to restore the facility to its original condition. The preventive maintenance activity takes a random amount of time, V, having a general distribution, $F(t)$. The cost of a preventive maintenance activity that takes time x is $K_2 + C_p x$.

Since preventive maintenance usually takes a considerable amount of time and cost, it is reasonable to carry out this activity during a non-production period after the facility has deteriorated to a point where it is not cost-effective to continue operation.

The control policy we consider, thus, includes a number, N, which denotes the number of production runs that take place before the next preventive maintenance overhaul. The facility is, therefore, operated as follows. As soon as the inventory level reaches a predetermined value, S, the facility is turned off. If the number of production runs since the last preventive maintenance activity is less than N, then a non-production period begins and continues until the inventory level is completely depleted. On the other hand, a preventive maintenance overhaul is carried out on the facility if the number of production runs is equal to N. This takes a random time, and so the inventory level at the end of this activity could be zero (a stock-out), in which case the facility is immediately set up for another production run. We assume that stock-outs result in lost sales, with a cost C_ℓ for each unit of demand that is lost. If the inventory level at the end of the preventive maintenance overhaul is greater than 0, the facility waits until the level drops to 0 before the next production run is initiated. See Figure 4.1.

Note that two control variables, S and N, are used to operate the production/inventory system. The objective of this study is to determine the optimal values of S and N which minimize the expected cost per unit time in the long run. Since the time at which a production run starts, following a preventive maintenance activity, is a regeneration point, the time interval between two successive regeneration points is a regenerative cycle. Let $L(S,N)$ denote the expected length of a cycle, and let $C(S,N)$ denote the expected cost incurred during a cycle when S and N are used as control values. From the renewal reward theorem [13], the expected cost per unit time, $TC(S,N)$, is then expressed as

$$TC(S,N) = \frac{C(S,N)}{L(S,N)}. \tag{4.1}$$

The expected length of a cycle, $L(S,N)$, is calculated as follows. Let B_S denote the time it takes to build up the inventory level to S during a production run, and let I_s denote the time it takes for the inventory level to drop from S down to 0. Since the production rate is p and the

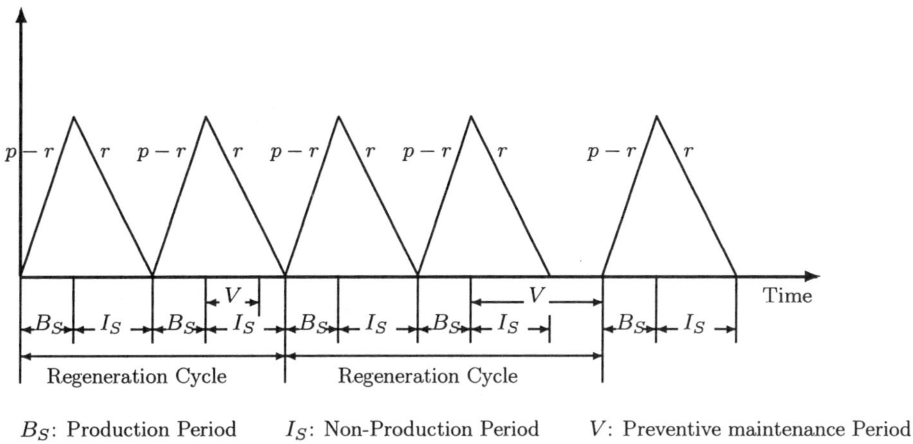

Figure 4.1 The Production/Inventory System Where $N = 2$.

demand rate is r, inventory grows at a rate $p - r$ during the production period, and so $B_S = S/(p - r)$. Similarly, since inventory depletes at a constant rate r, $I_S = S/r$. If the preventive maintenance activity is completed before the inventory is fully depleted, the length of a cycle is exactly $N(B_S + I_S) = NS\,p/((p - r)r)$. However, if the preventive maintenance duration, V, exceeds I_S, then the expected length of the cycle is larger by an amount $V - I_S$. Thus,

$$L(S, N) = N\frac{S\,p}{(p - r)r} + \int_{I_S}^{\infty} (v - I_S)\,dF(v). \tag{4.2}$$

To determine $C(S, N)$, note that each cycle consists of N setups having a cost $N\,K_1$ and a preventive maintenance activity that has an expected cost of $K_2 + C_p\,E[V]$. Let H_S denote the average holding time for all the units held in inventory, during a production period and the following non-production period. Note that H_S represents the area of a triangle in Figure 4.1. Therefore, we have

$$H_S = (B_S + I_S)\frac{S}{2} = \frac{p\,S^2}{2(p - r)r}.$$

Since there are N production runs in a cycle, the inventory holding cost per cycle is NH_SC_h. If the preventive maintenance duration, V, is greater than I_S, then a total of $r\,(V - I_S)$ units of demand are lost, and the expected cost of lost sales during a cycle is $r\, C_\ell \int_{I_S}^{\infty} (v - I_S)dF(v)$. Finally, since failure rates do not change due to minimal repairs, the failure arrival process follows a non-homogeneous Poisson process with rate function $h(t)$. Note that the total production time during a cycle is NB_S; the expected number of minimal repairs performed during a cycle is given by $\int_0^{NB_S} h(t)dt$. The expected cost incurred during a cycle is obtained as follows:

$$\begin{aligned} C(S,N) &= NK_1 + NH_sC_h + K_2 + C_pE(V) + C_m \int_0^{NB_S} h(t)dt \\ &+ r\, C_\ell \int_{I_S}^{\infty} (v - I_S)dF(v). \end{aligned} \tag{4.3}$$

In the following section, we develop some properties of the cost function that are used to characterize the optimal policy. We present an efficient procedure, based on these properties, for finding the optimal values of S and N.

3. PROPERTIES OF THE COST FUNCTION

To find the optimal policy (S^*, N^*), the function $TC(S,N)$ must be minimized over a two dimensional parameter space. This search can be performed quite efficiently if some properties of the cost function are exploited. For ease of presentation, we define

$$C'_N(S) = \frac{\partial C(S,N)}{\partial S},$$

$$L'_N(S) = \frac{\partial L(S,N)}{\partial S},$$

and

$$h'(t) = \frac{dh(t)}{dt}.$$

Thus, from (4.2) and (4.3),

$$C'_N(S) = \frac{N\, p\, S}{(p-r)r} C_h + \frac{N\, h(NB_S)}{p-r} C_m - C_\ell \int_{I_S}^{\infty} dF(v), \tag{4.4}$$

$$L'_N(S) = \frac{N\, p}{(p-r)\, r} - \frac{1 - F(I_S)}{r}. \tag{4.5}$$

From (4.1),

$$\frac{\partial TC(S,N)}{\partial S} = \frac{C'_N(S)L(S,N) - L'_N(S)C(S,N)}{L(S,N)^2} = \frac{q_N(S)}{L(S,N)^2}, \tag{4.6}$$

where

$$\begin{aligned} q_N(S) &= C_N'(S)L(S,N) - L_N'(S)\, C(S,N) \\ &= C_N'(S)\, L(S,N) - L_N'(S)\, TC(S,N)\, L(S,N) \\ &= \left[\frac{N\,p\,S}{(p-r)r}\, C_h + \frac{N\,h\,(NB_S)}{p-r}\, C_m \; - \; C_\ell(1 \; - F(I_S)) \right. \\ &\quad \left. - \left\{\frac{N\,p}{(p-r)\,r} - \frac{1-F(I_S)}{r}\right\} TC(S,N)\right] L(S,N). \end{aligned}$$

Rearranging the terms,

$$\begin{aligned} q_N(S) \;&=\; \left[\left\{\frac{N\,p\,S}{(p-r)\,r} C_h + \frac{N\,h(NB_S)}{p-r} C_m - C_\ell\right\}\right. \\ &\quad - \left\{\frac{N}{p-r} + \frac{N-1}{r}\right\} TC(S,N) \\ &\quad \left. + \frac{F(I_S)}{r}\{r\, C_\ell - TC(S,N)\}\right] L(S,N). \qquad (4.7) \end{aligned}$$

It is easily verified that $\lim_{S\to 0} q_N(S) < 0$. It can also be shown that $\lim_{S\to\infty} q_N(S) > 0$. Since the denominator in (4.6) is positive, $\frac{\partial TC(S,N)}{\partial S}$, thus, has at least one local minimum as S ranges from 0 to ∞. Theorem 1 now presents an important property of $TC(S,N)$; it shows that in order to find the optimum value of S that minimizes $TC(S,N)$ for a given N, we need look no further than the first local minimum value of $TC(S,N)$.

Theorem 1 *For a given N, suppose the first local minimum of $TC(S,N)$ occurs at S^*.*

a) If $TC(S^*,N) \geq r\; C_\ell$, then $TC(S,N) \geq r\; C_\ell$ for all S.

b) If $TC(S^*,N) < r\; C_\ell$, then S^* is a global minimum.

Proof: Note first, that since S^* is the first local minimum, $q_N(S^*) = 0$.

To prove part a), suppose that $TC(S^*,N) \geq r\; C_\ell$, but that there is another local minimum, S_b, at which $TC(S_b,N) < r\; C_\ell$. Thus, there exists a point, S_a, such that $S^* < S_a < S_b$; $TC(S_a,N) \leq r\; C_\ell$; and $q_N(S_a) < 0$. There are three sets of terms on the right hand side of (4.7), all enclosed in braces. The first set is clearly increasing in S. Since $TC(S_a,N) \leq r\; C_\ell$, and $F(I_S)$ is increasing in S, the value of the third set of terms, evaluated at S_a, is greater than the corresponding value, evaluated at S^*. Thus, since $TC(S_a,N) \;\leq\; TC(S^*,N)$, and

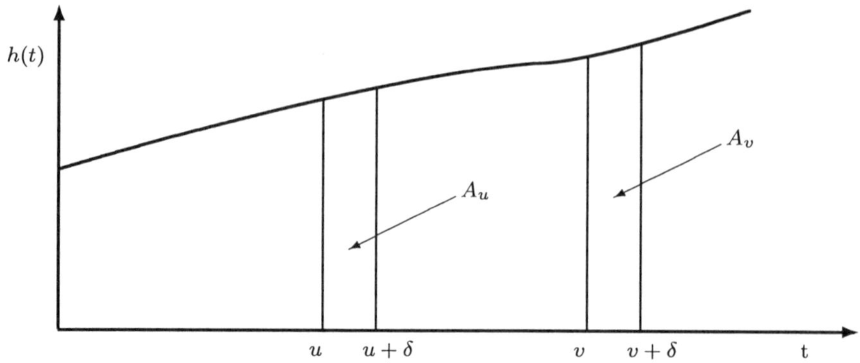

Figure 4.2 A_u and A_v for an increasing function, $h(t)$, with $v > u$.

$q_N(S^*) = 0$, we must obtain $q_N(S_a) > 0$, leading to a contradiction. Therefore, if $TC(S^*, N) \geq r\, C_\ell$, then $TC(S, N) \geq r\, C_\ell$ for all S.

To prove part b), suppose $TC(S^*, N) < r\, C_\ell$. Assume, as before, that there is another local minimum, S_b, at which $TC(S_b, N) < TC(S^*, N)$. Thus, there exists a point, S_a, such that $S^* < S_a < S_b$, $TC(S_a, N) \leq TC(S^*, N)$, and $q_N(S_a) < 0$. Since $S_a > S^*$ and $q_N(S^*) = 0$, following a similar approach as that used in part a), we see from (4.7) that $q_N(S_a)$ must be greater than zero, again leading to a contradiction. Hence, if $TC(S^*, N) < r\, C_\ell$, then S^* is a global minimum.

Theorem 1 implies that, in order to find the global minimum of $TC(S, N)$ for a given N, we have only to find the first local minimum of $TC(S, N)$. Note that if, at the first local minimum, S^*, $TC(S^*, N) \geq r\, C_\ell$, then it is better not to operate the system at all since $TC(S, N) \geq r\, C_\ell$ for all S. If, however, the first local minimum satisfies $TC(S^*, N) < r\, C_\ell$, then it is a global minimum from Theorem 1.

Property 1 *Let $a, b, c, d > 0$. Then $\frac{c}{d} > \frac{a}{b}$ if $(c - a)b - (d - b)a > 0$.*

Property 2 *If $h(t)$ is increasing in t, then for any arbitrary interval of length δ, and for $v > u$ (see Figure 4.2),*

$$A_v = \int_v^{v+\delta} h(t)\, dt > A_u = \int_u^{u+\delta} h(t)\, dt.$$

Let

$$\Delta C(S, N) = C(S, N+1) - C(S, N), \tag{4.8}$$

$$\Delta L(S,N) = L(S,N+1) - L(S,N). \tag{4.9}$$

Lemma 1 *For a given S,*

a) $C(S,N)$ is a convex function of N, and

b) $\Delta C(S,N+1)\ L(S,N+1) - \Delta L(S,N+1)\ C(S,N+1) > \Delta C(S,N)\ L(S,N) - \Delta L(S,N)\ C(S,N)$.

Proof: The proof for part a) follows from (4.3) and Property 2. We have:

$$\Delta C(S,N+1) - \Delta C(S,N) = C_m \int_{(N+1)B_S}^{(N+2)B_S} h(t)dt - C_m \int_{NB_S}^{(N+1)B_S} h(t)dt > 0.$$

To prove part b), the convexity of $C(S,N)$ is used:

$$\begin{aligned} L(S,N+1)\Delta C(S,N+1) &> L(S,N+1)\Delta C(S,N) \\ &= L(S,N)\Delta C(S,N) \\ &+ (L(S,N+1) - L(S,N))\Delta C(S,N) \\ &= L(S,N)\Delta C(S,N) \\ &+ \Delta L(S,N)(C(S,N+1) - C(S,N)). \end{aligned}$$

The result now follows from the fact that $\Delta L(S,N+1) = \Delta L(S,N)$.

Theorem 2 *$TC(S,N)$ is unimodal with respect to N.*

Proof: Suppose N^* is the smallest N for which $TC(S,N^*+1) > TC(S,N^*)$. To prove unimodality, we only need to show that $TC(S,N^*+k+1) > TC(S,N^*+k)$, for all $k \geq 0$. Since $TC(S,N^*+1) > TC(S,N^*)$, from Property 1, for $k = 0$, we have

$$\Delta\ C(S,N^*)\ L(S,N^*) - \Delta\ L(S,N^*)\ C(S,N^*) > 0.$$

Hence, from Lemma 1b), $\Delta\ C(S,N^*+1)\ L(S,N^*+1) - \Delta\ L(S,N^*+1)\ C(S,N^*+1) > 0$, and using Property 1 once again, we must obtain $TC(S,N^*+2) > TC(S,N^*+1)$. Applying Property 1 repeatedly, we see that $TC(S,N^*+k+1) > TC(S,N^*+k)$, for all $k \geq 0$.

We have conducted numerous experiments on a variety of test problems. These experiments suggest that the cost function has other useful properties. For instance, if we let $S^*(N)$ denote the optimal value of S for a given value of N, then $S^*(N+1)$ is always less than or equal to $S^*(N)$

Table 4.1 Optimal policy for Example 1

N	$S^*(N)$	$L(S^*(N), N)$	$TC(S^*(N), N)$
1	32.07	5.35	38.91
2*	25.45	8.48	36.31
3	22.31	11.15	36.58
4	20.28	13.52	37.58
5	19.46	16.27	38.89
6	18.10	18.29	40.29
7	17.01	20.14	41.69
8	16.10	21.86	43.06

(*: represents the optimal policy)

for any value of N. We also observed that $TC(S^*(N), N)$ is unimodal in N. These two observations, together with Theorems 1 and 2, are used to devise an effective search procedure. To find the optimal pair of control values, (S^*, N^*), we start with $N = 1$ and find $S^*(1)$ using Theorem 1. We then succesively increment N and find $S^*(N)$. Here we use the observation that $S^*(M) \leq S^*(N)$ for $M \geq N$. Based on the observation that $TC(S^*(N),\ N)$ is unimodal in N, we can use a search technique such as the Fibonacci method or the golden section method to find the minimum value. The search continues until we find the value of N that satisfies $TC(S^*(N-1), N-1) \geq TC(S^*(N), N) \leq TC(S^*(N+1), N+1)$.

4. NUMERICAL EXAMPLES

To verify the efficiency of our algorithm and to check the unimodality of $TC(S^*(N), N)$, we carried out extensive numerical tests. These tests also reveal how the algorithm facilitates answers to "what-if" type questions. Some of the numerical results are presented below.

Example 1. In the first example, the maintenance time is assumed constant, that is, $V = 2$ with probability 1. The other parameters are as follows: $p = 25, r = 10, h(t) = 4t, K_1 = 50, C_h = 1, C_\ell = 5, C_m = 2, K_2 = 50$, and $C_p = 2$. The results obtained by the algorithm are presented in Table 4.1.

Table 4.2 Optimal policy for Example 2

N	$S^*(N)$	$L(S^*(N), N)$	$TC(S^*(N), N)$
1	33.63	5.58	39.64
2*	26.73	9.43	37.11
3	23.23	12.24	37.30
4	20.94	14.66	38.19
5	19.28	16.83	39.35
6	17.98	18.79	40.61
7	16.93	20.61	41.91
8	16.05	22.30	43.22

(*: represents the optimal policy)

Example 2. For this example, we retain all parameter values in Example 1. The maintenance time, however, is assumed to follow an exponential distribution with mean 2. The results obtained by the algorithm are presented in Table 4.2.

The above two examples show that the system with deterministic maintenance times incurs a smaller expected cost for the optimal policy, compared with the equivalent system with exponential maintenance times. This is probably because the system with deterministic preventive maintenance times never "stocks out," even though its maximum inventory level, S^*, is lower than the corresponding S^* for the system with exponential maintenance times. (Note that with deterministic V, for $S^* = 2$, we have $I_S = S/r = 2.545$ which is greater than $V = 2$). Therefore, a system with deterministic maintenance produces saving in both the expected holding cost and the penalty cost of lost sales, compared with the system with exponential maintenance times, thereby making the case for reduction in the variance of preventive maintenance times. The next example demonstrates a case in which it is advisable to let the facility run for many cycles before it is set aside for preventive maintenance.

Example 3. For this example, we retain most of the parameter values in Example 2. Thus, the maintenance times follow an exponential distribution with mean 2, and p, r, C_h, C_m, K_2, and C_p are the same as before.

Table 4.3 Optimal policy for Example 3

N	$S^*(N)$	$L(S^*(N), N)$	$TC(S^*(N), N)$
1	37.61	6.57	35.74
2	29.26	10.21	30.68
3	24.97	13.06	28.88
4	22.18	15.45	28.09
5	20.17	17.54	27.76
6*	18.61	19.40	27.68
7	17.37	21.10	27.74
8	16.34	22.67	27.89

(*: represents the optimal policy)

The values of $h(t), K_1$, and C_ℓ are different. These values are $2t, 20$, and 10, respectively (compared with corresponding values of $4t, 50$, and 5 in Example 2.). The results obtained by the algorithm are provided in Table 4.3 and show that N, the optimal number of cycles the facility is run before it is set aside for preventive maintenance, is now $N^* = 6$, (compared with $N^* = 2$ from Example 2).

Again, this example shows how the algorithm could be used to answer what-if questions. For instance, it demonstrates that the presence of a smaller failure rate allows the facility to run, in the optimal policy, for a larger number of cycles before preventive maintenance is carried out. Comparing $TC(S^*(N), N)$ for each value of N, it also appears that a smaller setup cost (K_1) does not necessarily result in a smaller average inventory level. Note, from Tables 4.2 and 4.3, that $S^*(N)$ is larger in Table 4.3 for each value of $N = 1, \cdots, 8$, compared to Table 4.2. Thus, the effect of a 50% reduction in the failure rate, $h(t)$, outweighs the effect of a 60% reduction in the setup cost in this example.

5. CONCLUSIONS

We modeled a facility that manufactures a single type of part and considered a joint production/inventory and preventive maintenance policy that specifies both the maximum inventory level, S, and how often preventive maintenance is carried out. The objective was to determine the optimal control policy (S, N) that minimizes the average cost of operating the facility per unit time. A fairly general cost structure was assumed, and the properties developed for the various cost functions involved resulted in a very simple algorithm to find the optimal control policy. The search procedure was found to be very efficient; consequently, the algorithm determines the optimal values very quickly.

The model is, therefore, a valuable decision-making tool that will enable the analyst to answer what-if questions very easily. For instance, it can be used to analyze the tradeoffs involved in balancing preventive maintenance durations and costs against the cost of lost sales, setup costs, and holding costs. A number of examples demonstrate the efficiency of the algorithm as well as its ability to answer some of these what-if questions.

Acknowledgement: The research of the first author was supported through KOSEF Grant 981-1014-076-2.

References

[1] Barlow, R.E. and Proschan, F. *Mathematical Theory of Reliability*, John Wiley & Sons Inc., New York, 1965.

[2] Berg, M., M. J. M. Posner, and Zhao, H. Production-Inventory Systems with Unreliable Machines. *Operations Research*, **42**(1) (1994) 111-118.

[3] Groenevelt, H., L. Pintelon, and Seidmann A. Production Lot Sizing with Machine Breakdown. *Management Science*, **38**(1) (1992a) 104-123.

[4] Groenevelt, H., L. Pintelon, and Seidmann A. Production Batching with Machine Breakdowns and Safety Stocks. *Operations Research*, **40**(5) (1992b) 959-971.

[5] Hsu, L. and Tapeiro, C.S. Maintenance of an Unreliable $M/G/1$ Queueing Job Shop. *Queueing Systems*, **2** (1987) 333-350.

[6] Jorgenson, D.W., J.J. McCall, and Radner, R. *Optimal Replacement Policy*, Rand McNally & Co., Chicago, 1967.

[7] Lee, H.L. and Rosenblatt, M.J. Simultaneous Determination of Production Cycle and Inspection Schedules in a Production System. *Management Science*, **33**(9) (1987) 1125-1136.

[8] Meller, R.D. and Kim, D.S. The Impact of Preventive Maintenance on System Cost and Buffer Size. *European Journal of Operational Research*, **95** (1996) 577-591.

[9] Parthasarathy, P.R. and Sharafali, M. On the Unutilized Capacity of a Production-Storage System. *Operations Research Letters*, **6** (1987) 233-238.

[10] Peterson, R. and Silver, E. A. *Decision Systems for Inventory Management and Production Planning*, John Wiley & Sons, Inc., New York, 1979.

[11] Posner, M. J. M. and Berg, M. Analysis of a Production-Inventory System with Unreliable Production Facility. *Operations Research Letters*, **8** (1989) 339-345.

[12] Rahim, M. A. and Ben-Daya, M. A Generalized Economic Model for Joint Determination Production Run, Inspection Schedule and Control Chart Design. *International Journal of Production Research*, **36**(1) (1998) 277-289.

[13] Ross, S.M. *Applied Probability Models with Optimization Applications*, Holden-Day, Inc., San Francisco, 1970.

[14] Sharafali, M. On a Continuous Review Production-Inventory Problem. *Operations Research Letters*, **3** (1984) 199-204.

[15] Sherif, Y. and M.L. Smith, M. L. Optimal Maintenance Models for Systems subject to Failure - Review. *Naval Research Logistics Quarterly*, **28**, (1981) 47-74.

[16] Srinivasan, M.M., and Lee, H.S. Production-Inventory systems with preventive maintenance. *IIE Transactions*, **28** (1996) 879-890.

[17] Tseng, S. T., Yeh, R. H. and Wen-Tsung, Ho. Imperfect Maintenance Policies for Deteriorating Production Systems. *International Journal of Production Economics*, **55** (1998) 191-201.

[18] Valdes-Flores, C. and Feldman, R.M. A Servey of Preventive Maintenance Models for Stochastically Deteriorating Single Unit Systems. *Naval Research Logistics*, **36** (1989) 419-446.

Chapter 5

INTEGRATING MAINTENANCE LOT SIZING AND PRODUCTION PLANNING FOR A SINGLE MACHINE MULTI-PRODUCT SYSTEM

Suresh K. Goyal and
Department of Decision Sciences and MIS
Concordia University
1455 De Maisonneuve Blvd., West
Montreal, Qucbcc II3G 1M8 Canada.

Yasemin Kahyaoglu
Department of Management
Bogazici University
80815 Bebek, Istanbul, Turkey

Abstract In this article, we present a simple approach for determining a practical operating policy regarding scheduling of maintenance work and lot sizing of several products produced on a single machine. The article highlights the problems in determining a repetitive production schedule and in carrying out maintenance work on the machine at regular time intervals. A numerical example is provided for three products on a single machine to demonstrate the applicability of the alternative operating policies.

Keywords: Maintenance scheduling, lot-sizing, multi-product, single machine.

1. INTRODUCTION

The problem of determining an economic maintenance frequency of a machine has been dealt with extensively in the works of Christer and Doherty [1], Goyal and Kusy [5], Luss and Zvi [7], and Sule and Harmon [8]. In this paper, we consider the case of a single machine which produces several products over an infinite time horizon. The maintenance and operating costs of the machine are linear. The demand for each product is assumed to be constant. The problem, therefore, is to determine

1. the economic time interval between successive maintenance operations on the machine;
2. the lot size of each product; and
3. a feasible repetitive maintenance schedule as well as the production schedule. First, we will formulate a general total cost model, then point out the problems of optimizing such a model. We will also suggest a practical approach for solving the decision problem.

2. ASSUMPTIONS

The formulation of the total cost model is based on the following assumptions.

1. The maintenance work on the machine is carried out at equal time intervals.
2. The cost of carrying out maintenance on the machine is assumed to be linear with respect to the time 't' of machine operation.
3. The operating cost of the machine per unit of time after time 't' of operation is assumed to be linear.
4. The time required for maintenance is assumed to be constant.
5. The demand for each product is assumed to be constant over time.
6. The products are manufactured in equal lot sizes.
7. Shortages are not allowed.
8. The time horizon is infinite.

3. NOTATION

The following notation is used in the model.

$Z(T)$	total cost per unit of time
T	time interval between successive maintenance operations
t	machine operation time during successive maintenance operations
x	time required for maintenance

$a + bt$	linear maintenance cost (a and b are constants)
$f + vt$	linear operating cost of the machine per unit of time (f and v are constants)
n	number of products produced on the machine and for the ith product ($i = 1, 2, ..., n$)
D_i	demand per unit of time
P_i	rate of production
S_i	setup cost per setup
h_i	unit inventory cost per unit of time
w_i	setup time per setup
K_i	number of times the product is manufactured in equal batch sizes during time 'T'

The elements of the total cost per unit of time, $Z(T)$, are discussed below.

1. The cost of maintenance setups is given by the following:

$$\frac{a + bt}{T}.$$

2. Since the maintenance work is carried out on the machine after operating for time 't', the total operating cost during the time interval 0 to t equals

$$\int_0^t (f + v\tau)d\tau = ft + \frac{vt^2}{2}$$

Hence, the operating cost per unit of time is given by

$$\frac{t}{T}(f + \frac{vt}{2}).$$

3. The setup costs for the products can be evaluated as follows:

$$\frac{1}{T}\sum_{i=1}^{n} S_i K_i.$$

4. The inventory carrying cost is given by

$$\frac{T}{2}\sum_{i=1}^{n} D_i h_i (1 - D_i/P_i)/K_i.$$

Therefore, the total cost per unit of time is given by

$$Z(T) = \frac{1}{T}\left[a + \sum_{i=1}^{n} S_i K_i + t\left(b + f + \frac{vt}{2}\right)\right] + \frac{T}{2}\sum_{i=1}^{n} \frac{D_i h_i}{K_i}(1 - D_i/P_i). \tag{5.1}$$

In the above total cost model, the total operating time, t, for the machine between successive maintenance operations is obtained as follows:

$$t = T\sum_{i=1}^{n} D_i/P_i = T\sum_{i=1}^{n} r_i$$

$$t = rT \tag{5.2}$$

where

$$r_i = D_i/P_i, \tag{5.3}$$

and

$$r = \sum_{i=1}^{n} r_i \tag{5.4}$$

On substituting the value of t as obtained from (5.2) in (5.1), we get

$$Z(T) = r(b+f) + \frac{1}{T}\left(a + \sum_{i=1}^{n} S_i K_i\right) + \frac{T}{2}\left(vr^2 + \sum_{i=1}^{n} \frac{D_i h_i}{K_i}(1 - D_i/P_i)\right) \tag{5.5}$$

A necessary, but not sufficient condition for the feasibility of the maintenance and production of the lot sizes is obtained as follows:

$$\left(x + \sum_{i=1}^{n} (w_i K_i + T r_i)\right) \leq T$$
$$\frac{x + \sum_{i=1}^{n} w_i K_i}{(1-r)} \leq T \tag{5.6}$$

At a given value of T, the economic value of K_i equal to $K_i(T)$, for the ith product, can be obtained by satisfying the following condition, see Goyal [2,3]:

$$K_i(T) \cdot (K_i(T) - 1) \leq (m_i T)^2 \leq K_i(T) \cdot (K_i(T) + 1) \tag{5.7}$$

where for the ith item

$$m_i = \sqrt{\frac{D_i h_i (1 - r_i)}{2 S_i}} \tag{5.8}$$

By substituting $K_i = K_i(T)$, obtained from (5.7) in (5.5), and equating the first derivative of $Z(T)$ with respect to T equal to zero, we get

$$T = \sqrt{\frac{2\left(a + \sum_{i=1}^{n} S_i K_i(T)\right)}{vr^2 + \sum_{i=1}^{n} \frac{D_i h_i (1-r_i)}{K_i(T)}}} \tag{5.9}$$

To determine T, we need to know the K_i value of each product, which in turn requires us to know the value of T. Therefore, a search procedure can be adopted for determining the economic operating policy. Essentially, the problem is to obtain a feasible repetitive production schedule for n products and to find a free time equal to 'x' for carrying out maintenance.

The problem of finding a feasible production schedule for several products manufactured on a single machine can be partially resolved by restricting the value of K_i to powers of 2 (see Goyal [4] and Haessler [6]).

However, when the maintenance work requires a finite time period, the problem of obtaining a repetitive production schedule becomes more acute. Two simple alternatives are possible.

Alternative 1 : Assume $K_i = K_i(T) = 1$ for each item, and obtain the economic value of T from (5.9) and the value of $Z(T)$ from (5.5). If the condition given by (5.6) is not satisfied, crash the maintenance work, and add the additional cost due to the crash working to $Z(T)$.

Alternative 2 : At first, ignore the maintenance problem. Obtain a repetitive production schedule for the products. Schedule maintenance during the lengthiest idle time in the production schedule. If this idle time is less than the time required to carry out maintenance, then crash the maintenance work. Determine the total cost of this alternative by adding the additional crash cost of maintenance work to the $Z(T)$ obtained from (5.5).

The alternative providing the lower cost is selected. Once the economic value of T and $K_i(T)$ for each product is known, the lot size for each product is easily computed by evaluating $TD_i/K_i(T)$.

Table 5.1 Parameters for the example problem.

i	S_i	h_i	w_i	D_i	P_i	r_i
1	50	1.0	10	12,000	40,000	0.30
2	100	1.2	8	4,000	30,000	0.13
3	200	1.3	12	20,000	100,000	0.20

4. NUMERICAL EXAMPLE

For illustrative purposes, we will use the data in Table 5.1 to demonstrate the applicability of each proposed solution alternative. We assume that there are three products (denoted by i) to be produced on a single machine. The time unit is taken to be one year with 250 working days, 8 hours of regular shift per working day.

Linear maintenance cost $= a + bt = 450 + 1000t$
Linear operating cost $= f + vt = 1550 + 250t$.

$$r = \sum_{i=1}^{n} r_i = 0.63$$

In Table 5.1 above, parameters a,b,f,v,S_i and h_i are specified in terms of dollars, whereas w_i is specified in terms of hours. Demand rate, D_i, and production rate, P_i, are given in terms of units per year. In order to demonstrate the impact of maintenance time on the feasibility of the proposed solution alternatives, we provide a numerical example.

Maintenance time, x, is assumed to be 60 hrs per cycle. To obtain the solution for Alternative 1, assume $K_i(T) = 1$ for each product. The cycle time is computed according to equation (5.9) as

$$T_1 = \sqrt{\frac{2(450 + (50 + 100 + 200))}{(250(0.63)^2 + (8,400 + 4,160 + 20,800))}} = 0.22$$

where $T_1 = 0.22$ years is equivalent to 55 working days or 440 hrs. The production and maintenance schedule for the solution to Alternative 1 with maintenance time of 60 hrs/cycle is provided below both in the form of a table, Table 5.2, and a Gantt chart, Figure 5.1. Such a schedule results in an idle time of 131 hrs. Since the maintenance time (x=60 hrs) is less than the idle time, the schedule given in Figure 5.1 is a feasible

Table 5.2 Production and maintenance schedule in tabular format for the solution to Alternative 1 with maintenance time of 60 hrs/cycle.

	S_1 (P_1)	O_1 (P_1)	S_2 (P_2)	O_2 (P_2)	S_3 (P_3)	O_3 (P_3)	M	I	Cycle time
Time (hrs)	10	132	8	59	12	88	60	71	440

O_i: Operation i, S_i: Setup i, M: Maintenance, I: Idle time, P_i: Product i

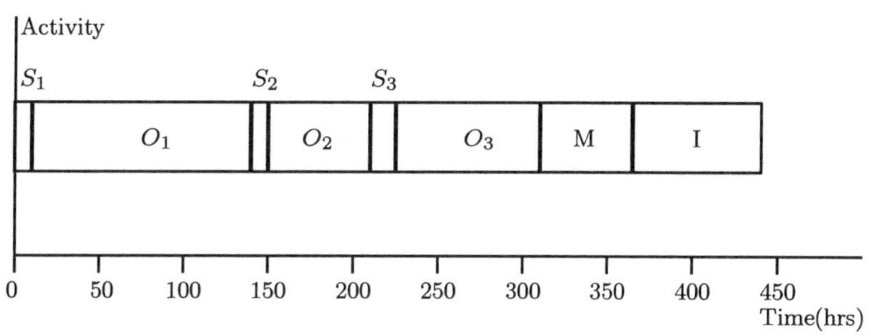

Figure 5.1 Production and maintenance schedule for Alternative 1

schedule. The schedule yields a total annual cost of $Z(T_1)$ =\$8,932 computed according to equation (5.5).

For the solution to Alternative 2, $K_i(T)$ is determined with respect to equation (5.7) and presented in Table 5.3.

The new cycle time as computed according to equation (5.9) is $T_2 = 0.24$ years and is equivalent to 60 days or 480 hours. In this case, the cycle time is to be divided in half since product 1 has to be split into two lots. The production and maintenance schedule for the solution to Alternative 2 with maintenance time of 60 hrs/cycle is provided both in the form of a table, Table 4, and a Gantt chart, Figure 5.2. The total

Table 5.3 Determination of $K_i(T)$ values.

i	Lot Size*	m_i	m_iT**	$K_i(T)$
1	1,309	9.16	2.02	2
2	877	4.56	1.00	1
3	2,774	7.21	1.58	1

* LotSize = $\sqrt{2D_iS_i/h_i(1-r_i)}$
** T_1=0.22 at $K_i(T) = 1$

Table 5.4 Production and maintenance schedule in tabular format for solution to Alternative 2 with maintenance time of 60 hrs/cycle.

	S_1 (P_1)	O_1 (1/2P_1)	S_2 (P_2)	O_2 (P_2)	M	I	Half Cycle
Time (hrs)	10	72	8	64	60	26	240

	S_3 (P_1)	O_3 (1/2P_1)	S_4 (P_3)	O_4 (P_3)	I	Full Cycle
Time (hrs)	10	72	12	96	50	480

O_i: Operation i , S_i: Setup i, M: Maintenance, I: Idle time, P_i: Product i.

annual cost of this alternative is $Z(T_2) = 8,660$, which is computed according to equation (5.5).

In this example, Alternative 2 is preferable with a total annual cost of $8,660 compared to Alternative 1 with a total annual cost of $8,932.

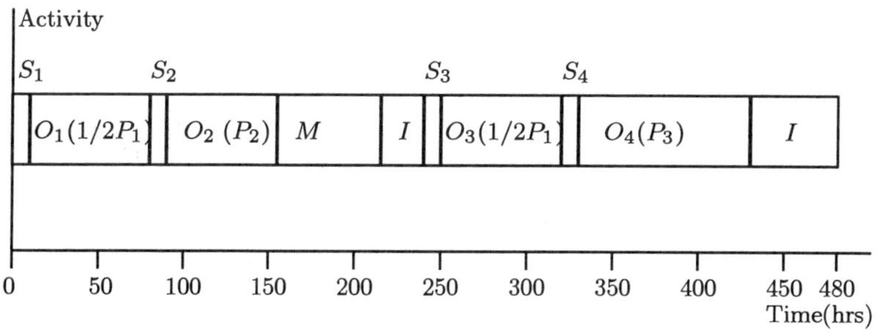

Figure 5.2 Production and maintenance schedule for Alternative 2

References

[1] Christer A., and Doherty D. Scheduling Overhauls of Soaking Pits. *Operational Research Quarterly*, **28** (1977) 915-926.

[2] Goyal S.K. Scheduling a Multi-Product Single Machine System - A New Approach. *International Journal of Production Research*, **13** (1975) 487-493.

[3] Goyal S.K. Determination of Economic Packaging Frequency for Items Jointly Replenished. *Management Science*, **20** (1973) 293-298.

[4] Goyal S.K. Scheduling a Multi-product Single Machine System. *Operational Research Quarterly*, **24** (1973) 261-269.

[5] Goyal S.K., and Kusy M.I. Determining Economic Maintenance Frequency for a Family of Machines. *Journal of the Operational Research Society*, **36** (1985) 1125-1128.

[6] Haessler R., and Hogne S.L. A Note on the Single Machine, Multi-Product Lot Scheduling Problem. *Management Science*, **22** (1976) 909-912.

[7] Luss H., and Zvi K. A Preparedness Model Dealing with N Systems Operating Simultaneously. *Operations Research*, **22** (1974) 117-128.

[8] Sule D., and Harmon B. Determination of Coordinated Maintenance Scheduling Frequencies for a Group of Machines. *American Institute of Industrial Engineers (Transactions)*, **11** (1979) 48-53.

Chapter 6

OPTIMAL MODELS OF PREVENTIVE MAINTENANCE AND REPLACEMENT POLICIES

Eli Shemesh, Abraham Mehrez and Gad Rabinowitz

Department of Industrial Engineering and Management,
Ben Gurion University of The Negev,
Beer Sheva, 84105, Israel

Abstract In this paper, we solve a family of problems related to machine replacement and preventive maintenance. We consider the optimal way of skipping the available technology and be ready for the next technological breakthrough. Skipping the existing technology may have a crucial impact on the firm's performance in a highly competitive world where the up-to-date technology affects both revenues and prestige. We investigate two cases of technological replacement problem for a discrete time finite horizon. The first deals with two available machines: the one-in-use and a more improved existing one that can be purchased to replace the first. The second case is a consideration of a third machine type that will be available in future. The maintenance level of each machine is chosen for each period in order to control performance deterioration. We show that the optimal maintenance policy for a single machine life span has a bang-bang structure, and employ this property in a dynamic programming setting to formulate and solve the problem efficiently. Numerical example demonstrates that management may improve profitability by delaying (but not necessarily foregoing) replacement with a more improved existing machine that is currently available.

Keywords: Machine replacement policy, optimal maintenance, dynamic programming, future technology.

1. INTRODUCTION

We aim at developing a class of finite-horizon models for replacement and maintenance at new and existing technology. In this paper, we concentrate on the optimal replacement strategy, specifically, on the possibility of skipping from the available in-use machine via the existing more advanced machine type directly to the next generation technological breakthrough. Our strategy is to help manufacturers to reach the highest possible exploitation level and expected benefits in limited time horizon. Analytical and numerical results that may assist decision-makers in planning the transition from today's machinery to that of tomorrow are also presented. We define the replacement problem as a discrete time optimal control optimization model possessing two aspects: a continuous level of maintenance policy and a sequence of replacement decisions. The continuous-state variable, which represents the machine's availability state, depends on its entire preventive maintenance program.

The authors of this paper were motivated by an actual maintenance and replacement problem in a textile plant. In this plant, preventive maintenance occurs according to a predetermined schedule, the chosen level of maintenance partially restoring the machine capabilities. The plant's policy is to provide the highest possible level of maintenance, which, it is claimed, maintains each machine in the best possible condition (BPC), although the optimality of this policy is a subject for research. Breakdown maintenance and system reliability are not in the focus of our attention, we consider only age, technology replacement, and preventive maintenance strategies. We do not consider the issue of forecast or planning horizon, in which short-term decisions correspond with long-term decisions (see, Bhaskaran and Sethi, [1]).

For a simple, real-world representation of the machine replacement problem, three technological levels (TL) of machines are considered: the machine currently in use, denoted as technological level 1 (TL1), a more technologically-advanced machine which is available (TL2) and the machine of next-generation technology (TL3). Continuous state-variables represent the machine's expected availability and its obsolescence value; consequently, both of these state-variables depend on the maintenance process.

In this paper, we have used an approach similar to that of Kamien and Schwartz [6] to analyze the effect of an unknown introduction time of a new machine on the optimal replacement timings. Kamien and Schwartz discussed only optimal maintenance and age replacement of machines of a single technological level. In order to extend their perspective, we permit any sequence of three TLs and a multiplicity of machines for each technological level. Both their model and our model assume a continuous maintenance control variable and a quality state variable. However, time is discrete in our model while continuous in theirs. Moreover, we develop a model that provides dynamic programming (DP) solutions for the finite-horizon problems for both deterministic and stochastic cases.

Our model for preventive maintenance permits the attenuation of both the gradual loss of machine quality and the reduction of obsolescence value. Expenditures and profits within the considered horizon are also discounted to maximize the net present profit. Our findings indicate that, under reasonable parametric conditions, a non-increasing bang-bang preventive maintenance policy, i.e. a policy that switches from the BPC maintenance level to the minimum level and vice-versa, is optimal. Thus, we provide a closed-form solution for time switching from the BPC to the minimum preventive maintenance strategy.

The literature on machine replacement models reveals different approaches to optimal replacement decisions due to technological improvements. For example, Kamien and Schwartz [6] analyzed the effect of preventive maintenance on the optimal replacing time of two machines. They have shown an important though intuitive result: the rate of desired preventive maintenance declines over time as a result of machine aging. While considering the forecast horizon problem, Goldstein, Ladany and Mehrez [4] found the planning horizon for the first optimal replacement in an infinite horizon problem when two types of machines are considered; i.e. the first type is currently in operation while the second type, which is more advanced, appears. In their study, Goldstein *et al.* generated some conditional decision rules of optimal replacement by employing a DP model, which, in turn, assumed identical production capacity and quality characteristics for machines of different or identical technologies. An alternative formulation of a machine replacement problem by means of preventive maintenance and probabilistic technological breakthrough was suggested by Sethi and Thompson [16] in a discrete maximum principle, analyzed by Sethi and Morton [14] and Sethi and Thompson [15]. This approach has been used in the Wagner-Whitin [19] model to derive a bang-bang solution by employing a mixed optimization technique dealing with probabilistic technological breakthroughs.

Different models for preventive-maintenance strategy within the framework of sequential decision-making processes and optimal control are presented by Gaimon and Thompson [3], who employed a real-time solution for preventive and repair maintenance. Blaguire [2] extended the basic models of Thompson [17] by adding the option of impulsive machine replacement.

A related area is that of capacity acquisition and disposal. A recent article on this issue was published by Rajagopalan *et al.* [11]. Although they did not discuss the maintenance problem or the decision as to when to replace machinery, they used a DP algorithm to solve the problem of the amount of machine units to acquire and dispose of in order to meet future demand growth. Their work considers the benefits of purchasing new technology when there is a demand increment for more capacity, taking into consideration the uncertainty of successive technological breakthroughs leading to the equipment obsolescence.

The issue of treating a problem under discrete and continuous settings has been previously discussed in the literature. Mehrez [7] evaluated and compared two alternative discrete and continuous formulations of a risky research and development model, while Hopp and Nair [5] examined the interaction between deterioration and technological changes and assumed that all replacements are motivated by equipment deterioration.

In our paper, the replacement decision is considered in a state of uncertainty regarding the introduction time of the TL3-type machine. Examples include personal computers, military gadgets, agricultural equipment, and other systems that are costly to upgrade. The types of the probability mass functions for the introduction time of TL3-type machine include the geometric distribution type with a constant failure rate, and the uniform and normal type with an increasing failure rate. In our context, the increasing failure rate function implies the conditional increase probability of the TL3 introduction time. This feature can be justified if technological knowledge is accumulated over time to increase the conditional probabilities of technological success. We employ a similar formulation to that of Mehrez and Berman [8] and Nair and Hopp [9] in order to analyze the effect of unknown introduction time of future technology on an optimal machine replacement decision. A clear and full structure of costs and benefits of exploiting each machine in the machine replacement chain is also provided. Furthermore, Nair and Hopp [9] did not address the important issue of technological generation skipping, and Mehrez and Berman [8] neither proved concavity nor presented an interior solution that they thought existed. In their paper, the possibility of exchanging a TL2 machine for a TL1 machine was eliminated from the set of possible operations. We do not pose such a restriction and, in our study, the probability distribution of the arrival time of TL3 machines is general and not limited to exponential distribution function. Our main contributions include a more general model, which enables an integrated solution for machine maintenance and replacement. The model clearly differentiates between machine replacement and technological replacement and provides a closed-form solution for optimal maintenance policy in the deterministic case. This solution, in turn, leads to an efficient algorithm for solving the stochastic case as well. This study generalizes previous models and offers a new perspective for future research.

First, we aim at the derivation of an optimal preventive maintenance strategy; second, at the optimal solution of the machine replacement problem, including the above-mentioned three types of technologies under various conditions; and, third, at the proof of an efficient algorithm that provides an integrated solution for these decision problems under any general distribution of TL3.

Our paper is organized as follows. In Section 2, we define the notations and assumptions used throughout the paper. In Section 3, we formulate and solve the deterministic finite horizon model, which allows us to maintain dependent quality deterioration and multiple machine replacements. In section 4, a solution for the stochastic replacement is provided. Numerical examples with some sensitivity tests are given in section 5 and, finally, we propose directions for future research in Section 6.

2. THE MODEL PARAMETERS AND NOTATION

In this section we define the notation and basic assumptions of our project.

Q_j^t = A state variable that designates the level of quality of a TLj machine after t periods of use ($Q_j^t \geq 0$).

T = The length of the planning horizon in time units.

t_3 = The introduction time of TL3 type machine.

$t_{i,j}$ = The planned time of replacement of a TLi machine by a TLj machine.

m_j^t = The actual expenditure in preventive maintenance of a TLj machine if it is used at time t; $j = 1, 2, 3$. Also $\mathbf{m}_j = (m_j^1, m_j^2, ..., m_j^T)$.

$\bar{m}_j$ = The expenditure limit of preventive maintenance for a TLj machine.

Π_j = Revenue per quality unit of a TLj machine, for one time period.

a_j = Quality deterioration rate of a TLj machine attributable to aging $0 < a_j < 1$.

b_j = Quality increase per dollar of preventive maintenance expenditure on a TLj machine.

K_j	= Salvage value (in dollars) per quality unit of a TLj machine.
P_j	= Purchasing cost, including set-up, of TLj machine.
$H_j(t, \mathbf{m}_j)$	= Net present value (NPV) of a TLj machine, for t periods and a feasible maintenance strategy $\mathbf{m}_j$.
$J_j(t)$	= The maximum value of $H_j(t, \mathbf{m}_j)$ over m_j, where t is the machine's life-cycle.
r	= The discount rate.

It is assumed that profits for period t are generated at the beginning of the period, and that maintenance expenses for period t are invested at the end of period $t-1$ for $t = 1, ..., T$. Note that $P_j \geq K_j Q_j^0$ because the seller would not sell for less than the salvage value at the time of sale.

In our study, quality is represented by a single variable, as given by Ross [13]. There are several different methods to measure quality, subjective methods and objective ones. We adopted the Kamien and Schwartz [6] approach to define the *state* of a machine as the percentage of time when the machine is available. Like Virtanen [18], we consider the maintenance policy to be a deterministic optimal control problem. The machine's quality is assumed to follow a dynamic linear relationship (in this section we omit the machine index j):

$$Q^t = Q^{t-1} - aQ^{t-1} + bm^t. \tag{6.1}$$

Employing this assumption, the single-machine finite horizon NPV is given by

$$H(t, \mathbf{m}) = \sum_{s=0}^{t-1} (1+r)^{-s} \left(\Pi Q^s - m^{s+1}\right) - P + (1+r)^{-t} KQ^t, t \geq 0. \tag{6.2}$$

By employing (6.1) iteratively, we express Q^s as a linear expression of $\mathbf{m}$ only.

$$Q^s = (1-a)^s Q^0 + b \sum_{i=0}^{s-1} (1-a)^i m^{s-i}, \quad s = 1, ..., t. \tag{6.3}$$

By inserting equation (6.3) into equation (6.2) and setting $\beta = \frac{1-a}{1+r}$, $\theta = \Pi \frac{1+r}{a+r}$, the NPV becomes a linear expression of $\mathbf{m}$ (all proofs are in the Appendix):

$$H(t,\mathbf{m}) = Q^0\left(\theta + (K-\theta)\beta^t\right) - P$$

$$+\sum_{s=1}^{t}\left(m^s(1+r)^{-s}\left((\Pi b-(a+r))\frac{1+r}{a+r}+(K-\theta)b\beta^{t-s}\right)\right). \quad (6.4)$$

In the next section we solve the deterministic maintenance and replacement case where the TL3 type machine is available at some definite future time.

3. GENERAL PROBLEM SOLUTION OF THE OPTIMAL MAINTENANCE STRATEGY

For solving the optimal maintenance and replacement problem, first, we develop a closed-form solution for the maintenance policy. Then, we employ this solution to express the expected profit of t periods for a single machine under optimal maintenance strategy (OMS) for any period t and use this value for solving the DP problem for multiple machine types. The state variable indicates the condition of the equipment, which degrades with age, the control variables are the maintenance effort and replacement.

We formulate a DP recursive equation describing a replacement problem according to an optimal maintenance policy in a finite horizon of T periods, where machines of all three TLs are available at $t = 0$:

$$f(n) = \max_{k=n+1,\ldots,T}\left\{\begin{array}{l}(1+r)^{-n}J_1(k-n)+f(k);\\(1+r)^{-n}J_2(k-n)+f(k);\\(1+r)^{-n}J_3(k-n)+f(k)\end{array}\right\}. \quad (6.5)$$

We assume a machine replacement at period (n) and search for both a later period (k) and the TL type for the machine replacement that would maximize the NPV. Following Bellman's principle of optimality, the function $f(k)$ is the NPV for optimal functioning from period k to period T. Note, that $f(T) = 0$ and the NPV for an optimal policy for T periods is $f(0)$. In order to avoid triviality, we assume that machines of TL2 and TL3 types are available only at $t \geq 1$; hence, at time 0, we force the purchase of the TL1 machine. In general, if a TLj machine is not available at time n, then $J_j(k-n)$ is considered non-existent.

Using Bellman's theory, we can guarantee that this DP method will reveal all optimal replacement patterns. There can be more than a single optimal path, and, if so, the algorithm will derive all of them. It is clear, that the machine replacement strategy for the multiple machine case does not necessarily include identical machine life-cycles, which is the case for a single TL facing a long horizon. The OMS for every machine's life-cycle and the structure of $J_j(k-n)$ will be defined and studied later.

3.1 OMS analysis. In Lemma 1 introduced below, we analyze the structure of $\mathbf{m}^*$, the OMS. The linearity of the NPV in $\mathbf{m}$ guarantees the bang-bang property of this strategy. However, we need the monotone structure of the coefficients of $\mathbf{m}$ to establish a closed-form policy.

We denote the linear coefficient of m^s in (6.4) by as α^s. While examining (6.4), we consider two productivity aspects:

a) The increase in revenue per dollar spent on preventive maintenance should be greater than the rate of value deterioration of the machine due to both discount rate and aging (i.e., $\Pi b \geq r + a$). Without this condition, even maximum preventive maintenance would not be sufficient to increase productivity.

b) If no maintenance occurs during the last period in the machine's life-cycle, then it is economical to function in this period only if the discounted revenue for t periods is higher than the discounted revenue of $t-1$ periods, i.e., if $m^t = 0$, then $H(t, \mathbf{m}) \geq H(t-1, \mathbf{m})$ is reduced to $\Pi Q^{t-1} + (1+r)^{-1} K (1-a) Q^{t-1} - K Q^{t-1} \geq 0$ and, thus, to $K - \theta \leq 0$.

Four Cases of Maintenance Conditions:

Lemmas 1 and 2 employ the linearity of $H(t, \mathbf{m})$ with $\mathbf{m}$ to show that, on the one hand, a bang-bang maintenance program maximizes $H(t, \mathbf{m})$, and on the other hand, that its structure is characterized by four defined cases. For further analysis, we define and characterize the following four cases in terms of the structure of α, which are the coefficients of $\mathbf{m}$ in $H(t, \mathbf{m})$.

Case 1: $\Pi b - (r+a) \geq 0$ and $(K - \theta) \leq 0$, α are non-increasing and convex with s.

Case 2: $\Pi b - (r+a) \leq 0$ and $(K - \theta) \geq 0$, α are non-decreasing and concave with s.

Case 3: $\Pi b - (r+a) \geq 0$ and $(K - \theta) \geq 0$, α are positive and convex with s.

Case 4: $\Pi b - (r+a) \leq 0$ and $(K - \theta) \leq 0$, α are negative and concave with s.

Lemma 1: The single-machine finite-horizon NPV can be maximized by employing a monotone bang-bang preventive maintenance program. Depending on the problem parameters, there are four cases of program structures.

Lemma 2: Partial maintenance in any period is a non-optimal solution to $J(t)$.

While the formulation may yield a partial maintenance solution, we find out when using Lemma 2 that a bang-bang maintenance strategy is more optimal. Specifically, if as $\alpha^s \geq 0$, then setting $m^s = \bar{m}$ is an improvement over partial maintenance. Optimal maintenance program solving (6.4) has the following bang-bang structure in each previously defined case of α, with the coefficients of $\mathbf{m}$ in $H(t, \mathbf{m})$.

i) In Case 1, the optimal maintenance program is non-increasing (both productivity aspects formerly defined as $\Pi b - (r + a) \geq 0$ and $(K - \theta) \leq 0$ are true); i.e., k periods with full maintenance (starting at $s = 1$) and $t - k$ with no maintenance.

ii) In Case 2, the optimal maintenance program is non-decreasing; i.e., k periods with no maintenance (starting at $s = 1$) and $t - k$ with full maintenance.

iii) In Case 3, maintenance should be performed periodically; i.e., all periods with full maintenance.

iv) In Case 4, no preventive maintenance should be given at all; i.e., all periods with no maintenance.

The fact that the time, during which the user may benefit, diminishes toward the end of the planning horizon is evidence of the deteriorating maintenance policy. In their survey, Pierskalla and Voelker [10] included models involving the optimal maintenance decision and the systems subjected to non-decreasing service. The practice of maintenance indicates that both non-increasing and non-decreasing preventive maintenance efforts might be rational.

The optimal solution, $J(t)$, defined for a single machine operated during a horizon of t periods, is determined by the alternative that maximizes its discounted objective function. For a given machine and time horizon, the optimal maintenance program is non-alternative. If Cases 1 or 2 prevail, the program may seldom contain two alternative solutions, due to the discrete structure of the problem under consideration. The results of solving this problem set will be employed in the following sections in order to discover a DP solution for the multiple replacement problems.

3.2 The maintenance switching time for Case 1.

For an optimal maintenance program, the bang-bang switching time t', when the preventive maintenance strategy switches from the maximum to minimum level is the time when α, the coefficient of m, switches from a positive to a negative value. Setting this coefficient to zero, we get the length of time with no maintenance: $d = \left\lceil \frac{1}{\ln(\beta)} \ln\left(\frac{\theta - \frac{1+r}{b}}{\theta - K}\right)\right\rceil$. Thus, the maintenance switching time from maximum to minimum is derived as follows:

$$t' = \begin{cases} 0 & d \geq t \\ t & d \leq 0 \\ t - d, \text{otherwise} & . \end{cases} \tag{6.6}$$

Assuming the conditions of Lemma 1 and using equation (6.6) in equation (6.4), the objective function becomes a function of t only. Setting $J(t)$ as $H(t, \mathbf{m}^*(t))$, we get

$$J(t) = \left[\theta\left(Q^0 + \frac{\bar{m}b}{r}\right) - \frac{\bar{m}}{r}(1+r) - P\right] - \left[(\theta - K)\left(Q^0 - \frac{\bar{m}b}{a}\right)\right]\beta^t$$

$$-\left[\frac{\bar{m}b}{a}(\theta - K)(1-a)^d + \frac{\bar{m}}{r}(b\theta - (1+r))(1+r)^d\right](1+r)^{-t}, \tag{6.7}$$

where $d = t - t' \geq 0$.

3.3 The maintenance switching time for Case 2.

In Case 2, where the maintenance strategy switches from the minimum to the maximum level, the bang-bang switching time is the time when the coefficient of m, switches from a negative to a positive value. As in Case 1, the switching time derived is (6.6). Calculating $J(t)$ for Case 2, we get

$$J(t) = (\theta Q^0 - P) - (\theta - K)Q^0\beta^t -$$

$$\left[\frac{\bar{m}b}{a}(\theta - K)\left(1 - (1-a)^d\right) - \frac{\bar{m}}{r}(b\theta - (1+r))\left((1+r)^d - 1\right)\right](1+r)^{-t} \tag{6.8}$$

3.4 The maintenance switching time for Case 3.

In Case 3, with full preventive maintenance, we get

$$J(t) = \left[\theta\left(Q^0 + \frac{\bar{m}b}{r}\right) - \frac{\bar{m}}{r}(1+r) - P\right] - \left[(\theta - K)\left(Q^0 - \frac{\bar{m}b}{a}\right)\right]\beta^t$$

$$- \left[\frac{\bar{m}b}{a}(\theta - K) + \frac{\bar{m}}{r}(b\theta - (1+r))\right](1+r)^{-t}. \qquad (6.9)$$

3.5 The maintenance switching time for Case 4.

For Case 4 with no preventive maintenance we get

$$J(t) = \left(\theta Q^0 - P\right) - (\theta - K)\, Q^0 \beta^t. \qquad (6.10)$$

Note, that using the natural assumptions that $H(t, \mathbf{m}) \geq H(t-1, \mathbf{m})$ and those in case 1, we can conclude that $\frac{\Pi}{r+a} \geq \frac{1}{b} \geq \frac{K}{1+r}$. Mehrez and Berman [8] have introduced a non-increasing strategy with identical restrictions for continuous time optimal maintenance. The non-increasing property of the optimal maintenance policy coincides with that of Kamien and Schwartz [6] under a continuous time setting.

We use closed-form solutions for the maintenance program and the $J(t)$ properties (Lemma 2) in order to solve the deterministic replacement problem as presented in equation (6.5). However, the deterministic approach to the problem seems to be, to some extent, limited, if the benefits of the stochastic approach are not considered. Thus, we face some difficulties, since the time to replace the now-in-service machine depends on the non-identified TL3 machine arrival time. In this case, we cannot pre-determine the maintenance switching time for pre-defined cases 1 and 2.

4. STOCHASTIC MACHINE REPLACEMENT IN FINITE HORIZON

Let us discuss now the case in which a TL3 machine becomes available at some unknown time in the future. A TL1 type machine is purchased at $t_1 = 0$, because a new and more advanced TL2 machine will not be available until $t_2 = 1$ (note that a TL1 machine would not be purchased at $t_1 = 0$ if, at that time, a TL2 machine was available). TL3 type machines are expected in the future at some unknown period t_3, and after their appearance on the market, any of the three TLs can be used repeatedly within the planning horizon. The parameters of all technologies are known, and the introduction time t_3, though uncertain, exists according to some special probability distribution. If a TL3 machine is not yet available at time n, the functional machine (TL1 or TL2) will be used for at least one additional time period. In this case, it might be optimal to work without a TL3 machine for more than one period and to consider purchasing it at period k. As in the deterministic case, the bang-bang property of the OMS is reduced here to the set of the machine's states. This property limits the size of the problem to a manageable level and allows a DP solution to the stochastic maximization problem.

Depending on the information about the uncertainty in the model, we face three types of control methods - (a) open loop (where no information is collected for $T > 0$ and the optimal decision is based solely on the information at $t = 0$); (b) feedback (which treats it as a sequence of open loop problems in the sense that future information is ignored and decision is based on the current information which changes with time) and, (c) closed loop (which takes into account the future information to be collected and used in the optimization). In our case we use the open loop control method.

Let us define $g_t(.)$ as the NPV of replacing the now-in-service machine with that of type TL 1, 2, or 3 (if it is available) at period t, and then optimally exploit it. Thus, during the t time period, we may decide to continue using the current machine. The NPV, presenting the optimal replacement and maintenance from time t to the horizon T, can be formulated in the following way.

Let A_3^{t+1} be the availability of TL3 machine at time $t+1$.

If the machine is available, its value is set to 1; otherwise, its value is set to 0.

Let $Y^{t+1}\left(Q_j^k, j_t, j_{t+1}\right)$ be the state of the machine in use at the beginning of time period $t+1$.

If the machine used at time t is replaced ($j_t \neq j_{t+1}$), then $Y^{t+1} = Q_{j+1}^0$. Otherwise, Y^{t+1} is equal to Q_j^k , which is the state of a TLj machine after working k periods.

Let $R\left(j_t, m^t, Y^t\right)$ be the revenue of operating during period t with a TLj machine, while the preventive maintenance policy and the machine quality are taken into consideration.

Let p_t be the probability that a TL3 machine will be available at time t, supposing that it was not available before t.

We can now formulate the NPV as

$$g_t\left(j_t, Y^t \left| A_3^t = 0\right.\right) = \underset{j_{t+1}, j'_{t+1}, m^t}{\text{maximize}} \{X\} \tag{6.11}$$

where

$$X = R\left(j_t, m^t, Y^t\right) + p_{t+1} g_{t+1}\left(j_{t+1}, Y^{t+1} \left| A_3^{t+1} = 1\right.\right)/(1+r') +$$

$$(1 - p_{t+1})\, g_{t+1}\left(j_{t+1}, Y^{t+1} \left| A_3^{t+1} = 0\right.\right)/(1+r')$$

and

$$g_t\left(j_t, Y^t \left| A_3^t = 1\right.\right) = \underset{j_{t+1}, j'_{t+1}, m^t}{\text{maximize}} \{Y\}, \tag{6.12}$$

$$Y = R\left(j_t, m^t, Y^t\right) + g_{t+1}\left(j_{t+1}, Y^{t+1} \left| A_3^{t+1} = 1\right.\right)/(1+r'),$$

which enables us to solve this problem using DP technique.

The maximization is conducted over $j_{t+1} \in \{1,2\}$, when TL3 is not available at $t+1$; if it is available, then $j'_{t+1} \in \{1,2,3\}$, and $m^t \in \{0, \bar{m}\}$ for the maintenance policy of a TLj machine during period t. Notice that Y^{t+1} depends on j_t, j_{t+1} or $j'_{t+1}$and on m^t.

Now, let us concentrate on the last stage, when $g_T(.) = 0$, and use the backward recursive equations (6.11) and (6.12) in order to receive the final result. The objective function for the three machine types in finite horizon under optimal strategy is thus $Z_{1,2,3}(T) = g_0\left(j_0 = 1, Y_1^0 \left| A_3^0 = 0\right.\right)$. The solution algorithm is a simple DP algorithm, but, since the objective function is more complicated, the complexity level of the solution changes to $O(T^3)$. It is obvious that the optimal replacement policy depends on the actual realization of the TL3 type machine arrival time.

5. NUMERICAL EXAMPLES

In this section we present the results of our computations. First, we examine the deterministic solution to the problem i.e., how to use the various machines we have designated within a pre-decided horizon. Next, we illustrate the effect of the probability distribution on the NPV and perform sensitivity tests.

5.1 Optimal replacement strategy. In general, a TL3 type machine is expected to be available at some unexpected time $\tilde{t}_3$ with a known probability distribution. First, we solve the deterministic case, assuming t_3 is known; this case is presented for values of $t_3 = 1, ..., T$. Next, we investigate the stochastic case with a geometric probability distribution of $\tilde{t}_3$ and parameter p=0.07. Table 6.1 presents the data for the numerical experiments. To simplify the presentation, let us consider identical machine types except for the initial quality level Q_j^0 and purchase costs P_j.

Table 6.1 Raw Data for the Three-Machine Types.

		Units	Machine TL1	Machine TL2	Machine TL3
Maintenance expenditure limit	$\bar{m}_j$	\$	20	20	20
Revenue per quality unit	Π_j	\$ per quality per week	30	30	30
Purchase cost	P_j	\$	3900	4000	4100
Initial quality level	Q_0^j	quality units	100	108	110
Rate of decrease in quality	a_j	%	2%	2%	2%
Maintenance increase	b_j	quality per \$	0.01	0.01	0.01
Salvage value	K_j	\$ per quality	15	15	15
Discount rate	r	%	5%		
Horizon	T	Weeks	30		

Figure 6.1 presents the results of T deterministic cases of the realization of the optimal replacement times (vertical scale) versus the introduction time of a TL3 machine (horizontal scale) under optimal strategy in a finite horizon. Each column is actually a solution that uses the DP technique for the deterministic problem.

For example, if a TL3 machine is introduced at time $t_3 = 13$, the best replacement strategy is to purchase a TL2 machine at $t = 5$, then to purchase a TL3 machine at $t = 13$, and to replace it with a second TL3 machine at $t = 22$. If the TL3 machine is introduced at time $t_3 \geq 25$, then it is not beneficial to purchase it at any point within the horizon of 30 time-units.

Let us now examine the solution of the stochastic problem defined by equations (6.11) and (6.12). Presenting the solution requires a T level binary tree that determines which machine and maintenance policy to choose if a TL3 is available and if it is not. Instead, we present the probability of using the ith TLj type machine at every given t period

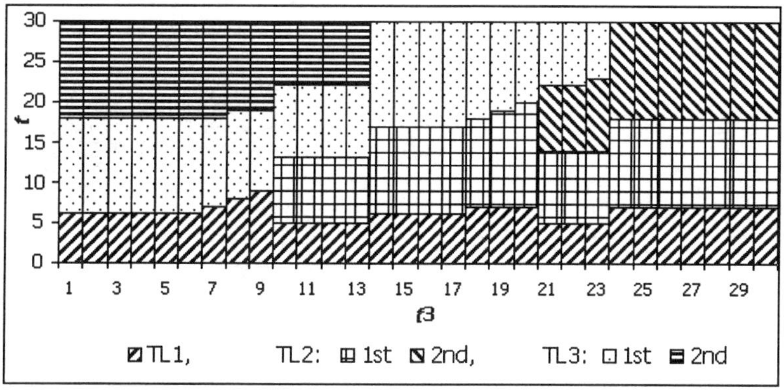

Figure 6.1 Optimal machine replacement sequences under deterministic cases of t_3.

within the problem horizon. For the following stochastic example, the probability of exploitation of all machine types is illustrated in Figure 6.2.

In this example for a horizon of 30 time-units at $t = 20$, there is a 1% probability that the machine-in-use is the first one of the TL2 type, a 16% probability that it is the first one of the TL3 type, a 46% probability that it is the second one of the TL3 type, and a 28% probability that it is the third one of the TL3 type. At time $t = 30$, there is a 8% probability that the now-in-use machine is the first one of the TL3 type, a 27% probability that it is the second one of the TL3 type, and 65% probability that it is the third one of the TL3 type.

5.2 Sensitivity Tests. We would like to use the machine parameters of the alternative TLs to derive a performance index for each of them, which would rank the expected machine performances. Unfortunately, it is impossible to derive such an index because the advantage of one machine over another depends on the machine life-cycle. Let us illustrate this issue with the following example, in which we compare TL1 machine and TL2 machine types (Table 6.1), with the following

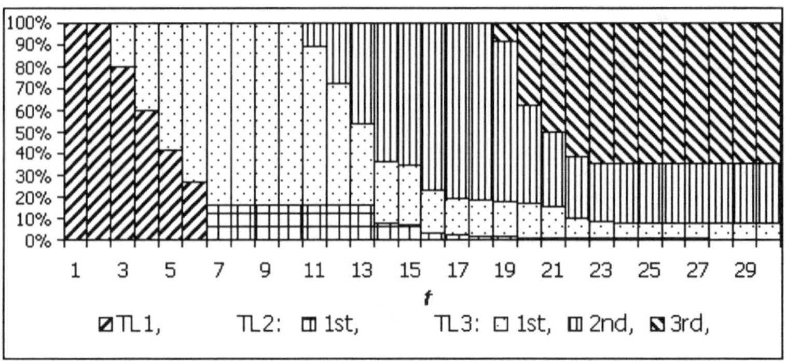

Figure 6.2 Probability of machine's use over time, under Geometric PDF with p=0.07

changes of TL2 parameters $Q_2^0 = 88$, $a_2 = 1\%$, and $b_2 = 0.05$. The results are presented in Figure 6.3.

It is clear that the horizon's effect on the preference of a single machine is relevant to the replacement sequence of different machine types. In the following example (Figure 6.4), a similar effect is demonstrated by the comparison between two variants of the finite horizon deterministic problem. The only difference between these variants is in the properties of the TL2 machine. The TL2 machine has the parameters given in Table 6.1, and the TL2* machine has nearly the same ones except $Q_2^0 = 85$, $a_2 = 1\%$, and $b_2 = 0.11$. The introduction time of a TL3 machine (t_3) affects the objective function in a significantly different pattern in each case. The results for a horizon of 30 time-units are presented in Figure 6.4.

The analysis of the phenomena demonstrated in Figures 6.3 and 6.4, clarifies that the term *machine quality* should be based not only on the parameters of the machines, but also on some additional information, e.g., horizon time, discount rate etc. Our result, which seems to agree with Rosenblatt and Carlson [12], in reality, emphasizes the trade-offs between efficiency and profitability. Dealing with n future technologies requires the definition of n dimensional probability distribution function

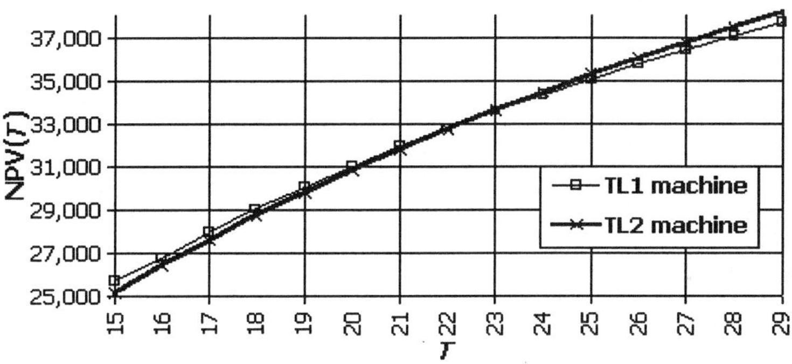

Figure 6.3 Deterministic NPV for TL1 and TL2 versus machine life-cycle.

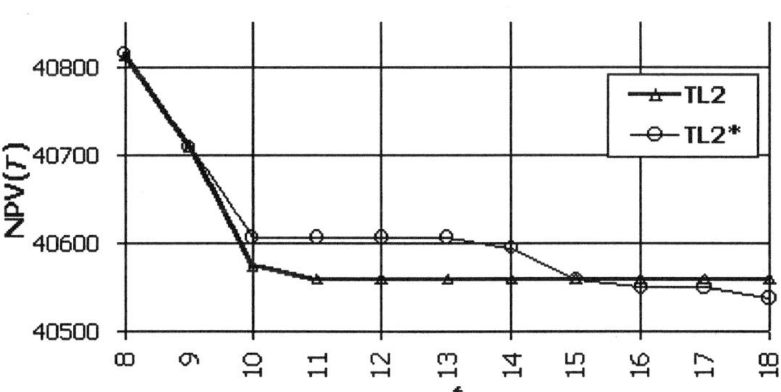

Figure 6.4 Deterministic NPV versus the introduction time of a TL3 machine.

(PDF) of their unknown introduction time. The dynamic programming node is then split into 2^n branches, so evidently our method may be applied successfully to a large number of industrial techniques. However, the mentioned above extension process influences the complexity of the problem solution.

In our study the probability of TL3 becoming available within the time horizon is not limited to a geometric or any other specific PDF. The effect of the PDF on the overall performance of the system requires special identification of the appropriate PDF to the particular technological development process in industry. Some industries, for example, are characterized by incremental technological improvements, while others make dramatic jumps in their development.

6. CONCLUSIONS AND APPLICATIONS FOR FUTURE RESEARCH

Our paper integrates technological changes in machine replacement, maintenance control under discrete time, and the stochastic introduction time of the new technology. Its main contributions are the definition and efficient solution of the integrated stochastic decision problem. The main finding that enables these results is that the optimal maintenance policy for a single machine's life-cycle is bang-bang (four cases defined by the machines parameters in Lemma 1). The numerical examples show that no simple index of the machine's parameters can be used for ranking them economically.

Frequently it is advantageous to purchase the TL2 type machine as a replacement for a TL1 machine immediately. However, as the gap between the TL parameters lessens, the purchase of TL2 might be delayed or foregone entirely. In this particular case, the discount rate, the other parameters of TL1 and TL2 or any of the parameters of TL3 have substantial impact on the outcome. Some combinations of parameters may also lead to interior solution, particularly, if TL2 is not evidently superior to TL1 in all respects.

Managers attempting to determine a policy for machine replacement under technological uncertainty face a complex problem, which we solved here. First, we can predict scientifically the maintenance and replacement strategy for the machine in use (Figure 6.1). Second, we can plan the machine's life-cycle, upgrade prospects, and get the best out of the machine (Figure 6.2). Extending this method to a larger number of technologies is also possible. The decision-maker could base an evaluation of the situation on additional related aspects such as the amount of time devoted to preventive maintenance or the prestige of having new technology. We will concentrate on these related aspects in our next paper.

Acknowledgements
The authors would like to thank Dr. Alex Tsookerman for his valuable comments and assistance. The authors also thank the referees for helpful comments and suggestions.

References

[1] Bhaskaran, S., and S.P. Sethi, Decision and Forecast Horizon in a Stochastic Environment: A Survey. *Optimal Control Application and Methods,* **8** (1987) 201-217.

[2] Blaguire, A. Necessary and Sufficient Condition for Optimal Strategies in Impulsive Control. In Liu, P.T., and E.O. Roxin, (Eds.). *Differential Games and Control Theory III.* Marcel Dekker, New York 1979.

[3] Gaimon, C., and G.L. Thompson, A Real-Time Solution for Preventive and Repair Maintenance. *Optimal Control Application and Methods*, **10** (1989) 211-228.

[4] Goldstein, T., S.P. Ladany, and A. Mehrez, A Discounted Machine Replacement Model with an Expected Future Technological Breakthrough. *Naval Research Logistics Quarterly,* **35** (1988) 209-220.

[5] Hopp, W.J., and S.K. Nair, (1994). Markovian Deterioration and Technological Change. *IIE Transactions*, **26**(6) 74-82.

[6] Kamien, N.L., and M.I. Schwartz, Optimal Maintenance and Sale Age for a Machine Subject to Failure. *Management Science*, **17** (1971) B495-B504.

[7] Mehrez, A. A Note on the Analysis of the Expected Value of Perfect Information with Respect to a Class of R&D Projects. *European Journal of Operations Research,* **19** (1985) 217-221.

[8] Mehrez, A., and N. Berman, Maintenance Optimal Control Three-Machine Replacement Model under Technological Breakthrough Expectations. *Journal of Optimization Theory and Applications,* **81**(3) (1994) 591-618.

[9] Nair, S.K., and W.J. Hopp, A Model for Equipment Replacement Due to Technological Obsolescence. *European Journal of Operational Research,* **63** (1992) 207-221.

[10] Pierskalla, W.P., and J.A. Voelker, A Survey of Maintenance Models: The Control and Surveillance of Deteriorating Systems. *Naval Research Logistics Quarterly,* **23** (1976) 353-388.

[11] Rajagopalan, S., M.R. Singh, and T.E. Morton, Capacity Expansion and Replacement in Growing Market with Uncertain Technological Breakthrough. *Management Science*, **44** (1998) 12-30.

[12] Rosenblatt, M.J., and R.C. Carlson, Designing a Production Line to Maximize Profit. *IIE Transactions*, **19**(2) (1985) 117-121.

[13] Ross, J.E., *Total Quality Management.* St. Lucie Press, St. Lucie, FL. 1993.

[14] Sethi, S.P., and T.E. Morton, A Mixed Optimization Technique for the Generalized Machine Replacement Problem. *Naval Research Logistics Quarterly*, **19** (1972) 471-481.

[15] Sethi, S.P., and G.L. Thompson, Christmas Toy Manufacturer's Problem: An Application of the Stochastic Maximum Principle. *OpSearch*, **14** (1977) 161-173.

[16] Sethi, S.P., and G.L. Thompson, *Optimal Control Theory: Applications to Management Science.* Martinus Nijhoff, Amsterdam, NL, 1981.

[17] Thompson, G.L., Optimal Maintenance Policy and Sale Date of a Machine. *Management Science*, **14** (1968) 543-550.

[18] Virtanen, I., Optimal Maintenance Policy and Planned Sale Date for a Machine Subject to Deterioration and Random Failure. *European Journal of Operational Research*, **9** (1982) 33-40.

[19] Wagner, H.M., and T.M. Whitin, Dynamic Version of Economic Lot Size Model. *Management Science*, **5** (1958) 89-96.

Appendix

Proof of Equation (6.4)
Since $Q^1 = (1-a)\,Q^0 + bm^1$, $Q^2 = (1-a)\,Q^1 + bm^2 = (1-a)^2\,Q^0 + (1-a)\,bm^1 + bm^2$, etc., then $Q^t = (1-a)^t\,Q^0 + \sum_{s=1}^{t}(1-a)^{t-s}\,bm^s = (1-a)^t\left(Q^0 + b\sum_{s=1}^{t}(1-a)^{-s}\,m^s\right)$.

We use this structure to re-write equation (6.2) as:

$$\begin{aligned} H(t,\mathbf{m}) &= \Pi\sum_{s=0}^{t-1}(1+r)^{-s}\,Q^s - P - \sum_{s=0}^{t-1}(1+r)^{-s}\,m^{s+1} \\ &\quad (1+r)^{-t}\,K\,(1-a)^t\left(Q^0 + b\sum_{s=1}^{t}(1-a)^{-s}\,m^s\right) \end{aligned}$$

$$
\begin{aligned}
&= \Pi Q^0 + \Pi \sum_{s=1}^{t-1} (1+r)^{-s} (1-a)^s \left[Q^0 + b \sum_{i=1}^{s} (1-a)^{-i} m^i \right] \\
&- P - \sum_{s=1}^{t} (1+r)^{-s+1} m^s + \beta^t K Q^0 + \beta^t K b \sum_{s=1}^{t} (1-a)^{-s} m^s \\
&= \Pi \frac{1-\beta^t}{1-\beta} Q^0 + \Pi b \sum_{s=1}^{t-1} \beta^s \sum_{i=1}^{s} (1-a)^{-i} m^i \\
&- P - \sum_{s=1}^{t} (1+r)^{-s+1} m^s + \beta^t K Q^0 + \beta^t K b \sum_{s=1}^{t} (1-a)^{-s} m^s .
\end{aligned}
$$

Using $\frac{\Pi}{1-\beta} = \theta$ and $\beta^t (1-a)^{-s} = \beta^{t-s} (1+r)^{-s}$, we obtain:

$$
\begin{aligned}
H(t, \mathbf{m}) &= Q^0 \left[\theta \left(1 - \beta^t\right) + \beta^t K \right] - P + \sum_{s=1}^{t-1} \left[\Pi b (1-a)^{-s} m^s \sum_{i=s}^{t-1} \beta^i \right] \\
&\quad - \sum_{s=1}^{t} (1+r)^{-s+1} m^s + \beta^t K b \sum_{s=1}^{t} (1-a)^{-s} m^s \\
&= Q^0 \left[\theta \left(1 - \beta^t\right) + \beta^t K \right] - P + \sum_{s=1}^{t-1} \Pi b (1-a)^{-s} m^s \frac{\beta^s - \beta^t}{1-\beta} \\
&\quad + \sum_{s=1}^{t} m^s \left[\beta^t K b (1-a)^{-s} - (1+r)^{-s+1} \right] \\
&= Q^0 \left[\theta \left(1 - \beta^t\right) + \beta^t K \right] - P + \sum_{s=1}^{t} \theta b (1+r)^{-s} m^s \left(1 - \beta^{t-s}\right) \\
&\quad + \sum_{s=1}^{t} m^s (1+r)^{-s} \left[\beta^{t-s} K b - (1+r) \right] \\
&= Q^0 \left[\theta \left(1 - \beta^t\right) + \beta^t K \right] - P \\
&\quad \sum_{s=1}^{t} m^s (1+r)^{-s} \left[\theta b - \theta b \beta^{t-s} + K b \beta^{t-s} - (1+r) \right],
\end{aligned}
$$

thus,

$$
H(t, \mathbf{m}) = Q^0 \left[\theta - (\theta - K) \beta^t \right] - P
$$

$$
+ \sum_{s=1}^{t} m^s (1+r)^{-s} \left[b\theta - (1+r) - (\theta - K) b \beta^{t-s} \right].
$$

Proof of Lemma 1:
The solution structure in each of the above four cases is straight forward.

Case 1: A non-increasing bang-bang maintenance program maximizes NPV, beginning with full maintenance in periods of positive coefficients and switching to no maintenance in periods of negative coefficients.

Case 2: A non-decreasing bang-bang maintenance program maximizes NPV, beginning with no maintenance in periods of negative coefficients and switching to full maintenance in periods of positive coefficients.

Case 3: Full maintenance in all periods maximizes NPV.

Case 4: No maintenance in any period maximizes NPV.

Thus, there exist four cases of monotone bang-bang maintenance programs that maximize NPV.

Proof of Lemma 2:
Suppose $\mathbf{m^*}$ is an optimal solution for (6.3) and $0 < m^{s^*} < \bar{m}$, where m^{s^*} is the sth component of $\mathbf{m^*}$. We will show that either $m^s = 0$ or $m^s = \bar{m}$, with no change in the other components, so that we will yield a solution with the same or even higher objective value. By the definition of x_2 [see (6.4)], and since we assumed that $m^{s^*} > 0$, no modification of m^s would change x_2 but setting $m^s = 0$; consequently, x_2 would be increased. If the coefficient of m^s in $x_1 =$ NPV [as defined by (6.7)] is negative, than setting $m^s = 0$ would increase x_1. In this case, since u_i is an increasing function of x_i, setting $m^s = 0$ would produce a higher objective value. If, on the other hand, the coefficient of m^s in x_1 is positive, then setting $m^s = \bar{m}$ would increase it and, thus, produce a higher objective value. In any case, therefore, setting either $m^s = 0$ or $m^s = \bar{m}$ would increase the objective value.

III

PRODUCTION/INVENTORY & QUALITY

Chapter 7

AN INTEGRATED ECONOMIC MODEL FOR INVENTORY AND STATISTICAL PROCESS CONTROL

Enrique Del Castillo and Isabel T. Salcedo
Department of Industrial and Manufacturing Engineering
The Pennsylvania State University
University Park, PA 16802

Abstract Determining the optimal design of a quality control chart has been the subject of study of many authors over the past four decades. In this paper, we present an optimization model for the design of a Shewhart $\overline{X}$ chart, taking into account both inventory and statistical process control considerations. The system consists of a single item produced at a fixed rate in finite-length production runs or batches. The production process experiences at random a single assignable cause that shifts the mean of the quality characteristic. Demand is assumed to follow a Poisson distribution. Search and repair times are assumed to be stochastic. The objective is to minimize the sum of quality-related and inventory-related expected costs per time unit. The model is based on a Markov-chain model for the analysis of an inventory system controlled with an $\overline{X}$ chart by Del Castillo [3] and a model for the economic design of a control chart in finite-length production runs by Ladany and Bedi [5]. A sensitivity analysis is performed on the optimization model based on results from a large experiment. Second order polynomial surfaces were fitted as smooth approximations of the actual responses of interest, some of which, although available in analytic form, were too complicated and non-linear to clarify the underlying relationships. Included is a discussion resolving the apparent trade-offs that exist between having increasingly better quality and improved production-inventory operations. It is shown that a continuous improvement program that reduces the rate of assignable causes and the time for repairs improves all performance metrics, both inventory-related and quality-related.

Keywords: Integrated model, inventory, statistical process control, Markov chain.

1. INTRODUCTION

In a recent paper, Del Castillo [3] proposed a Semi-Markov chain model to study the analytic relations of certain production control variables and certain quality control variables. The model integrates the design of a Shewhart $\overline{X}$ chart with the operation of a single stage, single-item stochastic inventory control model. Due to the exploratory nature of that work, the author did not propose any economic model. In this paper, we present an economic model based on the aforementioned semi-Markov chain approach for the integrated design of an $\overline{X}$ chart and the parameters of a stochastic inventory control model.

The purpose of the model presented here is to explain from an economic-analytic point of view a) the effects of inventory-related parameters on optimal design of an $\overline{X}$ chart and b) the effects of quality-related parameters on optimal production and inventory policy. This work and the work in [3] were motivated in part by the necessity to link, within an analytical framework, the conceptual ideas of the Just-in-Time (JIT) production techniques with the classical ideas of statistical process control (SPC). This includes the necessity to explain the apparent trade-off between a "continuous improvement effort" of the operations, including as an integral part, a reduction of inventories and the feasibility of accomplishing such a goal under the presence of "assignable causes of variation" and the concomitant production stoppages that may result in shortages. Recent publications along these lines are those by Rahim [19] and Ben-Daya [1]. Rahim [19] presents an economic model to determine a control chart design as well as a production schedule. Basic differences between his model and the model presented here are that in the former no reference is made to shortages and their cost; in other words, the service level of the inventory policy is assumed to be 100%. Furthermore, Rahim assumes a deterministic demand rate. Assignable causes occur according to an arbitrary probability distribution, probably with an increasing hazard rate. Whereas, in this paper, we restrict ourselves to exponential times between assignable causes, consider the possibility of shortages due to stoppages, and assume a stochastic (Poisson) demand. Ben-Daya's model [1] is based on Rahim's but incorporates the effect of a preventive maintenance policy on a increasing failure rate assignable cause process. For earlier literature on the subject of integrated SPC-inventory models, the reader is referred to [3], [19], [1] and the references therein.

The paper is organized as follows. Section 2 is a review of the Semi-Markov model, introduced in [3], which serves as the basis for the main contribution of this paper. Section 3 provides an economic optimization model for integrated inventory-SPC decisions. A sensitivity analysis is presented to study the effect of the various inventory-quality factors on the responses. A discussion is included on how to resolve the apparent trade-off that exists between a continuous improvement policy and the necessity of avoiding shortages and improving service level.

2. THE SEMI-MARKOV MODEL

2.1 Assumptions and notation. Consider a single item production system that experiences stochastic demand at a rate of v units per day and produces items at a fixed production rate of $P > v$. A single quality characteristic of the product is monitored using an $\overline{X}$ control chart. Following conventional recommendations in SPC and modern inventory management (e.g., the JIT system), each time there is a sample mean outside the control limits of the chart, the process is stopped and a search for assignable causes of variation is conducted. If a cause is found, it is removed from the system and the process re-started; otherwise, it is considered that a "false alarm" took place.

Shifts occur according to a Poisson Process with rate λ. Shifts in the mean quality level of size $\delta\sigma$ occur, where σ denotes the process standard deviation and $\delta\sigma$ denotes the size of the shift in the process expressed as a multiple of σ. Search and repair times are assumed to be exponentially distributed with mean r. It is assumed that false alarms and true signals take the same amount of time to search and correct.

Production variables include the *nominal* cycle time T, the lot size Q, the production rate P, and the demand of the item which is assumed Poisson with rate v. Quality control variables include the sample size n, the control limit width, k, and the time between samples, h.

The quality control responses we consider are as follow:

- the fraction of output produced while the process was running out-of-control, σ_2;
- the fraction of time the system is down for repairs, σ_3; and
- the fraction of time the system is down for false alarm searches, σ_4.

Production or inventory control responses of interest are as follow:

- the service level of the production system σ_1; and
- the average net inventory, $\overline{I}$.

Del Castillo [3] developed a semi-Markov chain model for the study of the following relations between the quality and inventory control systems:

1. the effect of the quality variables $n, k, h, \delta, \lambda, r$ on production control responses $\sigma_1, \overline{I}$; and

2. the effect of the production variables P, v, Q, T on the quality control responses $\sigma_2, \sigma_3, \sigma_4$.

2.2 The Markov chain model. If samples are taken every h units of time, then, under a Poisson assumption, there is a probability $P_f = 1 - e^{-\lambda h}$ that a process shift can occur during the interval between two samples. If there are f samples taken during the production time T_p, then $h = T_p/(f+1)$. The production time is such that $vT/P = T_p$. Note that in reality, the production time is the sum of times when production occurs. The sum being equal to that time which is needed to complete a batch. Several of such times may be needed to complete a batch if several stoppages occur. For this same reason, the nominal cycle time, T, differs from the actual cycle time, T_c, which is a random variable.

Figure 7.1 (adapted from [3]) shows a tree diagram of the different events that can take place in the time interval between any two samples t and $t+1$ when using a control chart for a process subject to random shocks. Ladany and Bedi [5] used this setup to develop an economic model for the design of p charts. The different transitions between the six states in Figure 7.1 have a simple Markov chain interpretation. Let $1-P_a$ be the power of the control chart, and let α be the probability of a false alarm. For an $\overline{X}$ chart, it is well-known that $1-P_a = \int_{-\infty}^{-k-\delta\sqrt{n}} \phi(z)dz + \int_{k-\delta\sqrt{n}}^{\infty} \phi(z)dz$ and $\alpha = 2\int_k^{\infty} \phi(z)dz$, where $\phi(z)$ is the standard normal density function.

The transition matrix that corresponds to the six states shown in Figure 7.1 is

$$\mathcal{P} = \begin{bmatrix} P_f(1-P_a) & P_fP_a & (1-P_f)\alpha & (1-P_f)(1-\alpha) & 0 & 0 \\ 0 & 0 & 0 & 0 & 1-P_a & P_a \\ P_f(1-P_a) & P_fP_a & (1-P_f)\alpha & (1-P_f)(1-\alpha) & 0 & 0 \\ P_f(1-P_a) & P_fP_a & (1-P_f)\alpha & (1-P_f)(1-\alpha) & 0 & 0 \\ P_f(1-P_a) & P_fP_a & (1-P_f)\alpha & (1-P_f)(1-\alpha) & 0 & 0 \\ 0 & 0 & 0 & 0 & 1-P_a & P_a \end{bmatrix}.$$

It is assumed that production resumes after a shift is detected, a search conducted and, if necessary, the process repaired. If a shift occurs, the process is assumed to return to the in-control state only through detection and correction. The stochastic process described by $\mathcal{P}$ is a semi-Markov process [7] since the length of time the system stays in each state is a function of the random repair time. Therefore, the long-run proba-

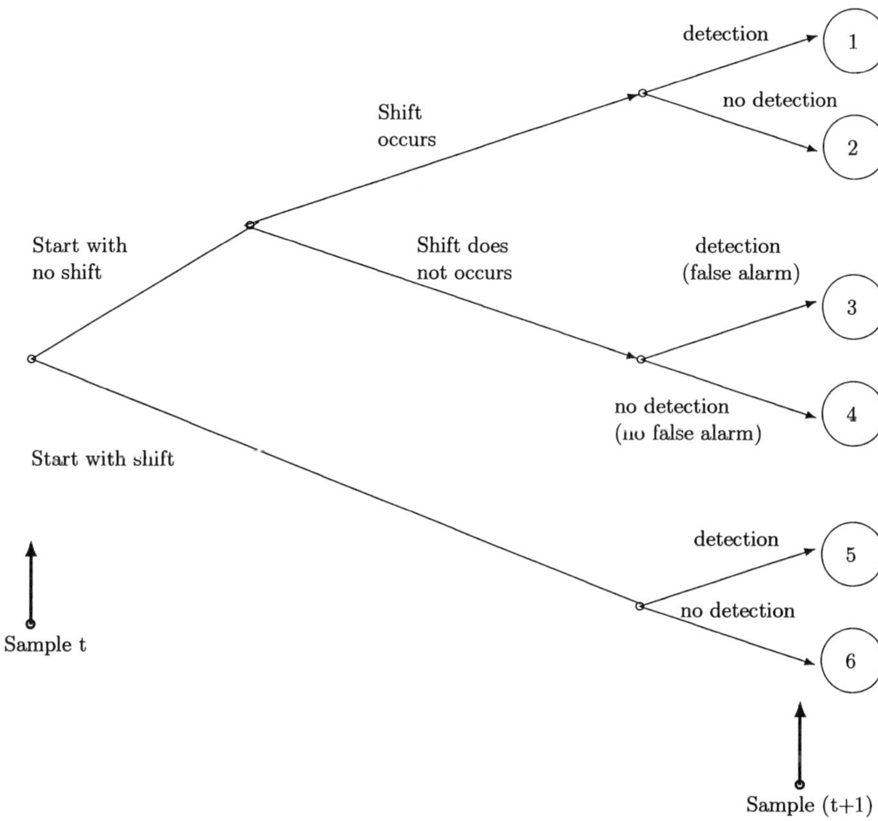

Figure 7.1 Transitions between two consecutive samples.

bilities $\pi_i (i = 1, ..., 6)$ cannot be interpreted as the long run proportion of time the system is at state i. It is easy to show that the matrix $\mathcal{P}$ is regular; therefore, the limiting distribution $\Pi = (\pi_1, \pi_2, ..., \pi_6)$ is obtained from $\Pi = \Pi\mathcal{P}$ and $\sum_{i=1}^{6} \pi_i = 1$. In this way, Del Castillo [3] obtained the following expressions.

$$\pi_2 = \frac{P_f P_a (1 - P_a + P_f P_a - P_f P_a^2)}{(1 - P_a + P_f P_a)(1 + P_f P_a)} \tag{7.1}$$

$$\pi_3 = \frac{(1 - P_f)\alpha(1 - P_a)}{1 - P_a + P_f P_a} \tag{7.2}$$

$$\pi_4 = \frac{(1 - P_f)(1 - \alpha)(1 - P_a)}{1 - P_a + P_f P_a} \tag{7.3}$$

$$\pi_5 = \frac{(1 - P_a) P_f P_a}{1 - P_a + P_f P_a} \tag{7.4}$$

$$\pi_6 = \frac{P_f P_a^2}{1 - P_a + P_f P_a} \tag{7.5}$$

$$\pi_1 = 1 - \sum_{i \neq 1} \pi_i \tag{7.6}$$

The long-run proportion of time the process is in state i is given by

$$t_i = \frac{\pi_i \gamma_i}{\sum_{\text{all } j} \pi_j \gamma_j}, \tag{7.7}$$

where γ_j is the mean time the process stays in state j. Clearly, $\gamma_1 = \gamma_3 = \gamma_5 = r + h$ and $\gamma_2 = \gamma_4 = \gamma_6 = h$.

For a batch-oriented production process, the inventory graph resembles Figure 7.2, where continuous demand during random stoppages (due to signals in the $\overline{X}$ chart) decrease the inventory level. The expected actual cycle time, $E[T_c]$, is given by $E[X] + E[Y]$, provided that I, the final inventory level after the batch is completed, is positive. X is the time needed to complete a batch of product, and Y is the time needed for the inventory to deplete from its final level. It was found in [3] that the average slope of the inventory graph, computed from

$$S = (1 - \sigma_3)(P - v) - \sigma_3 v = (1 - \sigma_3)P - v, \tag{7.8}$$

is vital in understanding the behavior of the process.

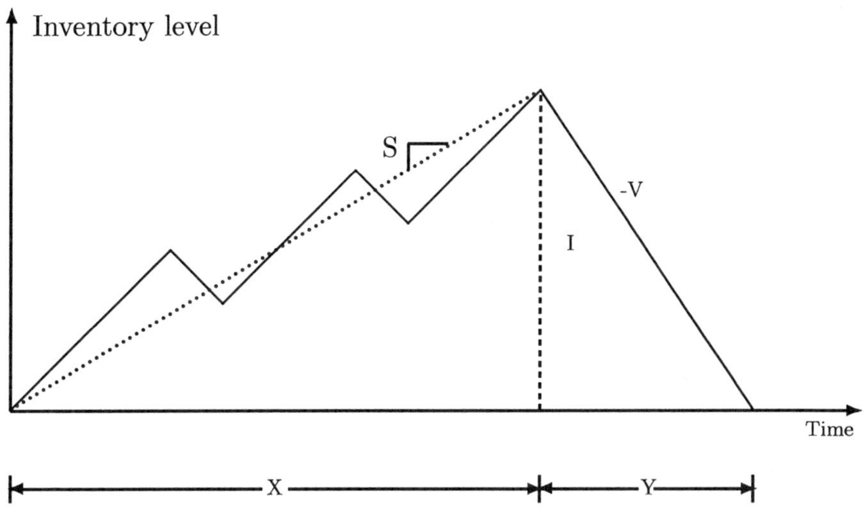

Figure 7.2 A typical inventory graph, showing the average inventory slope.

The following expressions were also derived in [3] and will be used in the next section:

$$E[I] = S^{+}E[X] = S^{+}f\sum \pi_i\gamma_i, \tag{7.9}$$

where $()^{+}$ denotes positive part and

$$E[Y] = E[I]/v. \tag{7.10}$$

The fraction of output produced while the system was running out of control is given by

$$\sigma_2 = \pi_2 + \pi_6. \tag{7.11}$$

The proportion of time the system is down for repairs is given by

$$\sigma_3 = \frac{fr(\pi_1 + \pi_3 + \pi_5)}{E[X]} = \frac{r(\pi_1 + \pi_3 + \pi_5)}{\sum \pi_i\gamma_i}. \tag{7.12}$$

The proportion of time the system is down due to false alarms is given by

$$\sigma_4 = \frac{fr(\pi_3)}{E[X]} = \frac{r\pi_3}{\sum \pi_i.\gamma_i}. \tag{7.13}$$

The average inventory, $\overline{I}$, is given by

$$\overline{I} = \frac{E[I]}{2}. \tag{7.14}$$

Whenever $S > 0$, there will be a steady-state average inventory given by (7.14) since $E[I] > 0$. Otherwise, inventory is continuously depleted (and eventually shortages are incurred) at an average rate of $-S$. Finally, the functional relation between σ_1, found experimentally for the service level, is:

$$\sigma_1 = \begin{cases} 0 & \text{if } \; S/(P-v) \leq 0 \\ 1 - e^{-3.934522 \, S/(P-v)} & \text{otherwise.} \end{cases} \tag{7.15}$$

Here, "service level" is defined as the proportion of time demand was satisfied with items in stock. With the formulation above, the average shortages per time unit are given by:

$$\overline{S} = (1 - \sigma_1)v.$$

3. ECONOMIC OPTIMIZATION MODEL

The economic model is partially based on the model by Ladany and Bedi, [5], who provided an economic model for a p chart for a finite production run system. This model was simplified and extended by Del Castillo and Montgomery [4] to the case of an $\overline{X}$ chart. We define the following costs per actual cycle, where an actual cycle is the random quantity T_c, previously defined as follows:

- A, sampling costs per cycle;
- B, cost per cycle of running the process in the out of control state;
- C, cost of false alarms;
- D cost of repairing and restoring the process.

From [4], expressions for these costs follow:

$$\begin{aligned} A &= (C_0 + C_1 n) f \\ B &= \frac{C_2 P_f T}{2(f+1)X_1} \{(1 + X_7)X_4 - X_7 X_5\} \\ C &= \frac{C_3 \alpha (1 - P_f)}{X_1} (X_3 + X_2) \\ D &= \frac{C_4 P_f}{X_1} (X_3 + X_2 - X_6), \end{aligned} \tag{7.16}$$

where

- C_0 is the fixed sampling cost
- C_1 is the variable sampling cost

- C_2 per time unit cost of running the process out of control
- C_3 cost per false alarm
- C_4 cost of detecting and correcting a false alarm,

and where the following relations define $X_1, \cdots, X_7$:

$$\begin{aligned}
X_1 &\equiv 1 - P_a(1 - P_f) \\
X_2 &\equiv \frac{P_a P_f (1 - P_a{}^f (1 - P_f)^f)}{X_1} \\
X_3 &\equiv f(1 - P_a) \\
X_4 &\equiv (f+1)(1 - P_a) + P_a{}^{f+1} P_f (1 - P_f)^f + X_2 \\
X_5 &\equiv P_a{}^{f+1}(1 - P_f)^f (P_f - 1) + 1 \\
X_6 &\equiv P_a - P_a{}^{f+1}(1 - P_f)^f \\
X_7 &\equiv \frac{2P_a}{1 - P_a}.
\end{aligned}$$

Using a renewal reward process justification, the optimization is based on minimizing the sum of expected costs incurred per time unit:

$$\min \ \mathrm{Cost}_{n,k,f,T} = \frac{C_6}{E[T_c]} + C_7 \overline{I} + C_8 \overline{S} + \frac{A + B + C + D}{E[T_c]}, \qquad (7.17)$$

subject to

$$\begin{aligned}
n &\leq Ph \\
2 \leq n &\leq 100 \\
1 \leq k &\leq 6 \\
f &\geq 1 \\
T &\geq 0.1 \\
P_a &\leq 0.99 \\
S &> 0 \\
\sigma_1 &\geq 0.6,
\end{aligned}$$

where C_6 is the setup cost (cost per batch), C_7 is the holding cost (per item per time unit), and C_8 is the cost of shortages (cost per item short per time unit). The constraints imply a service level of at least 60 %, a non-negative average slope, and bounds for the decision variables n, k, f, and T. LINGO v. 4.0 optimization software was used to carry out the optimization. The optimization model is solved for a latin hypercube grid of starting points (n, k, f, T), and the best solution is utilized (for details on the optimization, see Salcedo [8]).

3.1 Example. As an example of the solution of the optimization model, consider one of the cases described by Rahim [19]. Here, the cost parameters (in our notation) are $C_0 = 2, C_1 = 0.5, C_2 = 950, C_3 = 500, C_4 = 1100, C_6 = 20, C_7 = 0.1, C_8 = 0, \lambda = 0.050479, \delta = 0.5, P = 1500$, and $v = 1400$. Rahim assumes a Weibull distribution for the time between assignable causes, but a Weibull with shape paramater equal to one is equivalent to the exponential distribution used here for illustration purposes. The solution of the optimization model under the stated assumptions is as follow:

n	f	k	T	Cost
12	1	2.96	2.5332	22.825

σ_1	σ_2	σ_3	σ_4	$\overline{I}$	$\overline{S}$
0.875	0.3559	0.031	0.0015	63.6	174.0

Thus, only one relatively large sample ($n = 12$) must be used to assess control of the batch, a situation that resembles an acceptance sampling situation. The solution given by Rahim (under the assumptions of deterministic demand and no shortages) is $(n, k, f, T) = (54, 13, 2.54, 12.054)$. If the demand were Poisson with parameter v, this policy would provide a service level of 0.232, with shortages of 1075 items short per cycle. Although no shortage cost is specified, the service level constraint $\sigma_1 > 0.6$ results in a quite different policy.

3.2 Sensitivity Analysis. In order to analyze the behavior of the different responses of interest in the proposed optimization model with respect to variations in the parameters, second order polynomial approximations were fitted to the responses from a large experiment. The idea is that, even though we have available expressions for all the responses, these are quite complicated nonlinear functions of the decision variables. Therefore, simpler quadratic approximations allow us to observe relationships not easily detectable from the original formulation.

The parameters C_0 to C_8 (excluding C_5 which is the cost of running the process in control according to Ladany and Bedi [5] and is not used here), λ, δ, P, v, and r were varied according to a "faced-centered" central composite design consisting of 1051 runs, i.e., the optimization model was solved for 1051 combinations of the thirteen parameters. Regression techniques were used to fit a quadratic (second order) polynomial for the quantities n, f, k, T, Cost, σ_2, σ_3, $\overline{I}$, and $\overline{S}$. The regressor S was included to analyze the effect of the average slope on the responses, i.e., for each run, the average slope was computed and entered in the regression analysis as a regressor. Contrary to the case of other regressors, the correlation between S and the responses usually implies that the response affects the value of the slope, or that both responses "move" in the same direction. That is, the slope is not seen as a cause of the responses; nevertheless, the correlations explain a great deal of the underlying trade-offs involved.

In what follows, only the significant effects, defined as those with a P-value smaller than 0.05, are reported for each response.

Effects on n Positive:

$$C_1C_4, C_1C_7, SC_1, \delta C_1, \delta C_4, \delta C_7, \delta S, C_6, C_2.$$

Negative:

$$C_1C_2, C_1C_3, C_1C_6, \delta C_3, \delta C_6, \delta\lambda C_6C_4.$$

It is observed that the optimum sample size n is largely determined by the interaction of C_1 (variable sampling cost) and δ (shift size). According to the literature on economic design of control charts, as the cost of sampling increases and the shift size increases, sample sizes tend to decrease, an effect also observed here.

Effects on f Positive:

$$C_4C_0, SC_1SC_7, S\lambda, S^2.$$

Negative:

$$C_1C_2, C_1C_6, C_2\lambda, C_4\lambda S, SC_0, SC_2, SC_4, SC_6, SC_8.$$

In this case, the sampling costs have a strong effect on the optimal number of samples (f). An interesting correlation exists between f and S. As the number of samples decreases, S (the average slope) increases. This means that increasing the number of samples increases the number of opportunities to stop, decreasing the inventory level. Thus, the average slope goes down as the the number of samples goes up. Also of interest are the interactions of S with the costs of holding and shortages, C_7 and C_8. The SC_7 effect on f is positive, indicating that higher shortage costs and a steeper average slope correlates positively with the number of samples. Since stoppages reduce inventory, allowing shortages could be a way of saving money if the holding cost is high. Likewise, the SC_8 effect on f is negative: since shortages are very expensive, we want as few stoppages as possible.

Effects on k Positive:

$$S, C_7C_1, C_7^2, \lambda C_2, \lambda C_6, \delta C_3, \delta C_4, \delta C_7, \delta C_8, \delta\lambda, SC_1, SC_4, SC_7.$$

Negative:

$$C_7, C_3C_1, C_7C_2, \lambda C_8, \delta C_2, \delta C_6, SC_2, SC_8, S\lambda, S^2, C_8C_3, C_8C_4.$$

The control limits width, k, is affected by several factor combinations that can be classified in three groups: 1) those that contain C_7, 2) those that contain δ, and 3) those that contain S. Increases in shortage cost lead to narrower control limits for reasons similar to those explained in the previous paragraph: stopping too often makes the inventory decrease, but if holding inventory is expensive, stopping often due to a reduced control limits width is reasonable. Large shifts in the mean can be detected with wide control limits and vice versa, a common relation in the control chart design literature.

Effects on Cost Positive:

$$C_3C_2, C_6, C_4, C_7C_1, C_7C_6, C_7^2, \lambda_1C_1, \lambda C_2.$$

Negative:

$$C_7, C_7C_2, SC_2, SC_8, S\lambda.$$

The cost function being minimized is a very complicated function of the decision variables; therefore, using a simplified quadratic regression function allows us to observe some key relationships not readily observable. The levels in the experiment indicate that, the cost per time unit depends highly on the holding cost per unit. The optimal cost diminishes as the holding cost increases. When holding costs are higher, the resulting quality control policies are stricter, keeping inventory low (e.g., tighter control limits) and reducing the effect of holding costs on the optimal cost. The λC_2 interaction indicates that more frequent shifts increase cost since there are longer periods of time when the process is running out of control.

Effects on σ_2 Positive:

$$C_6C_1, C_6C_4, \lambda C_4, \lambda C_6, \delta C_4, \delta C_8, SC_7$$

Negative:

$$C_4C_1, C_4C_3, C_6C_2, C_7C_1, C_8C_2, C_8C_4, \delta C_1, \delta C_2, SC_8, S\lambda, S\delta.$$

The size of the shifts (δ) is an influential factor on the proportion of output produced while the system is running out of control. The smaller the shift, the harder it is to detect, and therefore, the larger σ_2 becomes. This general rule does not apply when there are interactions with shortage (δC_8) and detection (δC_4) costs. In both cases, the quality controls scheme is less strict to compensate for the high costs, so the time running out of control increases.

Effects on σ_3 Positive:

$$C_2C_1C_4C_0, C_4C_3SC_3, SC_4, SC_7S^2.$$

Negative:

$$S, C_3C_1, C_3C_2, C_4C_2, C_6C_1, SC_1, SC_2, S\lambda.$$

The proportion of time the system is down for repairs is highly dependent on the average slope. As the average slope increases, the time for repairs decreases since, clearly, there are fewer stoppages.

Effects on $\overline{S}$ Positive:

$$C_6, \lambda, C_0^2, C_4C_0, C_6^2, C_7C_6, C_8^2, \lambda C_8, SC_3, SC_8, S\delta, S^2.$$

Negative:

$$C_0, C_8, S, C_6^2, \lambda C_7, \lambda^2, \delta C_6, SC_7, S\lambda.$$

As expected, as the cost of shortages (C_8) increases, the average shortages decrease. In addition, shortages decrease as the fixed sampling costs increase since to reduce fixed sampling costs fewer samples are taken, reducing the opportunities for alarms and shortages. Shortages increase with an increasing assignable cause occurrance rate (λ), as expected. Finally, the following correlation between shortages and the average slope of the inventory graph is evident: as S increases, shortages decrease and vice versa.

Effects on T Whereas the quadratic fit for all responses so far average an adjusted R^2 value greater than 0.80, the quadratic approximation for T resulted in a very poor fit; thus, the individual effects are not reported. However, it was observed that C_6 (setup cost) and C_7 (holding cost) were the most significant regressors, with C_6 being positively correlated and C_7 being negatively correlated. This is a result in accordance with basic relations in inventory control theory, i.e., the economic order quantity model reveals that $T = \sqrt{2C_6/(vC_7)}$.

Effects on $\overline{I}$ The quadratic fit on the average inventory is poor; thus, its individual effects are not reported. It was observed, however, that $\overline{I}$ is positively correlated with the setup cost and negatively correlated with holding costs. Given that we have available a formula for average inventory, the regression analysis does not add any new insight.

Comparison with standard results on the economic design of control charts According to the classic paper by Duncan [2], the optimum sample size n is largely determined by the shift size δ. Also, n is affected by the variable sampling cost C_1. These results are similar to those found in the present study. In Duncan's model, the cost of false alarms (C_3) and the cost of running out of control (C_4) influence the control limits width the most. In the model presented here, these factors do affect k when interacting with the shift size δ, but are determinant relationships. The average slope and the holding cost have much more significant effect on the behavior of the optimal value of k. Thus, we see that inventory control parameters are of influence in the optimal determination of a control chart design.

3.3 Economic incentive and continuous improvement: a trade-off resolution analysis. As part of any continuous improvement program, companies are interested in reducing the rate at which shifts occur and the time it takes to search and repair the production process. To study this, the values of λ (rate of occurrence of shifts) and r (repair time) were varied, keeping all other parameters fixed, and the optimization model was solved.

It was observed that as the rate of shifts decreases, all performance measures improve, showing that there is no trade-off between having good quality and adequate inventory control policies. The same is valid for the time of search and repair. Thus, there is no disparity between the economic design of control charts and continuous improvement policies and their inventory-production control implications. The continuous use of a control chart is a powerful, yet simple, tool that can help to remove assignable causes of variation in a process, thereby improving it. Likewise, decreasing the time of searching yields better performance of the production process, including the reduction of shortages.

References

[1] Ben-Daya, M. Integrated Production Maintenance and Quality Model for Imperfect Processes. *IIE Transactions*, **31** (1999) 491-501.

[2] Duncan, A.J. The Economic Design of $\overline{X}$ Charts used to Maintain Current Control of a Process. *Journal of the American Statistical Association*, **51** (1956) 228-242.

[3] Del Castillo, E. Relations between $\overline{X}$ Control Chart Design Variables and Production Control. *International Journal of Production Research*, **33**(10) (1995) 2709-2721.

[4] Del Castillo, E. and Montgomery, D.C. Optimal Design of control Charts for Monitoring Short Production Runs. *Economic Quality Control*, **8** (1993) 225-240.

[5] Ladany, S.P., and Bedi, D.N. Selection of the Optimal Setup Policy. *Naval Research Logistics Quarterly*, **23** (1976) 219-233.

[6] Rahim, M.A. Joint Determination of Quantity, Inspection Schedule and Control Chart Design. *IIE Transactions*, **26** (1994) 2-11.

[7] Ross, S. *Stochastic Processes*, New York: John Wiley, 1983.

[8] Salcedo, Isabel, T. An Integrated Model for Inventory and Statistical Process Control *Master Thesis, Department of Industrial and Manufacturing Engineering*, Penn State University, 1999.

Chapter 8

A GENERALIZED INTEGRATED ECONOMIC MODEL FOR INVENTORY AND QUALITY CONTROL PROBLEMS

Hiroshi Ohta,
Department of Industrial Engineering
College of Engineering
Osaka Prefecture University
Sakai, Osaka 599-8531, JAPAN

Aritoshi Kimura and
Department of Management Systems Engineering
Faculty of Engineering
Kinki University, Kure
Hiroshima 737-0112, JAPAN

M. Abdur Rahim
Faculty of Administration
University of New Brunswick
Fredericton, N.B. E3B 5A3, CANADA

Abstract In this paper, we develop an integrated economic model for inventory and quality control problems, extending the work of Rahim (1994) and Rahim and Ben-Daya (1998). The production process is subject to an assignable cause which shifts the process from an in-control state to an out-of-control state. In the paper, we consider the shifts in both the process mean and the process variance. When a warning is triggered, a search for the assignable cause is initiated and is terminated upon finding the cause within a pre-specified target time. The process is then brought back to an in-control state by repair. However, if the assignable cause is not discovered within the pre-specified time, production is allowed to continue until the next sampling or warning, whichever occurs first. In this case, either the alarm is considered to be false with a probability of Type I error, or the assignable cause has not been eliminated with a probability of Type II error. In the latter case, the process produces products in an out-of-control state until the next sampling or warning, whichever occurs first. However, this state does not indicate any severe damage to the system. Joint $\bar{X}$ and R charts are used for process control. Under these conditions, a generalized economic model for the joint determination of production quantity, an inspection schedule, and the design of the $\bar{X}$ and R control charts are developed.

Keywords: Production processes, economic production quantity, inspection schedule, control charts design, integrated model.

1. INTRODUCTION

Control charts are widely used for monitoring production processes. The relevant literature has increased considerably since Duncan's (1956) pioneering work on the economic design of process control charts. Earlier work in this area is summarized by Montgomery (1980) and Vance (1985). Recently, Svoboda (1991) and Ho and Case (1994) have provided comprehensive reviews of the economic design of process control charts. In the economic design of the $\bar{X}$ chart, the objective is to determine the optimal design parameter values of the sample size n, the sampling (inspection) interval h, and the control limit coefficient k for the chart so as to minimize the expected cost per unit time of operation. The key assumption in the process model has always been that the process in-control period obeys a Markovian shock model. In Duncan's approach to this model, the length of the sampling intervals is kept constant and uniform. Using a renewal theory approach, Banerjee and Rahim (1987) derived economic models for some non-Markovian processes. However, the issue of a non-uniform sampling scheme had not been addressed until Banerjee and Rahim (1988) demonstrated that increasing the frequency of sampling as the system ages yields a lower operational cost per unit time for a Weibull distributed shock model.

The problem of determining the economic production quantity (EPQ) in production processes has also been well studied (see, for example, Silver *et al.* (1998)). In the development of EPQ models, controlling the quality of the product has not been a general consideration. Rather, product quality has been assumed to be perfect. As a result, the effects of a deteriorating process on the EPQ have been ignored. In many industrial situations, however, the quality and the quantity of the product(s) are equally important. Rosenblatt and Lee (1986) have studied the effects of an imperfect production process on the optimal production cycle. The problem of joint determination of the EPQ and inspection schedules for an imperfect production process has been studied by Lee and Rosenblatt (1987). In their studies, the state of the production process is monitored by an inspection scheme. The EPQ and inspection schedule are dependent on the trade-off among set-up cost, holding cost, inspection cost, and process deterioration. It is assumed that the process failure mechanism follows a Markovian shock model. This assumption may not be always applicable. In the determination of EPQ, a non-Markovian shock model has not been addressed. Furthermore, the effect of EPQ on the economic design of control charts has not been considered.

As mentioned above, traditionally, quality control and inventory control have been viewed as two separate problems. However, Rahim (1994) presented a model for integrating the EPQ, inspection schedule, and control chart design of an imperfect production process. Accordingly, the following assumptions are made. The process is subject to the occurrence of a non-Markovian shock having an increasing failure rate. The product quality of the process is monitored under the surveillance of an $\bar{X}$ chart. The objective is to determine the optimal control chart design parameters and EPQ so as to minimize the expected total cost (the quality control cost and the inventory control cost) per unit time. However, the time for searching for an assignable cause when the alarm is false is assumed to be zero. More recently, Rahim and Ben-Daya (1998) have introduced a more realistic assumption concerning the stoppage of equipment during the in-control phase of a false alarm. Nevertheless, following the traditional approach, an $\bar{X}$ chart is used to monitor the process where any changes have been observed in the process mean under the assumption that the process variance remains constant throughout the production cycle. In practice, however, it is possible that both the process mean and the process variance may change simultaneously during a production cycle. In that case, an R control chart is often used to monitor the process variance. Considerable attention has been devoted to the study of the joint economic design of $\bar{X}$ and R control charts in recent years. This growing interest is due to the greater power of joint $\bar{X}$ and R charts compared to that of the $\bar{X}$ or R chart used separately.

The purpose of this paper is to present a generalized integrated economic process model for inventory and quality control problems from these viewpoints.

1. When the process is out of control, we consider the shifts not only in the process mean but also in the process variance of the quality characteristic and use joint $\bar{X}$ and R charts for the process control.

2. Even though the warning is true, we consider the possibility that we cannot always identify the cause. The search for the assignable cause is initiated as soon as the warning is noticed. However, if the search time exceeds the maximum threshold value, we terminate the search and allow the process to continue. If the cause is found within the threshold time, repair is done to bring the process back under control.

Under these conditions, we developed a generalized economic model for the joint determination of EPQ and $\bar{X}$ and R control charts design. The paper was motivated by the idea of a repairable equipment system adherent to statistical process control. The problem can be viewed as a combination of the inspection policy and the control policy. The output of the product quality is assumed to be normally distributed and monitored by both $\bar{X}$ and R control charts. For example, consider the bottle filling process studied by Costa (1998), where the amount of liquid in each bottle is the quality characteristic of interest. The process fills the bottle with 600 ml ($\mu_0 = 600$). The process standard deviation is known ($\sigma_0 = 200$ ml). Occasionally, the liquid contains some impurities that partially clog the feeder line, thereby shifting the process mean. To return the mean (μ) to the target (600 ml), the feeder line must be cleaned. Since cleaning the feeder line requires a shut down of the process, unnecessary cleaning must be avoided. Historical data have shown that the process variance increases when the process mean shifts. The R chart has been used jointly with an $\bar{X}$ chart for monitoring the bottle contents. Consider another example of a similar type (see Rahim and Banerjee, 1988), a container filling process where the nozzle of the feeder line occasionally overflows or underflows due to overheating/overpressure or vibration and shifts the process mean, process variance, or both.

Let us assume that the occurrence time of any one of the above assignable causes follows a Weibull failure distribution. The manufacturer uses an $\bar{X}$ control chart to control the process mean and an R chart to control the process variance. The problem for the manufacturer is to find the optimal sample size, inspection intervals, control limit coefficients for both charts, and the economic production quantity which is measured in terms of number of bottles or containers filled per unit time.

The paper is organized as follows. The next section provides the notations and assumptions. Section 3 provides the general model formulation, and section 4 provides illustrative examples and the sensitivity analysis of the model. The final section offers some concluding remarks.

2. NOTATIONS AND ASSUMPTIONS

The following notations are used to develop the model related to the joint $\bar{X}$ and R control charts design.

Decision variables

m: number of inspection intervals
n: sample size
h_j : length of the jth sampling (inspection) interval
k_1 : control limit coefficient for the $\bar{X}$ chart
k_{2u} : upper control limit coefficient for the R chart
k_{2l} : lower control limit coefficient for the R chart
γ : time for searching the assignable cause under the condition that the process was stopped

Other notations

a: fixed sampling cost
b: cost per unit sampled
D_0: quality cost per unit time while producing a nonconforming item when the process is in control
D_1: quality cost per unit time while producing a nonconforming item when the process is out of control
S_U: upper specification limit of the product
p_0: fraction defective while process is in control
p_1: fraction defective while process is out of control
C_d: unit cost caused by producing a defective item
W_C: cost per unit time for searching the assignable cause
Y : set-up cost per false alarm
W : cost to repair the assignable cause
$\alpha_{\bar{X}}$: Type I error of the $\bar{X}$ chart
α_R: Type I error of the R chart
α: Type I error of the joint $\bar{X}$ and R charts
$P_{\bar{X}}$: power of the $\bar{X}$ chart
P_R: power of the R chart
P : power of the joint $\bar{X}$ and R charts
β: Type II error of the joint $\bar{X}$ and R charts
e_2: probability of missing the assignable cause in the searching time
δ: shift of the process mean
κ: ratio of the process standard deviations ($\kappa = \sigma_0/\sigma_1$)
ω_j: expected time at the end of the jth inspection interval
$f(t)$: probability density function of the process failure time
$F(t)$: cumulative distribution function of the process failure time
$r(t)$: process failure rate at time t

$S(t)$: salvage value for working equipment at time t

The following notations are used for the EPQ model.

r_d : demand rate
r_p : production rate
C_h : inventory holding cost per item per unit time
C_0 : set-up cost for each production cycle
T : total length of an inventory cycle
T_U : total length of production run in the inventory cycle
T_D : total length of non production time for searching the assignable cause in the inventory cycle
T_p : length of the production cycle including the mth inspection during which inventory is built up
Q^* : EPQ

The integrated model is based on the following assumptions.

1. The measurable quality characteristic is assumed to be normally distributed with mean μ_0 and standard deviation σ_0 while the process is in control.

2. The measurable quality characteristic is assumed to be normally distributed with mean $\mu_1(=\mu_0+\delta\sigma_0)$ and standard deviation $\sigma_1(=\sigma_0/\kappa)$ while the process is in an out-of-control state. In this case, shifts in both the mean and the standard deviation of the product quality characteristic are occurring simultaneously due to a single assignable cause.

3. The duration of an in-control period is assumed to follow a Weibull distribution having an increasing hazard rate $r(t) = \lambda\nu t^{(\nu-1)}$ and a cumulative distribution function $F(t) = 1 - \exp(-\lambda t^{\nu})$.

4. The process is inspected at times h_1, $h_1 + h_2$,... to determine its state, and the output quality of the product is monitored by the joint $\bar{X}$ and R control charts.

5. The production cycle ends either with a true alarm or at time ω_m, depending on which occurs first. In other words, if no true alarm is observed by time ω_{m-1}, then the cycle is allowed to continue for an additional time h_m. At time ω_m, necessary maintenance work is carried out. Thus, there is no cost of sampling and charting during the mth sampling interval.

6. If any inspection shows that the state of the process is out of control, production ceases during γ (unit time for searching an

assignable cause). If an assignable cause is found during γ, the process is brought back to an in-control state by repair. If an assignable cause is not found during γ, production is resumed. Note that in this case, the alarm is either false or the assignable cause cannot be eliminated with probability e_2 during the searching time.

7. The time to sample and chart one item is negligible.

8. A production cycle begins when a new component is installed and ends when repair is needed or after a specified number m of sampling intervals has passed, whichever occurs first. The process is brought back to an in-control state by repair. Thus, a renewal occurs at the end of each cycle.

Assumptions (2), (4) and (6) are original points in this paper.

3. GENERAL MODEL

In this section, we develop a generalized integrated cost model for the joint determination of EPQ, inspection interval schedules, and the design parameters of the joint $\bar{X}$ and R charts.

3.1 Inspection interval expression. Under assumptions (3) and (4) and following Rahim (1994), we propose that the length of sampling intervals should be chosen in such a way that the integrated hazard over each interval should be equal. That is,

$$\int_{h_j}^{h_j+h_{j+1}} r(t)\,dt = \int_0^{h_1} r(t)\,dt. \tag{8.1}$$

This can be derived from equation (8.1) by defining the length of sampling intervals $h_j (j = 1, 2, ...)$ in the following expression (Rahim 1994):

$$h_j = \left\{ j^{1/\nu} - (j-1)^{1/\nu} \right\} h_1 \tag{8.2}$$

3.2 Determination of quality costs D_0 and D_1 . Under assumption (2), fraction defectives p_0 and p_1 are given by

$$p_0 = \int_{\frac{S_U - \mu_0}{\sigma_0}}^{\infty} \varphi(u)\,du = \int_A^{\infty} \varphi(u)\,du = 1.0 - \Phi(A), \tag{8.3}$$

and

$$p_1 = \int_{\frac{S_U-\mu_1}{\sigma_1}}^{\infty} \varphi(u)\,du = \int_{\kappa(A-\delta)}^{\infty} \varphi(u)\,du = 1.0 - \Phi(\kappa(A-\delta)), \quad (8.4)$$

where

$$S_U = \mu_0 + A\sigma_0 \quad (A > 0), \quad (8.5)$$

and $\varphi(.)$ and $\Phi(.)$ are the probability density function and the cumulative distribution function of a standard normal variable, respectively. Using equations (8.3) and (8.4), we see that D_0 and D_1 can be derived as follows:

$$D_0 = C_d \times \lfloor r_p \times p_0 \rfloor, \quad (8.6)$$

and

$$D_1 = C_d \times \lfloor r_p \times p_1 \rfloor \quad (8.7)$$

where $\lfloor z \rfloor$ stands for the maximum integer that is less than or equal to z.

3.3 Determination of the missing probability.

The probability e_2 is assumed to be given by

$$e_2 = \exp(-\frac{1.0+\delta}{\kappa}\gamma). \quad (8.8)$$

Equation (8.8) indicates that e_2 decreases as the process mean and process variance increases (i.e., δ increases and κ decreases), and the time for searching the assignable cause, γ, increases.

3.4 Derivation of Type I errors and powers of joint $\bar{X}$ and R control charts.

Assume that the upper and lower control limits for the $\bar{X}$ chart are

$$\text{UCL}_{\bar{X}} = \mu_0 + k_1\sigma_0/\sqrt{n}, \quad (8.9)$$

and

$$\text{LCL}_{\bar{X}} = \mu_0 - k_1\sigma_0/\sqrt{n}. \quad (8.10)$$

The expressions for the Type I error and the power of the $\bar{X}$ chart are derived as follows:

$$\alpha_{\bar{X}} = 2[1 - \Phi(k_1)], \quad (8.11)$$

and

$$P_{\bar{X}} = 1 - \Phi\left\{(-\delta\sqrt{n} + k_1)\kappa\right\} + \Phi\left\{(-\delta\sqrt{n} - k_1)\kappa\right\}. \tag{8.12}$$

Denoting the upper and lower control limits, respectively, for the R chart as

$$\text{UCL}_R = k_{2u}\sigma_0 \tag{8.13}$$

and

$$\text{LCL}_R = k_{2l}\sigma_0, \tag{8.14}$$

we obtain expressions for the Type I error and the power of the R chart as follows:

$$\alpha_R = 1 - F_n(k_{2u}) + F_n(k_{2l}), \tag{8.15}$$

and

$$P_R = 1 - F_n(k_{2u}\sigma_0/\sigma_1) + F_n(k_{2l}\sigma_0/\sigma_1), \tag{8.16}$$

where $F_n(.)$, the cumulative distribution function for the standardized range $w = R/\sigma_0$, is given by

$$F_n(w) = \{\Phi(w/2) - \Phi(-w/2)\}^n + 2n\int_{w/2}^{\infty} \varphi(u)\{\Phi(u) - \Phi(u-w)\}^{n-1}\,du. \tag{8.17}$$

Thus, the joint Type I error and power, respectively, for the joint $\bar{X}$ and R control charts can be obtained as follows:

$$\alpha = \alpha_{\bar{X}} + \alpha_R - \alpha_{\bar{X}} \cdot \alpha_R, \tag{8.18}$$

and

$$P = P_{\bar{X}} + P_R - P_{\bar{X}} \cdot P_R. \tag{8.19}$$

3.5 Derivation of the production run time and production stop time. Under the assumptions (1) - (8), we obtain the expected total length of production run in the inventory cycle, $E(T_U)$, as

$$\begin{aligned} E(T_U) &= \sum_{j=1}^{m} h_j \overline{F}(\omega_{j-1}) + \{\beta + (1-\beta)e_2\} \\ &\quad \times \sum_{j=1}^{m-1} \Delta F(\omega_j) \sum_{i=j+1}^{m} \{\beta + (1-\beta)e_2\}^{i-(j+1)} h_i. \qquad (8.20) \end{aligned}$$

Furthermore, the expected total length of production stop for searching the assignable cause in the inventory cycle, $E(T_D)$, can be obtained by the following:

$$E(T_D) = \gamma\alpha \sum_{j=1}^{m-1} \bar{F}(\omega_j). \qquad (8.21)$$

3.6 Derivation of the quality control cost. Following Rahim (1994), the expected quality control cost per cycle, $E(C)$, can be described as the expected cost of operating while the process is initially in control and then out of control, subsequently triggering a true alarm. The $E(C)$ can be expressed as the expected cost of operating while in control with no alarm, the expected cost of false alarms, the expected cost of operating while out of control with no alarm, the repair cost, and the expected cost of sampling minus the salvage value for working equipment of age x:

$$\begin{aligned} E(C) &= D_0 \sum_{j=1}^{m-1} h_j \bar{F}(\omega_{j-1}) \\ &\quad + \left\{ (1-\alpha-\beta) \sum_{j=1}^{m-1} \Delta F(\omega_j) + \alpha \sum_{j=1}^{m-1} \bar{F}(\omega_{j-1}) \right\} \gamma W_c \\ &\quad + \left[\{(1-\beta)e_2 - \alpha\} \sum_{j=1}^{m-1} \Delta F(\omega_j) + \alpha \sum_{j=1}^{m-1} \bar{F}(\omega_{j-1}) \right] Y \\ &\quad + (D_1 - D_0) \sum_{j=1}^{m} \omega_j \Delta F(\omega_j) - (D_1 - D_0) \int_0^{\omega_m} x f(x) dx \\ &\quad + D_1[\beta + (1-\beta)e_2] \\ &\quad \times \sum_{j=1}^{m-1} \Delta F(\omega_j) \sum_{i=j+1}^{m} \{\beta + (1-\beta)e_2\}^{i-(j+1)} h_i \end{aligned}$$

$$+(a+bn)\left[1+\sum_{j=1}^{m-2}\bar{F}(\omega_j)+\{\beta+(1-\beta)e_2\}\sum_{j=1}^{m-2}\Delta F(\omega_j)Q_j\right]$$

$$-(1-\beta)(1-e_2)\sum_{j=1}^{m}S(\omega_j)\Delta F(\omega_j)+W, \tag{8.22}$$

where

$$Q_j=(1-\beta)\sum_{i=1}^{m-1-j}i\beta^{i-1}+(m-1-j)\beta^{m-1-j}. \tag{8.23}$$

3.7 An integrated cost model. The expected total cost of the model consists of the set-up cost, inventory holding cost, and quality control cost. The integrated production, inventory, and quality control cycle is depicted in Figure 8.1. The approximate expected total cost per unit time, ETC, is obtained as follows:

$$ETC=\left[\frac{r_d}{r_p}C_0+\frac{1}{2}E(T_U)\{r_pE(T_U)-r_dE(T_P)\}C_h+\frac{r_d}{r_p}E(C)\right]\Big/E(T_U), \tag{8.24}$$

where the relation between the expected total length of the inventory cycle, $E(T)$, and the expected total length of the production run in the inventory cycle, $E(T_U)$ is as follows:

$$E(T)=\frac{r_p}{r_d}E(T_U). \tag{8.25}$$

The problem of interest is to determine, simultaneously, the parameters m, n, k_1, k_{2u}, k_{2l}, h_1, and γ which minimize the ETC.

4. EXAMPLES

For illustrative purposes, we use the following values:
$a=2.0$, $b=0.5$, $Y=5.0$, $W=1100$, $W_c=400$, $A=2.0$, $r_d=1400$, $r_p=1500$, $C_d=2.5$, $C_h=0.1$, $C_0=20.0$, $\lambda=0.05$, $\nu=2.0$ and $S(t)=1100exp(-t)$.

The values of p_0, p_1, D_0 and D_1 under several sets of (δ,κ) are obtained from equations (8.3)-(8.7) as shown in Table 8.1. Table 8.2 provides the results of the near-optimal designs obtained by using the Simulated Annealing Method and illustrates the following.

1. When the process mean and the process variance increase, the sample size n, the sampling interval h_1, and the economic production quantity Q^*, decrease, and the expected total cost ETC increases.

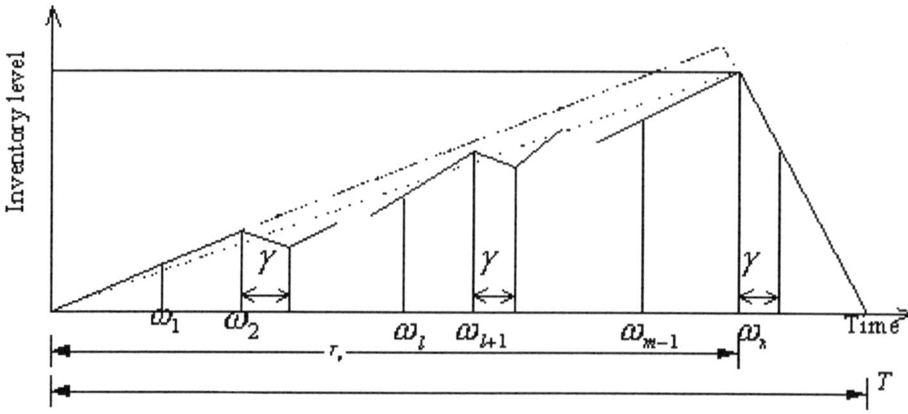

Figure 8.1 Quality and inventory control cycle

2. The time taken for searching for the assignable cause, γ, under the condition that the process is shut down, is not affected by changes in the process mean and the process variance since there is a trade-off between the cost for searching for the assignable cause and the quality cost due to the failure to locate the assignable cause.

Table 8.1 Fraction defectives and quality cost per unit time, where p_0= 0.0231 and D_0 = 85.0 ($\delta = 0, \kappa = 1$).

δ	0.5	0.5	1.0	1.0	1.5	1.5
κ	p_1	D_1	p_1	D_1	p_1	D_1
0.9	0.0887	332.5	0.1841	690.0	0.3264	1222.5
0.8	0.1152	430.0	0.2119	792.5	0.3446	1290.0
0.7	0.1469	550.0	0.2420	907.5	0.3632	1360.0
0.6	0.1841	690.0	0.2743	1027.5	0.3821	1432.5
0.5	0.2267	850.0	0.3085	1155.0	0.4013	1502.5

Table 8.2 Near-optimal designs obtained using the Simulated Annealing Method.

δ	κ	m	n	k_1	k_{2u}	k_{2l}	h_1	γ	Q^*	ETC
	0.9	57	38	1.67	6.60	2.32	3.25	0.20	10616	386.9
	0.8	54	40	1.77	5.93	2.00	2.78	0.30	8533	421.5
0.5	0.7	44	30	2.07	5.31	2.47	2.02	0.32	7295	439.1
	0.6	62	16	2.32	4.72	1.60	1.47	0.32	6650	441.3
	0.5	60	9	2.42	4.30	0.18	1.21	0.30	6386	440.0
	0.9	74	10	1.73	6.49	1.23	1.33	0.27	6679	436.8
	0.8	77	11	1.74	5.84	1.34	1.31	0.32	6521	444.2
1.0	0.7	56	12	1.72	6.11	1.58	1.28	0.35	6401	450.8
	0.6	62	11	2.13	4.95	0.90	1.19	0.32	6322	450.2
	0.5	71	10	2.31	5.20	0.86	1.09	0.32	6220	447.5
	0.9	78	7	2.36	6.27	0.78	0.96	0.36	6180	451.9
	0.8	79	7	2.36	6.62	0.64	0.94	0.33	6158	452.0
1.5	0.7	88	8	2.42	5.77	0.66	0.95	0.32	6164	452.8
	0.6	76	8	2.40	5.54	0.77	0.95	0.31	6121	451.7
	0.5	70	8	2.49	5.67	0.86	0.97	0.30	6100	448.8

5. CONCLUDING REMARKS

This paper extends the work of Rahim (1994) and Rahim and Ben-Daya (1998) and develops a generalized economic model for the problem of inventory and quality control. This model takes into account the simultaneous change in the process mean and the process variance; it includes an R chart, in addition to an $\bar{X}$ chart, thereby making it more realistic. To accomplish this, the paper makes an assumption that there is a pre-specified time to find an assignable cause when there is an out-of-control signal. If the source of the assignable cause cannot be detected within a pre-specified time, the process is allowed to continue until the next warning signal is triggered. This may lead to Type I or Type II errors depending on the actual existence or non-existence of an error. The time for searching an assignable cause is considered to be an additional decision variable. This assumption may invite some criticism and deserves some explanation and/or justification. Some researchers might interpret it as contradicting the modern view in quality management that quality is the number one priority or as supporting the traditional concept of quantity over quality. In other words, time is considered to be more important than quality; hence, time is not to be wasted in fixing a problem. The implication of this assumption is that, in the long run, resources may be wasted in producing inferior goods. Customers may end up receiving poor quality products that should have been detected in the process. Loss of the customer's goodwill may be incurred, and the manufacturer will eventually lose competitiveness.

The following explains and justifies the above assumption. To keep the process in control, the magnitude of the shift/drift must be maintained within a specified range. This can be accomplished by resetting the tool or cleaning the nozzle of the feeder. But this resetting operation may be costly and may involve loss of production. Some degree of deterioration of the quality of the product may be tolerable at a cost. However, as the production cycle increases, the cost of defective items will increase to the point where the cost of resetting may be more economical. What is needed is an optimal rule for resetting. Duncan's (1956) and Chiu's (1975) models for the economic design of control charts have been widely studied in the literature. According to Duncan, the process remains in operation during the search for the assignable cause. For Chiu, the process is shut down during the search for the assignable cause. Our assumption to allow the process to continue if the assignable cause has not been detected within the specified time adheres to the concepts of both Duncan and Chiu. Furthermore, in some canning processes, the products are very hot when they come off the production line, and the inspector has to wait until the product cools before taking a sample. In many industrial situations, the quality and quantity of the products are equally important to the producer. Nevertheless, it has been recognized that improvements in production efficiency, quality, and cost can be realized through a careful selection of the design parameters, and this has been accomplished by using an optimization method in the current paper. From our numerical studies, we found that the optimal value of Type II error lies, approximately, between 0.01 and 0.05. That is, the probability of catching the shift if there is one, lies between 0.95 and 0.99. This indicates that if the assignable cause occurs and is not detected immediately, it will be detected, on average, after one sampling interval.

The following problem may be worth investigating in future research. In our model we stop the process at ω_m or at failure (detection time), whichever occurs first. The inventory is depleted to zero, before we start a new cycle. In many production situations, manufacturers may not like to stop the process to reduce inventory to zero. Rather, they may prefer to continue production of a predetermined quantity q, then stop the process to deplete the inventory to zero. In this case, the following may arise:

1. q items are produced before the first failure, then the process is allowed to bring the inventory level to zero; or
2. q items are produced by ω_m, during which time there could be a number of failures requiring that the process be repaired each time. Then, the process is allowed to bring the inventory level to zero.

Each of these cases pose an interesting and important problem for future study.

Acknowledgements: The financial assistance of the Natural Sciences and Engineering Research Council (NSERC) of Canada in support of this collaborative research is gratefully acknowledged.

References

[1] Banerjee, P. K., and Rahim, M. A. Economic Design of Control Charts: A Renewal Theory Approach. *Engineering Optimization*, **12** (1987) 63-73.

[2] Banerjee, P. K., and Rahim, M. A. Economic Design of $\bar{X}$-Control Chart under Weibull Shock Models. *Technometrics*, **30** (1988) 407-414.

[3] Chiu, W. K. Economic Design of Attribute Control Charts. *Technometrics*, **17** (1975) 81-87.

[4] Duncan, A. J. The Economic Design of $\bar{X}$ Chart used to Maintain Current Control of a Process. *Journal of the American Statistical Association*, **51** (1956) 228-242.

[5] Ho, C., and Case, K. E. Economic Design of Control Charts: A Literature Review for 1981-1991. *Journal of Quality Technology*, **26** (1994) 39-53.

[6] Lee, H. L., and Rosenblatt, M. J. Simultaneous Determination of Production Cycles and Inspection Schedules in a Production System. *Management Science*, **33** (1987) 1125-1136.

[7] Montgomery, D. C., The Economic Design of Control Charts: A Review and Literature Survey. *Journal of Quality Technology*, **12** (1980) 75-87.

[8] Rahim, M. A. Joint Determination of Production Quantity, Inspection Scheduling, and Control Chart Design. *IIE Transactions*, **26**(6) (1994) 2-11.

[9] Rahim, M. A., and Banerjee, P. K. Optimal Production Run for a Process with Random Linear Drift. *Omega*, **16**(4) (1988) 347-351.

[10] Rahim, M. A., and Ben-Daya, M. A Generalized Economic Model for Joint Determination of Production Run, Inspection Schedule, and Control Chart Design. *International Journal of Production Research*, **36**(1) (1998) 277-289.

[11] Rosenblatt, M. J., and Lee, H. L. Economic Production Cycles with Imperfect Production Processes. *IIE Transactions*, **18**(1) (1986) 48-55.

[12] Silver, E. A., Pyke, D.F., and Peterson, R. *Inventory Management and Production Planning, and Scheduling*, New York: John Wily & Sons, Inc., 1998.

[13] Svoboda, L. Economic Design of Control Charts: A Review and Literature Survey (1979-1989). In *Statistical Process Control in Manufacturing*, J. B. Keats and Montgomery D. C. Editors, New York: Marcel Dekker, 1991.

[14] Vance, L. C. A Bibliography of Statistical Quality Control Chart Techniques: 1970-1980. *Journal of Quality Technology*, **15** (1985) 59-62.

Chapter 9

OPTIMAL INVENTORY ORDERING POLICIES FOR QUALITY-DEPENDENT MARKETS

M. Hariga and M. N. Azaiez
Industrial Engineering Program,
King Saud University,
P. O. Box 800,
Riyadh, 11421, Saudi Arabia

Abstract In this study, we investigate the single-period inventory problem when the lot received may contain defective items. We consider the case where good and poor-quality items are sold in primary and secondary markets, respectively. We determine the optimal ordering quantity as well as the corresponding expected profit in this context. We also establish comparisons with the results of the standard single-period problem with a single primary market and defect-free products. In addition, we provide tools to help the vendor select the appropriate arrangement for handling the defective items with the supplier. The analysis extends to investigate pricing policies in the different markets and their impact on the order size and the expected profit. We consider both cases of complete and partial information regarding the distributions of the price-dependent demands. Finally, we present a number of examples for illustration.

Keywords: Inventory, single-period, defective items, quality-dependent markets, ordering policy, pricing.

1. INTRODUCTION

It is frequently the case in many purchasing/production situations that items ordered/manufactured are not of perfect quality. Imperfection may vary in severity from minor to serious defects. Handling the defective items can be arranged through negotiations between the vendor/user and the manufacturer/supplier. The two parties may agree that defective items should be returned to the supplier at his/her own expenses to refund/replace them. Shih (1980) investigated ordering policies of the vendor that minimize his/her total (expected) inventory cost under this type of agreement. Alternatively, negotiations may lead to other types of arrangement including a reduction in the purchasing cost of the items. Meanwhile, the vendor will have to handle the defective items. This is frequently the case if the defects are minor and the defective items can be sold (perhaps after some repairs) at some lower price. Such a situation occurs quite often in practice. For instance, in the garment industry, it is common to find markets selling branded items at much lower prices than usual. These items usually contain minor defects. This is observed particularly in developing countries where large classes of consumers who can not afford to buy expensive branded items at regular prices are tempted by the low price in these markets.

If the vendor accepts a lot with defective units, then he/she has to deal with two types of stocks: a stock of good items referred to in this study as first class items, and a stock of defective items referred to as second class items. The two classes of items are displayed in two different stores, referred to as primary and secondary markets, respectively. Note that the secondary market can lead to a successful business in itself, transforming the burden of handling defective items into a profit-making activity.

Our primary purpose in this study is to determine optimal ordering policies of style goods (say fashionable items) in the presence of defective units, where the vendor has to handle both types of stock, first and second class items, at a primary and a secondary market respectively. We will also provide tools for the vendor to select the appropriate arrangement to handle defective units. We will start by investigating the case where the vendor has no control on the prices in each of the markets. Then, we will consider the integrated inventory-pricing policies to help the vendor decide on selling prices of both classes of products. We will formulate the models in a single-period framework since we are concerned with fashionable items that have to be sold during a short period of time.

The single-period problem consists on determining the optimal order quantity that maximizes the expected profit over a short period with stochastic demand. This problem has been extensively studied in the literature. It is particularly applicable for inventory management of "style goods". A product is referred to as a style good if it has a random demand over a short selling season and a relatively long production process so that only one order can be made for the season. Various extensions to the newsboy problem have been developed to include different objectives, pricing policies, substitutable products, multi-echelon systems, multi-levels of storage, and random yields, among others. Khouja (1999) provides a state-of-the-art review of the newsboy problem as well as its extensions.

Inventory problems with random yields constitute an important class of problems that have been investigated both for the single and multiple period cases. Randomness arises from the uncertainty in the production capacity or from the presence of defective units. Yano and Lee (1995) surveyed the literature of these problems. Shih (1980) considered the case where the yield uncertainty is due to defective units. He assumed that such units are unsaleable and are to be returned to the manufacturer at his/her own expenses. Shih investigated two models, the modified EOQ and the modified newsboy, and derived the optimal solutions. He also performed comparisons with the traditional models via numerical examples. Ehrhardt and Taube (1987) extended Shih model by dealing with general forms of holding and shortage cost instead of the linear case. They derived the necessary optimality conditions and provided closed-form expressions of the optimal ordering quantity when the demand is uniformly distributed. They also provided a heuristic in the case where the demand follows a negative binomial distribution and the defective ratc has a Beta distribution.

In traditional inventory models, demand is assumed to be price independent. More specifically, the standard single-period model assumes that the procurement decision is not affected by the level at which the price of the style good is set at the beginning of the season. However, as noted by Whitin (1955), inventory models that integrate pricing and procurement decision have a stronger managerial implication, compared to fixed price models. In the early days of inventory theory, Whitin was the first to link pricing theory and inventory control by assuming a price-dependent demand. After his pioneering work, several research studies extended some of the classical inventory models (economic order quantity, single-period, etc.) to include the effects of pricing policies. Interested readers can refer to Eliashberg and Steinberg (1991) for a survey on integrated marketing-production/inventory decision models and to Khouja (1999) for extensions to different single-period pricing policies. Recently, using different price-demand relationships, Federgruen and Heching (1999) and Polatoglu and Sahin (2000) integrated pricing and inventory control under demand uncertainty in a perfect single item, periodic review model for a single market. Also, Rahim et al. (2000) developed a single-period perishable inventory model where deterioration begins at a random point in time during the cycle. The seller may change the selling price only once after the deterioration starts. This selling price depends upon on-hand inventory level and the residual life of the item.

2. MODEL FORMULATION

The model is developed under the following assumptions:

1. A complete and perfect inspection is performed upon receipt of the lot. The unit procurement cost includes the cost of carrying out this type of inspection.

2. All defective units are salable in the secondary market. Note however that the model can easily be extended to the case where a fraction of defective units are to be scrapped.

3. The demands at the primary and secondary market are uncorrelated. This is justified by the fact that these demands are generated from different categories of customers.

4. The repair cost, if needed, of defective units before their transfer to the secondary market is negligible. However, the model can be easily extended to account for a non-zero repair cost.

5. Both markets operate simultaneously; that is, they share the same selling season.

Note that, from the above assumptions, upon completion of the inspection process at the beginning of the season, all defective units are transferred to the secondary market. In this section and the next one, we will also assume that the demands in both markets are price independent throughout the short selling period. The following notations, which are somewhat similar to the ones in Gallego and Moon (1993), are used to formulate the model:

c	: unit procurement cost
r	: defective rate
Q	: ordering quantity
i	: index referring to primary market for $i = 1$ and to secondary market for $i = 2$
p_i, $i = 1, 2$	: unit selling price in market i
s_i, $i = 1, 2$	: unit salvage value in market i
x_i, $i = 1, 2$	: random demand in market i
F_i, $i = 1, 2$	: CDF of the random demand in market i
μ_i, $i = 1, 2$	: expected demand in market i
σ_i, $i = 1, 2$	: standard deviation of the demand in market i

Additional notations will be introduced throughout the paper, whenever needed. The following relationships (for which the interpretations are straightforward) hold:

$$p_1 > c > s_1 > s_2; \quad s_2 < p_2; \quad \text{and} \quad r < 1/2.$$

Note that the item can be sold in the secondary market at a price below its procurement cost. In the single-period problem with zero defects (Gallego and Moon (1993)), the expected profit is given by

$$ETP_0(Q) = \{(p_1 - s_1)\mu_1 - (c - s_1)Q - (p_1 - s_1)\int_Q^{\infty} (x_1 - Q)dF_1(x_1)\}. \tag{9.1}$$

The ordering quantity maximizing this expected total profit is given by

$$Q_0 = \bar{F}_1^{-1}\left(\frac{c - s_1}{p_1 - s_1}\right), \tag{9.2}$$

where $\bar{F}_1^{-1}$ is the inverse of the complementary cumulative distribution function of the demand in the primary market.

For the case of the current study, there is the possibility of receiving a portion of defective units in the lot. These units are to be transferred at the beginning of the season to the secondary market and sold at a lower price. The total expected profit generated from both markets has the following expression:

$$ETP_r(Q) = -cQ + p_1E[Min((1-r)Q, x_1)] + s_1E[((1-r)Q - x_1)^+]$$
$$+p_2E[Min(rQ, x_2)] + s_2E[(rQ - x_2)^+],$$

where $E[.]$ is the expected value operator, and $(.)^+ = max(., 0)$. The last expression of the total expected profit can be rewritten as

$$ETP_r(Q) = \sum_{i=1}^{2}\{(p_i - s_i)\mu_i - (c_i - s_i)r_iQ\}$$

$$-\sum_{i=1}^{2}\{(p_i - s_i)\int_{r_iQ}^{\infty}(x_i - r_iQ)dF_i(x_i)\} \tag{9.3}$$

where $r_1 = 1 - r$ and $r_2 = r$.
Observe that when $r = 0$, the expression in (9.3) reduces to that in

(9.1). Moreover, it is evident that maximizing $ETP_r(Q)$ is equivalent to minimizing the expected total cost, $ETC_r(Q)$ given by

$$ETC_r(Q) = \sum_{i=1}^{2} \{(c - s_i)r_iQ + (p_i - s_i)\int_{r_iQ}^{\infty}(x_i - r_iQ)dF_i(x_i)\} \quad (9.4)$$

The first order condition for the optimal ordering quantity is given by

$$g(Q) = \frac{dETC_r(Q)}{dQ} = \sum_{i=1}^{2} r_i\{(c - s_i) - (p_i - s_i)\bar{F}_i(r_iQ)\} = 0 \quad (9.5)$$

It can easily be verified that ETC_r is convex and therefore a solution of (9.5) is indeed a minimum. Moreover, it can be shown that equation (9.5) has a positive and finite root if the expected mark-up, $(p_1 - c)r_1 + (p_2 - c)r_2$ and the expected discount $(c - s_1)r_1 + (c - s_2)r_2$ are positive. These two conditions can be equivalently stated as

$$r_1s_1 + r_2s_2 < c < r_1p_1 + r_2p_2 \quad (9.6)$$

Next, for $i = 1, 2$ let Q_i be solution to

$$g_i(Q) = c - s_i - (p_i - s_i)\bar{F}_i(r_iQ) = 0 \quad (9.7)$$

Note that when the mark-up in the secondary market is negative (i.e., if $p_2 - c < 0$), then (9.7) has no solution for $i = 2$. In this case, we set $Q_2 = 0$.

In the following proposition, we provide a search interval for the optimal ordering quantity Q^* satisfying (9.5).

Proposition 1
$Q^* \in [Min(Q_1, Q_2), Max(Q_1, Q_2)]$.
Proof: We will prove the result when $p_2 - c > 0$. The case when $p_2 - c \leq 0$ can be treated easily. We will also assume that $Q_1 \leq Q_2$. The proof of the other case can be carried out in a similar way.

Note that g, g_1 and g_2 are non-decreasing functions of Q from a negative value at $Q = 0$ to positive values for sufficiently large values of Q and that $g = r_1g_1 + r_2g_2$. Therefore, $Q_1 \leq Q_2$ implies that $g_2(Q_1) \leq 0$, and since $g_1(Q_1) = 0$, it follows that $g(Q_1) \leq 0$. Similarly, we show that $g(Q_2) \geq 0$. Given that $g(Q^*) = 0$, it follows that $Q_1 \leq Q^* \leq Q_2$.

In earlier research related to imperfect procurement systems (see Yano and Lee (1995)), it was observed that the optimal ordering quantity is larger than the corresponding quantity when there is no defect. However, this may not be the case when defective units can be sold at a secondary market and generate revenues. In the next proposition, for which the proof is straightforward, we present conditions under which the optimal imperfect lot size Q^* can be larger than the perfect one, Q_0.

Proposition 2
$Q^* > Q_0$ under each of the following conditions:
1. $Q_1 < Q_2$.
2. $Q_0 \leq Q_2 \leq Q_1$.
Note that $Q^* > Q_0$ does not necessarily hold if $Q_2 < Q_0$. This can easily be shown by a counter-example.

Next, we will try to determine when it is more profitable for the vendor to order a lot having defective units. After substituting (9.5) into (9.4) and (9.3), the minimum expected total cost and the corresponding maximum expected total profit are given, respectively by

$$ETC_r^* = \sum_{i=1}^{2} (p_i - s_i) \int_{r_iQ^*}^{\infty} x_i dF_i \tag{9.8}$$

$$ETP_r^* = \sum_{i=1}^{2} (p_i - s_i) \int_{0}^{r_iQ^*} x_i dF_i \tag{9.9}$$

Note that the optimal expected profit in the case of zero defects is given by

$$ETP_0^* = (p_1 - s_1) \int_{0}^{Q_0} x_1 dF_1 \tag{9.10}$$

Prposition 3
If condition 1 of Proposition 2 holds, then it is more profitable to order a lot with defective units that can be sold in the secondary market than to have a perfect lot and limit the business to the primary market.
Proof: It follows from condition 1 of Proposition 2 (using Proposition 1) that Q^* is larger than Q_1. Also, from (9.9), (9.10) and $r < 1/2$, we obtain

$$ETP_r^* - ETP_0^* \quad = \quad (p_1 - s_1) \left[\int_{0}^{r_1Q^*} x_1 dF_1(x_1) - \int_{0}^{Q_0} x_1 dF_1(x_1) \right]$$

$$+(p_2 - s_2) \int_0^{r_2 Q^*} x_2 dF_2(x_2)$$

$$> \quad (p_1 - s_1) \left[\int_0^{r_1 Q_1} x_1 dF_1(x_1) - \int_0^{Q_0} x_1 dF_1(x_1) \right]$$

$$+(p_2 - s_2) \int_0^{r_2 Q^*} x_2 dF_2(x_2)$$

Finally, using $Q_0 = rQ_1$ and from $p_1 - s_1 > 0$ and $p_2 - s_2 > 0$, the result follows. Note that Proposition 3 holds in a special case where selling the defective items in the secondary market is in itself a profit-generating activity ($P_2 > c$, which implies in particular that $P_2 > s_1$). However, it is often the case where the secondary market is used only to reduce the loss due to the presence of defective items in the ordered lot. In this case (where $c > p_2$), the following intuitive result holds:

Proposition 4
If $c > p_2$, then it is more profitable to have a perfect (defect-free) lot.
Proof: Set for $i = 1, 2$
$ETP_{ri}(w) = (p_i - s_i)\mu_i - (c - s_i)w + (p_i - s_i) \int_w^\infty (x_i - w) dF_i(x_i)$.
Thus, $ETP_r^* = ETP_{r1}(r_1 Q^*) + ETP_{r2}(r_2 Q^*)$
Note that $ETP_{r1}(w)$ reaches its maximum, ETP_0, at Q_0. Therefore, $ETP_{r1}(r_1 Q^*) \leq ETP_0$. Also, when $c > p_2$, it can be easily shown that $ETP_{r2}(w)$ is non-increasing and vanishes at zero. The result follows.

Example 1
Jawda-Fashion is a chain of stores specialized in ladies fashionable products. A particular dress is offered every summer with the modifications proposed by the new fashions. The procurement cost of the dress is 1 MU (Monetary Unit). The selling unit price is 1.5 MU. If minor defects are detected, then the corresponding dresses are transferred to some secondary stores, where they are sold at 1.2 MU, each. At the end of the summer, the dress is cleared at 0.6 MU in the primary stores and at 0.1 MU in the secondary stores. The defective rate is approximately equal to 0.1. The demand is estimated to be normal with mean and standard deviation of 500 and 100 respectively in the primary stores and 100 and 20 respectively in the secondary stores.

The optimal ordering quantity is $Q^* = 577$ (rounded to the nearest integer), and the corresponding expected profit is 225.78 MU. If perfect dresses are to be ordered and sold in the primary stores, then the optimal ordering quantity would be $Q_0 = 514$ and the corresponding expected profit is 214.38 MU. This is a case where Q^* is larger than Q_0 and when it is more profitable to order an imperfect lot rather than to order a perfect one and limit the business to the primary stores. This case is consistent with the conclusion of Proposition 3. In fact, the condition is satisfied since we have $Q_1 = 571.045$ and $Q_2 = 818.33$.

We next perform some sensitivity analyses on the defective rate to observe the behavior of the optimal ordering quantity and the corresponding expected profit, compared with the case of ordering perfect dresses. The results are given in Table (9.1).

From the table above, ETP_r increases in r for $0.02 \leq r \leq 0.14$, after which the expected profit starts to decay. Also, ETP_r remains larger than ETP_0 for all defective rates $r \leq 0.18$. When $r = 0.14$, Q^* reaches its highest level of 596.328 and starts decreasing in r. In addition, Q^* remains larger than Q_0 for all $r \leq 0.32$. For $r \leq 0.12$, $Q_2 > Q_1$ and the result of Proposition 3 applies. The result still applies for $r = 0.14, 0.16$ and 0.18 while $Q_2 < Q_1$, showing that the condition of Proposition 3 is sufficient, but not necessary.

3. SELECTING AMONG ALTERNATIVE ARRANGEMENTS

Shih (1980) developed models for optimal ordering quantities in the presence of defective units while assuming that the supplier would refund each defective unit detected by the vendor. Alternative arrangements between the supplier and the vendor when dealing with defective units may include in practice the following two policies:

Policy 1: *The supplier is responsible for taking all the necessary measures so that the ordered quantity will be perfect (zero defects).*
Policy 2 *The supplier provides a reduction in the procurement cost in order to compensate for the presence of defective units.*

The new discounted procurement cost will be decided through negotiations between the two parties. We assume that the vendor does not share the cost reduction with his/her customers due to the discount offered by the supplier. In other words, The new procurement cost will not have any effect on the selling prices of good and defective units.

In this section, we will try to determine when policy 2 is more beneficial for the vendor than policy 1. This would serve as a tool for the vendor to negotiate the new procurement cost.

Table 9.1 Effect of the defective rate on the optimal ordering policy

$Q_0 = 513.941$ and $ETP_0 = 214.381$

r	Q_1	Q_2	Q^*	ETP_r
0.02	524.429	4091.65	525.627	216.525
0.04	535.355	2045.825	537.844	218.75
0.06	546.746	1363.883	550.636	221.08 1
0.08	558.631	1022.913	563.902	223.381
0.1	**571.045**	**818.33**	**577.385**	**225.78**
0.12	584.024	681.942	589.379	227.869
0.16	611.834	511.456	596.052	226.709
0.18	626.757	454.628	590.449	221.447
0.2	642.426	409.165	582.801	212.975 1
0.22	658.899	371.968	575.608	202.034 1
0.24	676.238	340.971	570.157	189.642 1
0.26	694.515	314.742	565.822	176.282
0.28	713.807	292.261	560.667	162.487
0.3	734.201	272.777	551.479	148.372
0.32	755.795	255.728	533.598	134.226
0.34	778.698	240.685	493.404	120.366
0.36	803.033	227.314	392.608	108.111
0.38	828.937	215.35	337.438	98.365
0.4	856.568	204.582	304.625	90.178

Let the new procurement cost per unit be τc, for some $0 < \tau \leq 1$. Note that from (9.6) the procurement cost must always be at least as large as $r_1 s_1 + r_2 s_2$. Therefore, a tighter lower bound of τ will be $\tau_{min} = [r_1 s_1 + r_2 s_2]/c$. Next, set $Q^*(\tau)$ to be the optimal ordering quantity for the new procurement cost. Thus, $Q^*(\tau)$ must satisfy:

$$\sum_{i=1}^{2} r_i((c\tau - s_i) - (p_i - s_i)\bar{F}_i(r_i Q)) = 0 \tag{9.11}$$

The corresponding optimal expected profit is given by:

$$ETP_r(Q^*(\tau)) = \sum_{i=1}^{2} (p_i - s_i) \int_0^{r_i Q^*(\tau)} x_i dF_i \tag{9.12}$$

The largest value of τ, τ_{max}, for which it is still worth opting for policy 2 is the solution of the following equation:

$$G(\tau) = ETP_r(Q^*(\tau)) - ETP_0 = 0 \tag{9.13}$$

Note that $Q^*(\tau)$ increases as τ decreases. Therefore, the following algorithm is proposed to find τ_{max}, and to determine the appropriate policy for the vendor.

Step 1 If $G(1) > 0$, then $G(\tau) > 0$ for any τ ((9.13) has no solution. In this case, policy 2 is always more beneficial to the vendor than policy 1. It is better off accepting any discounted procurement cost than having a perfect lot). Select policy 2. Stop.
Else, goto step 2.

Step 2 If $G(\tau_{min}) < 0$, then $G(\tau) < 0$ for any τ ((9.13) has no solution. In this case, policy 1 is always more beneficial to the vendor than policy 2, independently of the size of the discount).
Select policy 1. Stop.
Else, goto step 3.

Step 3$G(\tau_{min}) > 0, G(1) < 0$ and G is non-increasing. Use a line-search method to find τ_{max}, solution of $G(\tau) = 0$.
For $\tau > \tau_{max}$, select policy 1. Else, select policy 2.

Example 2
In example 1, $ETP_r > ETP_0$ for $0.02 \leq r \leq 0.18$. Therefore, it should be clear that policy 2 is always better, and $\tau_{max} = 1$. If $r \geq 0.20$, then a discount is needed in order for the vendor to keep with policy 2. For $r =$

0.2, τ_{max} was found to be 0.99758 suggesting that any discount offered by the supplier would support selecting policy 2. When $r = 0.4$, a 3% discount would be enough for the vendor to opt for policy 2. The complete set of computations is displayed in Table (9.2) below. Note that in all cases, the discount needed in order to favor policy 2 is very small. This is because the secondary market is in itself rewarding since $p_2 > c$.

4. INTEGRATED INVENTORY-PRICING MODEL

In the model developed above, we assume that the ordering decision is not affected by the price-demand relationship in both markets. In this section, we extend this model to incorporate the impacts of considering the selling prices as decision variables. At the beginning of the season, the problem consists of determining the procurement and pricing policies that simultaneously maximize the expected total profit in the primary and secondary markets. The following notation is used.

$x_i(p_i)$	random demand in market i ($i = 1, 2$) when the price level is p_i,
$F_i(x_i; p_i)$	cumulative distribution function of $x_i(p_i)$,
$\mu_i(p_i)$	expected demand in market i as function of its price.

In addition, we assume that the price at each market is bounded from below and above by p_{Li} and p_{Ui}, respectively.

The expected total profit is then given by:

$$ETP_r(Q, p_1, p_2) = \sum_{i=1}^{2} (p_i - s_i)\mu_i(p_i)$$

$$-\sum_{i=1}^{2} \{(c - s_i)r_iQ + (p_i - s_i) \int_{r_iQ}^{\infty} (x_i(p_i) - r_iQ)dF_i(x_i; p_i)\} \qquad (9.14)$$

The joint inventory-pricing optimization problem to solve can be stated as

JIP: Maximize $ETP_r(Q, p_1, p_2)$
Subject to
$c < \sum_{i=1}^{2} r_i p_i$
$p_1 > p_2$
$p_{Li} \leq p_i \leq p_{Ui}, \quad i = 1, 2$
$Q > 0.$

For the perfect single period problem with a single market, Lau and Lau (1988) and Polatoglu (1991) noted the difficulties in obtaining an

Table 9.2 Effect of the defective rate on selecting the appropriate policy

$Q_0 = 513.941$ and $ETP_0 = 214.381$

r	τ_{min}	τ_{max}	Q^*	ETP_r
0.02	0.5941	1	525.627	216.525
0.04	0.5842	1	537.844	218.75
0.06	0.5743	1	550.636	221.08
0.08	0.5644	1	563.902	223.381
0.1	**0.5545**	**1**	**577.385**	**225.78**
0.12	0.5446	1	589.379	227.869
0.14	0.5347	1	596.328	228.624
0.16	0.5248	1	596.052	226.709
0.18	0.5149	1	590.449	221.447
0.2	0.505	0.99758	583.645	214.35
0.22	0.49628	0.98131	584.562	214.295
0.24	0.49543	0.97807	592.208	214.386
0.26	0.49597	0.97662	605.823	214.406
0.28	0.49702	0.97593	622.257	214.351
0.3	0.49818	0.97525	639.994	214.329 1
0.32	0.49944	0.97458	658.882	214.391
0.34	0.50079	0.97416	678.784	214.332
0.36	0.50209	0.97301	700.154	214.474
0.38	0.5037	0.97262	722.734	214.47
0.4	0.50522	0.97198	746.664	214.334

analytical solution to the joint inventory-pricing problem with general price-demand relationship. These difficulties are mainly due to the fact that the optimal price and ordering quantity cannot be explicitly evaluated. To overcome these analytical difficulties, several authors considered special forms for the relationship between demand and price. In most joint inventory-pricing papers, the random demand is expressed as a combination of its expected demand and a price-independent random term. This form of demand-price dependency allows the isolation of the effects of price and uncertainty. It also makes the developed mathematical models more tractable. The additive, multiplicative, hybrid and riskless forms are the most used relationships in demand modeling. Under the additive model, the random demand at price p is $x(p) = \mu(p)+\varepsilon$, where $\mu(p)$ is a decreasing function of p and ε is a random variable with a known distribution having zero mean and price independent variance. In the multiplicative model, random demand is expressed as $x(p) = \varepsilon\mu(p)$ where the mean of the random term is one. In the hybrid model, demand is represented by a combination of the additive and multiplicative forms, i.e., $x(p) = \varepsilon\delta_1(p) + \delta_2(p)$, where $\delta_1(p)$ and $\delta_2(p)$ are decreasing functions of p. Note that for multiplicative and hybrid demand models, both the mean and the variance of the random demand are price-dependent. In the riskless model (also called deterministic model), it is assumed that the price-demand relationship can be perfectly estimated; that is, $x(p) = \mu(p)$. Using linear expected demand, $\mu(p) = a - bp$, where $a, b > 0$, Lau and Lau (1988) and Polatoglu (1991) showed the unimodality of the expected profit for uniformly and normally distributed additive demands, respectively. Polatoglu (1991) provided similar proofs for the case of an exponentially distributed multiplicative demand.

In this paper, we consider additive demand models for both markets, i.e., $x_i(p_i) = \mu_i(p_i) + \varepsilon_i$. This price-demand relationship is the most commonly used form in the joint pricing-inventory literature. We also consider both cases, of complete and partial information, regarding the distribution of the price-independent random term. In the complete information case, we assume that the random term is normally distributed in both markets. However, under the partial information approach (also called distribution-free approach, see Gallego and Moon (1993)), the probability distribution of the random term is not specified.

4.1 Complete information approach. Under the assumption of normally distributed $\varepsilon_i (i = 1, 2)$ with zero mean and σ_i as standard deviation, the expected total profit is

$$ETP_r(Q, p_1, p_2) = \sum_{i=1}^{2} (p_i - s_i)\mu_i(p_i)$$

$$-\sum_{i=1}^{2} \{(c - s_i)r_i Q + (p_i - s_i)\sigma_i[\phi(z_i) - z_i\bar{\Phi}(z_i)]\}, \qquad (9.15)$$

where

$$z_i = \frac{r_i Q - \mu_i(p_i)}{\sigma_i}, \qquad (9.16)$$

ϕ is the density function of the standard normal distribution, and $\bar{\Phi}$ is the complementary cumulative distribution of the standard normal. The first order conditions of the unconstrained version of (JIP) are

$$\frac{\partial ETP_r(Q, p_1, p_2)}{\partial Q} = \sum_{i=1}^{2} r_i[-(c - s_i) + (p_i - s_i)\bar{\Phi}(z_i)] = 0 \qquad (9.17)$$

$$\frac{\partial ETP_r(Q, p_1, p_2)}{\partial p_i} = \mu_i(p_i) + (p_i - s_i)\frac{d\mu_i(p_i)}{dp_i}\Phi(z_i) - \sigma_i[\phi(z_i) - z_i\bar{\Phi}(z_i)] = 0 \qquad (9.18)$$

It can also be verified that

$$\frac{\partial^2 ETP_r(Q, p_1, p_2)}{\partial Q^2} = -\sum_{i=1}^{2} r_i^2 \frac{p_i - s_i}{\sigma_i}\phi(z_i) < 0.$$

Therefore, the expected total profit is unimodal over Q. Moreover, $ETP_r(Q, p_1, p_2)$ is also unimodal over each of the market prices since

$$\begin{aligned}\frac{\partial^2 ETP_r(Q, p_1, p_2)}{\partial p_i^2} &= 2\frac{d\mu_i(p_i)}{dp_i}\Phi(z_i) + (p_i - s_i)\frac{d^2\mu_i(p_i)}{dp_i^2}\Phi(z_i) \\ &\quad - \frac{(p_i - s_i)}{\sigma_i}\left(\frac{d\mu_i(p_i)}{dp_i}\right)^2\phi(z_i) < 0,\end{aligned}$$

given that $\mu_i(p_i)$ is assumed to be strictly decreasing and concave in p_i.

It should be noted that solving the system of equations (9.17) and (9.18) is analytically difficult even using linear functions $\mu_i(p_i)$. Similarly, solving (JIP) using any analytical optimization technique will also be difficult. Instead, in the sequel, we determine ETP_r^*, given by (9.9) for different values of p_1 and p_2 which satisfy the constraints of (JIP). The values of p_1 are taken over the range $[p_{L1}, p_{U1}]$ with step size α_1 and the values of p_2 over the range $[p_{L2}, \min(p_1, p_{U2})\ ]$ with step size α_2 such that $c < r_1 p_1 + r_2 p_2$. The optimal ordering quantity, Q^* and prices, p_1^* and p_2^* are the ones that result in the largest expected total profit.

Example 3
Continuing with the same illustrative example, we assume that the linear demand model provides good estimates for the demand as a function of price in each market. Other authors in the literature have also used this type of demand-price function. Moreover, using real data collected from a specialty retailer of high-end women apparel, Federgruen and Heching (1999) found that the standard deviation of the weekly sales is independent of the price charged, justifying the use of a linear function. Suppose that the sales people of the company feel that the price-demand function can be well fitted by the following simple regression equations for each market.

$$x_1 = 800 - 200p_1 + \varepsilon_1 \quad 1.2 \le p_1 \le 3 \quad \sigma_1 = 100$$

$$x_2 = 220 - 100p_2 + \varepsilon_2 \quad 0.5 \le p_2 \le 1.8 \quad \sigma_2 = 20$$

Keeping the other input parameters of the example unchanged, the optimal policy is to set the prices at $p_1^* = 2.5$ and $p_2^* = 1.6$ and to order $Q^* = 430$ units which will result in an expected total profit of $ETP_r^* = 417.571$. Comparing this value of the expected total profit with the one obtained in example 1, one can see the financial importance of considering the prices as decision variables.

For the standard single-period problem, Lau and Lau (1988) observed that the lower slope, b, for the demand-price function results in a smaller ordering quantity and a higher expected profit at a lower selling price. They also noticed that an increase in demand uncertainty σ, would slightly decrease the price and increase considerably the ordering quantity. Our computational study revealed the same effects of the slope b and the standard deviation on the ordering policy and the pricing polices in both markets. We also investigated the effect of the defective rate on the optimal prices and ordering quantity. Table (9.3) shows that the price in the primary market increases as the defective rate increases, whereas the price in the secondary market is decreasing. This is intuitive since the defective rate has an increasing effect on the secondary market supply and a decreasing impact on the primary market supply. In fact, it has been observed by marketing researchers/practitioners that a large

Table 9.3 Effect of the defective rate on the integrated ordering-pricing policy

r	p_1^*	p_2^*	Q^*	ETP_r^*
0	2.5		380.440	395.129
0.02	2.5	1.7	390.799	400.42
0.04	2.5	1.7	401.378	405.747
0.06	2.5	1.7	411.522	410.723
0.08	2.5	1.7	420.081	414.886
0.1	**2.5**	**1.6**	**429.764**	**417.571**
0.12	2.5	1.6	434.562	419.407
0.14	2.5	1.5	442.962	419.669
0.16	2.5	1.5	443.774	418.602
0.18	2.5	1.4	450.865	416.153
0.2	2.5	1.4	449.424	411.821
0.22	2.5	1.3	455.628	406.497
0.24	2.6	1.3	437.689	399.302
0.26	2.6	1.2	443.422	391.44
0.28	2.6	1.2	442.376	382.748
0.3	2.6	1.2	443.218	372.615
0.32	2.6	1.2	446.165	361.454
0.34	2.7	1.2	429.095	349.69
0.36	2.7	1.2	433.486	337.797
0.38	2.7	1.2	438.41	325.388
0.4	2.7	1.1	443.298	312.371

amount of displayed goods (supply) generates a high rate of demand (Levin et al. (1972) and Silver et al. (1998)). Therefore, the price will increase when the supply decreases and vice-versa, given the positive correlation between the demand and supply.

4.2 Partial information approach. Practically speaking, obtaining a probability distribution that accurately fits the stochastic price-demand function is a very difficult task. Moreover, as we mentioned above, for a general distribution of the random term representing the demand uncertainty, the integrated procurement-pricing policy is difficult to determine in which case the problem should be addressed heuristically. Without specifying the form of the random term distributions, the partial information approach determines the ordering quantity and the prices that maximize the expected total profit against the worst possible distributions with mean $\mu_i(p_i)$ and standard deviation $\sigma_i (i = 1, 2)$. This approach is referred to in the inventory literature (Scarf (1958)) as the min-max distribution free approach. For more exposure to this approach to other production and inventory problems, we refer interested readers to Gallego (1992), Gallego and Moon (1993), Hariga (1998) and Hariga and Ben-Daya (1999). For the classical single-period problem, Gallego and Moon (1993) used the following important inequality

$$\int_{Q}^{\infty} (x - Q) dF(x, \mu, \sigma) \leq \frac{\sqrt{\sigma^2 + (Q-\mu)^2} - (Q - \mu)}{2}$$

to provide a tighter lower bound to the expected profit. Using this inequality, a lower bound to the expected total profit is given by

$$ETP_r^L(Q, p_1, p_2) = \sum_{i=1}^{2} (p_i - c)\mu_i(p_i)$$

$$- \sum_{i=1}^{2} \frac{\sigma_i}{2} [(p_i - s_i)\sqrt{1 + z_i^2} - ((p_i - c) - (c - s_i))z_i] \qquad (9.19)$$

where z_i is defined as above in (9.16). Using the same analysis as for the complete information case, it can be shown that the lower bound expected profit is unimodal over Q and $p_i (i = 1, 2)$. Therefore, we can use the same search technique by maximizing $ETP_r^L(Q, p_1, p_2)$ over Q for different values of the selling prices. For fixed p_1 and p_2, the ordering quantity maximizing ETP_r^L is the solution to

$$\frac{\partial ETP_r^L(Q, p_1, p_2)}{\partial Q} = h(Q) = \sum_{i=1}^{2} r_i h_i(Q) = 0 \qquad (9.20)$$

where

$$h_i(Q) = (p_i - s_i)\frac{z_i}{\sqrt{1 + z_i^2}} - ((p_i - c) - (c - s_i)).$$

Letting Q_i^L be the solution to $h_i(Q) = 0$, which is given by

$$Q_i^L = \mu_i(p_i) + \frac{\sigma_i}{2}\left[\sqrt{\frac{p_i - c}{c - s_i}} - \sqrt{\frac{c - s_i}{p_i - c}}\right], \tag{9.21}$$

it can be shown that, $Q^{L^*}(p_1, p_2)$, the solution to equation (9.20), belongs to the interval

$$[Min(Q_1^L, Q_2^L), Max(Q_1^L, Q_2^L)]. \tag{9.22}$$

The optimal procedure to find a lower bound for the expected total profit is as follows.

Step 1. For $p_1 = p_{L1}$ to p_{U1} step α_1 do:
For $p_2 = p_{L2}$ to p_{U2} step α_2 do:
If p_1 and p_2 satisfy the constraints of (JIP) do:
Use any one-dimensional search technique to find $Q^{L^*}(p_1, p_2)$.
Compute $ETP_r^L(Q^{L^*}(p_1, p_2), p_1, p_2)$.
Step 2. Set $(Q^{L^*}, p_1^{L^*}, p_2^{L^*}) \leftarrow Arg\ max\ ETP_r^L(Q^{L^*}(p_1, p_2), p_1, p_2)$.

Example 4
Using the same data as in the above example, the joint inventory-pricing policy under the partial information approach is to order 439 units and to set the prices at 2.4 and 1.6 in the primary and secondary markets, respectively. The expected total profit of this approximate solution is 417.113. Note that $ETP_r(Q^*, p_1^*, p_2^*) - ETP_r(Q^{L^*}, p_1^{L^*}, p_2^{L^*}) = 0.458$ is the expected value of additional information (Gallego and Moon (1993)). It is the largest amount that one would be willing to pay for the knowledge of the random term distribution. We have also solved the same example with different values of the defective rate to assess the performance of the partial information approach. The results of these numerical runs are summarized in Table (9.4). In the last column of this table, we reported the penalty ratio, $ETP_r(Q^*, p_1^*, p_2^*)/ETP_r(Q^{L^*}, p_1^{L^*}, p_2^{L^*})$, of using the approximate solution, $(Q^{L^*}, p_1^{L^*}, p_2^{L^*})$, instead of the optimal one. It can be noticed that the maximum penalty ratio is 0.4%, which confirms the robustness of the partial information approach. The good performance resulting from this approach has also been observed by other authors in different production/inventory contexts (see for example Hariga (1998)).

5. CONCLUSION

In this study, we investigated the single-period inventory problem when the lot received may contain defective items. We analyzed the

Table 9.4 Effect of the defective rate on the performance of the approximate solution

r	p_1^{L*}	p_2^{L*}	Q^{L*}	ETP_r	Ratio
0.02	2.4	1.6	396.963	398.967	1.004
0.04	2.4	1.7	408.029	405.151	1.001
0.06	2.4	1.7	418.716	410.318	1.001
0.08	2.4	1.7	428.035	414.416	1.001
0.1	**2.4**	**1.6**	**437.881**	**417.113**	**1.001**
0.12	2.4	1.6	442.452	418.619	1.002
0.14	2.4	1.5	450.813	418.737	1.002
0.16	2.5	1.5	434.064	418.381	1.001
0.18	2.5	1.4	441.19	415.928	1.001
0.2	2.5	1.4	438.327	411.502	1.001
0.22	2.5	1.3	444.782	406.159	1.001
0.24	2.5	1.3	444.183	398.903	1.001
0.26	2.5	1.2	450.588	390.57	1.002
0.28	2.6	1.2	433.762	382.568	1
0.3	2.6	1.2	437.149	372.537	1
0.32	2.6	1.2	442.117	361.422	1
0.34	2.6	1.2	448.02	349.525	1
0.36	2.6	1.2	454.461	337.006	1.002
0.38	2.7	1.2	436.802	325.375	1
0.4	2.7	1.2	443.151	312.371	1

case where good and defective items are sold in a primary and secondary market, respectively. The problem occurs quite often in practice, particularly in the garment industry in developing countries as witnessed by the authors themselves. We developed results for the optimal ordering policies. We also established comparisons with the case of a 100% perfect lot. We showed in particular that the order size may be smaller in the presence of defective items. This result is not consistent with the findings of other studies on lot sizing with random yields, where only a primary market is considered (see for instance Shih (1980)). We also showed that ordering a lot with defective items may even be more profitable than ordering a perfect lot. This can be the case when the secondary market is in itself profitable. In addition, we provided tools for the vendor to negotiate alternative arrangements with the supplier regarding defective items. Finally, we included the pricing decisions in our model in order to determine the procurement and pricing policies that simultaneously maximize the expected total profit in the primary and secondary markets. We considered both cases of complete and partial information regarding the distributions of the price-dependent demands. Our results are fully illustrated by hypothetical examples throughout the paper.

While we restricted the analysis to the deterministic case of a failure rate, our empirical results show that the order size does not vary widely for a reasonably small range of defective rates. Therefore, the deterministic approach may still apply if the variance of the defective rate is not very large in a stochastic framework. As an extension to the current study, some research work is underway by the authors of this work to develop integrated polices for both supplier and vendor in the presence of defective items. The objective of this supply chain system is to determine the optimal procurement cost, ordering quantity, and selling prices in both markets.

Acknowledgements-The authors would like to thank Prof. Rahim and the anonymous referees for their valuable comments that improved the presentation of the paper.

References

[1] Ehrhardt, R. and Taube, L. An inventory model with random replenishment quantities. *International Journal of Production Research*, **25** (1987) 1795-1804.

[2] Eliashberg, J. and Steinberg, R. Marketing-production joint decision making. *Management Science in Marketing*, Volume 5, J. Eliashberg,, J. D. Lilien, eds., of *Handbooks in Operations Research and management Science*, North Holland, Amsterdam, 1991.

[3] Federgruen A. and Heching A. Combined pricing and inventory control under uncertainty. *Operations Research*, **47** (1999) 454-475.

[4] Gallego, G. A min-max distribution free procedures for the (Q, r) inventory model. *Operations Research Letters*, **11** (1991) 55-60.

[5] Gallego, G. and Moon, I. The distribution free newsboy problem: review and extensions. *Journal of Operational Research Society*, **44** (1993) 825-834.

[6] Hariga, M. A single period, multi-echelon stochastic model under a mix of assemble to order and assemble in advance policies. *Naval Research Logistics*, **45** (1998) 599-614.

[7] Hariga, M. and Ben-Daya, M. Some stochastic inventory models with deterministic variable lead time. *European Journal of Operational Research*, **113** (1999) 42-51.

[8] Khouja, M. The single-period (news-vendor) problem: literature review and suggestions for future research. *Omega: The International Journal of Management Science*, **27** (1999) 537-553.

[9] Levin, R. I., McLaughlin, C. P., Lamone, R. P., and Kottas, J. F., *Production/operations management: Contemporary policy for managing operating systems.* McGraw Hill, New York, 1972.

[10] Polatoglu, H. Optimal order quantity and pricing decisions in single period inventory systems. *International Journal of Production Economics*, **23** (1991) 175-185.

[11] Polatoglu, H. and Sahin I. Optimal procurement policies under price-dependent demand. *International Journal of Production Economics*, **65** (2000) 141-171.

[12] Rahim, M. A., Kabadi, S. N. and Banerjee, P.K. A single period perishable inventory model where deterioration begins at a random point in time. *International Journal of Systems Science*, **31**(1) (2000) 131-136.

[13] Scarf, H. A min-max solution of an inventory problem. In *Studies in the Mathematical Theory of Inventory and production*, Stanford University Press ,K. Arrow, S. Karlin, and H. Scarf, *eds.*, Stanford, CA, 201-209, 1958.

[14] Shih, W. Optimal inventory policies when stockouts result from defective products. *International Journal of Production Research*, **18** (6) (1980) 677-686.

[15] Silver, E. A., Pyke, D. F. and Peterson R., *Inventory management and production planning and scheduling*, 3rd Ed., John Wiley & Sons, New York 1998.

[16] Yano, A. R. and Lee, H. L. Lot sizing with random yields: A review. *Operations Research*, **43**(2) (1985) 311-334.

[17] Whitin T. M. Inventory control and price theory. *Management Science*, **2** (1955) 61-68.

Chapter 10

MANUFACTURING SYSTEM MODELING AND CONTROL: PRODUCTION QUALITY & MARKETING

E. K. Boukas and Z. K. Liu
Mechanical Engineering Department,
École Polytechnique de Montréal
P.O. Box 6079, station "centre-ville"
Montréal, Québec, H3C 3A7
Canada

Abstract This chapter extends the continuous flow model of stochastic manufacturing system to include deteriorating items, quality and marketing control. The production quality varies with the change of the system mode. To control the production quality, the jump rates from working mode to repair are considered as control variables. To model the marketing problem, the system is assumed to have many demand rates. The jump rates from one level to another are considered as controller variables. Numerical techniques are developed to synthesize the optimal control.

Keywords: Manufacturing system, production system, piecewise deterministic system, dynamic programming, numerical methods, Markov decision process, maintenance control, inventory control, quality control, marketing, viscosity solution.

1. INTRODUCTION

Manufacturing systems are often complex and large scale stochastic systems. This complexity makes their modeling and control a challenging problem and has attracted a lot of authors from control system and operations research communities. Some interesting models have been developed, among which we quote the ones based on queuing theory, control system, Petri nets, discrete event system, etc. For more details about these techniques, we refer the readers to Sethi and Zhang

[38], Gershwin [24], Buzacott and Shanthikumar [15], Viswanadham and Narahari [45] and references therein.

Manufacturing systems, where each machine switches from producing one part-type to another, require some setup, which may consume time or money or both. If the system is setup at no cost, it is called a flexible manufacturing system (FMS). In the field of FMS, the continuous flow model is of particular interest. This model is, in fact, a hybrid system. That is, the system is comprised of two components, a discrete part denoted by $\xi(t)$ and a continuous part denoted by x_t. The discrete part denotes the working mode of the machine and the continuous part represents the inventory level surplus/backlog. The mode switching is governed by a continuous-time Markov process taking values in a finite state space. When the system mode is fixed, the continuous state evolves as a deterministic dynamical system. For more information on the continuous flow model of a manufacturing system, the reader is referred to Akella and Kumar [2], Olsder and Suri [35], Gershwin [24], Sethi and Zhang [38] and references therein.

The current trend in manufacturing systems is to produce multiple part types in small batch so that the system can respond quickly to the changes in the market. To handle multiple part types in small batch, the production has to be well scheduled. In this respect, Connolly *et al.* [17], Gershwin *et al.* [25], Hu *et al.* [28], and Sharifnia *et al.* [40] have developed various possible heuristic policies and have carried out numerical computations and simulations. Sethi and Zhang [38] considered the asymptotically optimal policy using a hierarchical control. Yan and Zhang [48] established an approximate optimality condition and developed a computation algorithm for solving the optimal production control and setup scheduling problem.

Variants of this model including preventive maintenance and corrective maintenance, have also been proposed (see Boukas and Liu [10] and references therein). This chapter will consider some variants of the continuous flow model of manufacturing with the following factors considered:

- A continuous flow model with deteriorating terms;
- The quality control problem;
- The marketing and production control problem.

The remainder of this chapter is organized as follows. Section 2 introduces the basic continuous flow model of failure-prone manufacturing system. Section 3 addresses the definition, the optimal control problem of a piecewise-deterministic process (PDP), and its numerical method.

Section 4 studies the manufacturing system model with deteriorating terms. Section 5 and Section 6 develop a continuous flow model for a manufacturing system with quality control and marketing control problems respectively.

2. BASIC PRODUCTION MODEL

One of the most important achievements in the field of continuous flow modeling has been to identify the structure of the optimal feedback control policy. Kimemia and Gershwin [31] showed that the optimal control for such systems has a special structure, called the *hedging point policy.* The key point of this policy is that the optimal controlled system should have a non-negative production surplus (the hedging level) of parts to hedge against future possible shortages resulting from machine failure. When inventory deviates from this point, the controller should replenish it as soon as possible. For a one machine one part-type system with a constant demand rate, the optimal hedging point of the discounted cost model was first identified by Akella and Kumar [2]. The setting of average cost was first solved by Bielecki and Kumar [6]. In this section, their results will be outlined.

Let $x(t)$, $t \geq 0$ be the inventory when positive or the backlog when negative of the system and $\xi(t) \in \mathcal{S} = \{1, 2\}$ be a Markov process, representing the state of the machine, i.e. $\xi(t) = 1$ indicates that the machine is working and $\xi(t) = 2$ indicates that the machine is under repair. Let

$$Q = \begin{pmatrix} -q_1 & q_1 \\ q_2 & -q_2 \end{pmatrix}$$

be the generator of the Markov process $\{\xi(t), t \geq 0\}$. Let $u(t)$ be the production rate at time t, which is a control variable and satisfies the following constraints:

$$0 \leq u(t) \leq \bar{u} I_{\{\xi(t)=1\}}, \tag{10.1}$$

where $\bar{u}$, a constant, is the production capacity of the machine, and $I_{\{\cdot\}}$ is the indicator function defined by

$$I_{\{\xi(t)=1\}} = \begin{cases} 1, & \text{if } \xi(t) = 1, \\ 0, & \text{otherwise.} \end{cases}$$

The dynamics of the stock level, $x(t)$, is described by the following differential equations:

$$\dot{x}(t) = u(t) - d,$$

where d is the demand rate which is assumed to be constant. In the sequel, for notational simplicity $x(t), u(t), \xi(t)$ are also written as x_t, u_t, ξ_t so that no confusion is caused. Consider the following discounted cost function:

$$\begin{aligned} J(x,\alpha,u(\cdot)) &= \mathbb{E}\left[\int_0^\infty e^{-\rho t}[c^+x^+(t)\right. \\ &\qquad \left. +c^-x^-(t)]dt|x(0)=x,\xi(0)=\alpha\right], \qquad (10.2) \end{aligned}$$

where c^+, c^- are positive constants, $x^+ = max(x,0), x^- = max(-x,0)$, $\rho > 0$ is the discounted rate, and $\mathbb{E}$ is the expectation operator. The optimization problem is to seek a control law satisfying constraint (10.1) that minimizes the cost $J(\cdot,\cdot,\cdot)$. This model falls into the framework of Rishel's control system with a Markov jump disturbance.

Let $v(x,\alpha) = \inf_{u(\cdot)} J(x,\alpha,u(\cdot))$ be the value function corresponding to initial condition $\xi(0) = \alpha, x(0) = x$, thus it can be shown that $v(x,\alpha), \alpha = 1,2$ is the unique solution of the following Hamilton-Jaccobi-Bellman (HJB) equations (see Akella and Kumar [2]):

$$\begin{aligned} \begin{pmatrix} \min_{u\in[0,\bar{u}]}(u-d)\frac{dv(x,1)}{dx} \\ -d\frac{dv(x,2)}{dx} \end{pmatrix} &= \begin{pmatrix} \rho+q_1 & -q_1 \\ -q_2 & \rho+q_2 \end{pmatrix} \begin{pmatrix} v(x,1) \\ v(x,2) \end{pmatrix} \\ &- \begin{pmatrix} 1 \\ 1 \end{pmatrix} c^+x^+ + c^-x^-, \qquad (10.3) \end{aligned}$$

from which it follows that the optimal production rate is of the following form

$$u(x(t),\xi(t)) = \begin{cases} \bar{u}, & \text{if } \frac{dv(x,1)}{dx} < 0 \\ d, & \text{if } \frac{dv(x,1)}{dx} = 0 \\ 0, & \text{otherwise,} \end{cases}$$

where the purpose of letting $u(t) = d$ when $\frac{dv(x,1)}{dx} = 0$ is to prevent the chattering phenomenon. In [2], it is proved that $v(x,\alpha)$ is convex and continuously differentiable, so the optimal control is

$$u(x(t),\xi(t)) = \begin{cases} \bar{u}, & \text{if } x(t) < x^* \text{ and } \xi(t) = 1 \\ d, & \text{if } x(t) = x^* \text{ and } \xi(t) = 1 \\ 0, & \text{otherwise.} \end{cases} \qquad (10.4)$$

Note that the expression of the optimal control law depends on the hedging point, x^* which is given by

$$x^* = \max\left\{0, \frac{1}{\lambda_-}log\left[\frac{c^+}{c^+ + c^-}\cdot\left(1+\frac{\rho d}{q_1 d - (\rho + q_2 + d\lambda_-)(\bar{u}-d)}\right)\right]\right\},$$

where λ_- is the unique negative eigenvalue of the matrix

$$A_1 \triangleq \begin{pmatrix} \frac{\rho+q_1}{\bar{u}-d} & \frac{-q_1}{\bar{u}-d} \\ \frac{q_2}{d} & -\frac{q_2+\rho}{d} \end{pmatrix}.$$

Bielecki and Kumar [6] showed that for a manufacturing system with an unreliable machine, the zero-inventory policy can still be optimal in some cases. Their optimization problem was to seek a feedback control policy that minimized the following long run average expected cost incurred per unit time:

$$\lim_{T\to\infty} \frac{1}{T}\mathbb{E}\left[\int_0^T [c^+x_t^+ + c^-x_t^-]dt\right].$$

They proved, in this case, that policy (10.4) is still optimal with an optimal inventory level given by

$$x^* = \begin{cases} 0, \quad \text{if } \frac{\bar{u}q_1(c^+ + c^-)}{c^+(\bar{u}-d)(q_1+q_2)} \le 1 \text{ and } \frac{\bar{u}-d}{q_1} > \frac{d}{q_2} \\ \infty, \quad \text{if } \frac{\bar{u}-d}{q_1} < \frac{d}{q_2} \\ \frac{d(\bar{u}-d)}{q_2(\bar{u}-d)-q_1 d} log\left[\frac{\bar{u}q_1(c^+ + c^-)}{c^+(\bar{u}-d)(q_1+q_2)}\right], \quad \text{otherwise}. \end{cases}$$

For the average cost problem, using the discounted vanishing approach, Sethi *et al.* [37] developed the existence condition for the optimal control and established the verification theorem in the case of one machine having multiple states.

The above models, the jump rates of the Markov process governing the system mode switching of the system are assumed to be constant. However, it is a common knowledge that the faster the machine processes the parts, the more it is prone to fail. Based on this observation, Boukas and Haurie [7] and Hu *et al.* [29] studied the production control problem with a production rate dependent jump rate. In the case of one machine with two modes, Hu *et al.* [29] show that the linearity of the jump rate function is both necessary and sufficient for the optimality of the hedging point policy.

Since the hedging point policy was introduced, although in general its optimality can not be proved, it has received considerable attention due to its simplicity. Some structural properties of the hedging points for single part type and multiple machine state systems were established in [30, 51]. Sharifnia [39] showed how the optimal hedging point in the case of one part type and multiple machine states can be calculated. Liberopoulos and Caramanis [33] showed that the technique developed by Sharifnia can be applied to compute the hedging point policy even when the transition rates of the machine states depend on the production rate.

Since solving the optimal production rate control problem often involves solving either a set of partial differential equations or dynamic programming equations, it is difficult to find an analytical solution for the multiple mode or multiple part-type systems. However, when restricted to the subset of hedging point policies, the optimal control problem is, in fact, a parameter optimization problem. Therefore, the parameter optimization techniques can be applied to get sub-optimal controllers. In particular, the perturbation analysis techniques developed by Ho and his co-workers have been shown to be very efficient in finding the optimal hedging points (see Bremaud *et al.* [14], Caramanis and Liberopoulos [16] and Tu, Song and Lou [43] and references therein).

In general, for multiple machines and multiple part types, it is impossible to get the closed-form solution of the optimal control problem, and the only available recourse is to use numerical techniques. The next section will be devoted to controlled PDP and its numerical technique, which will be used to solve the optimal control problem in the sequel of this chapter.

3. CONTROLLED PIECEWISE-DETERMINISTIC PROCESSES AND NUMERICAL TECHNIQUE

Piecewise-deterministic processes include a variety of stochastic processes arising from engineering, economics, and inventory systems etc. Examples are queuing systems, insurance analysis (Dassios and Embrechts [18]), capacity expansion (Davis, Dampster, Sethi and Vermes [22]), permanent health insurance models (Davis [19]), inventory control models (Sethi and Zhang [38]), and production and maintenance models (Boukas and Haurie [7]).

3.1 The Optimal Control Problem of PDP.

Due to its extensive applications, this optimal control problem of PDP has received considerable attention. For more information on the optimal control of PDPs, the reader is referred to Davis [19], Soner [42], Boukas [5] and references therein.

In this chapter, the models of the manufacturing system will be presented as a special class of piecewise deterministic Markov processes without active boundary points in the state space and without jump in the continuous state. The model can be described as follows:

$$\dot{x}(t) = f(x(t), \xi(t), u(t)), x(0) = x_0, \xi(0) = \alpha \in \mathcal{S}, \tag{10.5}$$

where $x(t) \in \mathbb{R}^n$ is the continuous state, $u(t) \in \mathbb{R}^m$ is the control input, and $\xi(t)$ is a continuous-time Markov process with finite state space $\mathcal{S} = \{1, 2, \cdots, N\}$. The generator of the process $\{\xi(t), t \geq 0\}$ may be dependent on the control variable $u(t)$, denoted by $Q = (q_{\alpha\beta}(u(t))), \alpha, \beta \in \mathcal{S}$ with $q_{\alpha\beta}(u(t)) \geq 0, \beta \neq \alpha, q_\alpha(u(t))$ and $\sum_{\beta \in \mathcal{S}} q_{\alpha\beta}(u(t)) = 0$. Let $q_\alpha(u(t)) = -q_{\alpha\alpha}(u(t))$.

Suppose for each $\alpha \in \mathcal{S}$, $f(x, \alpha, u)$ is a bounded continuously differentiable function with bounded partial derivatives in x. Let $U(\alpha), \alpha \in \mathcal{S}$ (a closed subset of $\mathbb{R}^m$) denote the control constraints. Any measurable function defined on $[0, \infty)$ taking value in $U(\xi(t))$ such that (10.5) has a unique solution is called an admissible control. Let $\mathcal{A}$ be the class of admissible control. The control $u(x(t), \xi(t), t)$ is supposed to be piecewise continuous in t and continuously differentiable with bounded partial derivatives in x. Thus, each control law $u(\cdot) \in \mathcal{A}$ is associated with a probability measure P_u on the probability space $(\Omega, \mathcal{F})$ on which our discussion is based, such that the process $\{(x(t), \xi(t)), t \geq 0\}$ is well defined and the performance index defined by

$$J(x, \alpha, u(\cdot)) = \mathbb{E}_u \left[\int_0^\infty e^{-\rho t} c(x(t), \xi(t), u(t)) dt | x(0) = x, \xi(0) = \alpha \right] \tag{10.6}$$

is finite. The optimization problem is to seek a control law in $\mathcal{A}$ such that $J(x, \alpha, u(\cdot)$ is minimized. Let the value function $v(x, \alpha)$ be defined by

$$v(x, \alpha) = \min_{u(\cdot) \in \mathcal{A}} J(x, \alpha, u(\cdot)).$$

Thus, we have the following theorem.

Theorem 3 A necessary and sufficient condition for a control $u(\alpha, x) \in U(\alpha)$ to be optimal is that, for each $\alpha \in \mathcal{S}$, its performance function satisfies the coupled set of the following partial differential equations:

$$\begin{aligned} \rho v(x, \alpha) &= \min_{u \in U(\alpha)} \left\{ c(x, \alpha, u) + \sum_{i=1}^m v_{x_i}(x, \alpha) f_i(\alpha, x, u) \right. \\ &\quad \left. + \sum_{\beta \in \mathcal{S}} q_{\alpha\beta} v(x, \beta) \right\}, \quad \forall \alpha \in \mathcal{S}. \end{aligned} \tag{10.7}$$

Proof: See Theorem 6 in Boukas [5] for the proof.

In fact, the conditions of Theorem 3 are too strict to be useful. In a later, these restrictions will be loosened by introducing the viscosity solution of the HJB equation.

3.2 Numerical Technique. Theorem 3 indicates that to solve the optimal control problem of PDP, one has to solve the partial differential equations (10.6) since these are usually intractable, the numerical technique is applied.

To find an approximate solution to the Hamilton-Jacobi-Bellman (HJB) equation, corresponding to the deterministic or the stochastic optimal control problem, many approaches have been proposed. For more information on this subject, we refer the reader to Boukas [4] and Kushner and Dupuis [32].

In this section, we will give an extension of some numerical approximation techniques which were used by Kushner and Dupuis [32] to solve an elliptic and parabolic partial differential system associated with a stochastic control problem with diffusion disturbances. Our aim is to use these approaches to solve a set of coupled partial differential equations (10.6) representing the optimality conditions of the optimization problem. The idea behind these approaches consists, within a finite grid G_z^h with unit cell of lengths $(h_1, ..., h_n)$ for the state vector and a finite grid G_u^h with unit cell of lengths $(y_1, ..., y_m)$ for the control vector, of using an approximation scheme for the partial derivatives of the value function $v(x, \beta)$ which will transform the initial optimization problem to an auxiliary discounted Markov decision problem. This will allow us to use the well-known techniques used for this class of optimization problems such as successive approximation or policy iteration.

Since our problem has a continuous state vector z and a continuous control vector u, we need first to choose an appropriate discretization of the state space and the control space. Let G_z^h and G_u^h denote, respectively, the corresponding discrete state space and discrete control space and let us assume that they have finite elements with n_z points for G_z^h and n_u points for G_u^h respectively.

For the mode of the piecewise deterministic system, we do not need any discretization. Let $\mathcal{E}$ denote the global state space $\mathcal{E} = \mathcal{S} \times G_z^h$ and N the number of its elements. As we will see later, the constructed approximating Markov process X_t will jump between these states, $(s = (z, \alpha) \in \mathcal{E})$, with the transition probabilities $P_{ss'}(a)$, when the control action a is chosen from G_u^h. These transition probabilities are defined as

$$P_{ss'}(a) = \begin{cases} p_h^\beta(z, z + h; a), & \text{if } z \text{ jumps} \\ \tilde{p}_h^\beta(z, \beta; \beta', a), & \text{if } \alpha \text{ jumps} \end{cases},$$

where $p_h^\beta(z, z + h; a)$ and $\tilde{p}_h^\beta(z, \beta; \beta', a)$ are the transition probabilities between state s when the action a is used. The corresponding instantaneous cost function $c(s, a)$ and the discount factor δ of the approximating

DMDP depend on the used discretization approach. Their explicit expressions will be defined later.

Let h_i denote the finite difference interval in the coordinate i, and e_i denote the unit vector in the ith coordinate direction. The approximation that we use for $\frac{\partial}{\partial z_i}v(z,\beta)$ for each $\beta \in \mathcal{S}$, will depend on the sign of $f_i(z,\beta,u)$. Let G_z^h denote the finite difference grid which is a subset of $\mathbb{R}^n$.

This approach was used by Kushner to solve some optimization problems, and it consists of approximating the value function $v(z,\beta)$ by a function $v_h(z,\beta)$ and to replacing the first derivative of the value function, $\frac{\partial}{\partial z_i}v(z,\beta)$, by the following expressions:

$$\frac{\partial}{\partial z_i}v(z,\beta) = \begin{cases} \frac{1}{h_i}[v_h(z+e_ih_i,\beta) - v_h(z,\beta)], & \text{if } f_i(z,\beta,u) \geq 0 \\ \frac{1}{h_i}[v_h(z,\beta) - v_h(z-e_ih_i,\beta)], & \text{otherwise.} \end{cases} \tag{10.8}$$

For each β, define the functions $p_h^\beta(.;.,.)$, $\tilde{p}_h^\beta(.,.;.,.)$ and $Q_h^\beta(.,.)$, respectively, as follows:

$$\begin{aligned} Q_h^\beta(z,u) &= q_\beta(u) + \sum_{i=1}^{p}[|f_i(z,\beta,u)|/h_i], \\ p_h^\beta(z; z \pm e_ih, u) &= f_i^\pm(z,\beta,u)/[h_iQ_h^\beta(z,u)], \\ \tilde{p}_h^\beta(z,\beta;\beta',u) &= q_{\beta\beta'}(u)/Q_h^\beta(z,u). \end{aligned}$$

Let $p_h^\beta(z; z \pm h, u) = 0$ for all points z not in the grid.

Substituting the finite difference approximation of the partial derivatives as defined by (10.8) into (10.6) and collecting coefficients of the terms $v_h(z,\beta)$, $v_h(z \pm e_ih_i,\beta)$, yields for a finite difference interval h applying to z, the following:

$$v_h(z,\beta) = \Bigg\{\frac{c(z,\beta,u)}{Q_h^\beta(z,u)\left[1+\frac{\rho}{Q_h^\beta(z,u)}\right]} + \frac{1}{\left[1+\frac{\rho}{Q_h^\beta(z,u)}\right]}\Bigg[$$

$$\sum_{z' \in G_h} p_h^\beta(z;z',u)v_h(z',\beta) + \sum_{\beta' \in \mathcal{S}-\{\beta\}} \tilde{p}_h^\beta(z,\beta;\beta',u)v_h(z,\beta')\Bigg]\Bigg\}. \tag{10.9}$$

Let us define $c(s,u)$ and δ as follows:

$$\begin{aligned} c(s,u) &= \frac{c^\beta(z,u)}{Q_h^\beta(z,u)\left[1+\frac{\rho}{Q_h^\beta(z,u)}\right]}, \\ \delta &= \frac{1}{1+\frac{\rho}{Q_h^\beta(z,u)}}. \end{aligned}$$

A careful examination of (10.9) reveals that the coefficients of $v_h(.,.)$ are similar to transition probabilities between points of the finite set $\mathcal{E}$ since they are nonnegative and sum to, at most, unity. The term $c(s,u)$ is also nonnegative and bounded. As defined, δ actually represents a discount factor with values in $(0,1)$. Then, the Equation (10.9) has the basic form of the cost equation of the discounted Markov decision process optimization for a given control action. The approximating optimization problem built on the finite state space $\mathcal{S}$ has then the following cost equation:

$$v_h(z,\beta) = \min_{u \in G_u^h} \Bigg\{ \frac{c(z,\beta,u)}{Q_h^\beta(z,u)[1+\frac{\rho}{Q_h^\beta(z,u)}]} + \frac{1}{[1+\frac{\rho}{Q_h^\beta(z,u)}]} \Bigg[\sum_{z' \in G_h} p_h^\beta(z;z',u) v_h(z',\beta) + \sum_{\beta' \in \mathcal{S}-\{\beta\}} \tilde{p}_h^\beta(z,\beta;\beta',u) v_h(z,\beta') \Bigg] \Bigg\}. \quad (10.10)$$

Therefore, it is plausible that the algorithms used in the discounted Markov process optimization would be helpful in computing this solution.

To obtain an approximation of this solution, we can use the successive approximation technique or policy iteration technique. In the following, we present the policy iteration technique.

For a given policy $u(.)$, let us introduce the mappings T_u and T^*, acting on $v_h = (v_h(\beta))_{\beta \in \mathcal{E}}$ and defined by

$$\begin{aligned} T_u(v_h(z,\beta)) &= \frac{c(z,\beta,u)}{Q_h^\beta(z,u)\left[1+\frac{\rho}{Q_h^\beta(z,u)}\right]} + \frac{1}{\left[1+\frac{\rho}{Q_h^\beta(z,u)}\right]} \Bigg[\sum_{z' \in G_h} p_h^\beta(z,z',u) \\ &\qquad v_h(z',\beta) + \sum_{\beta' \in \mathcal{S}-\{\beta\}} \tilde{p}_h^\beta(z,\beta,\beta',u) v_h(z,\beta') \Bigg], \quad (10.11) \\ T^*(v_h(z,\beta)) &= \min_{u \in G_u^h} \{T_u(v_h(z,\beta))\}. \quad (10.12) \end{aligned}$$

The policy iteration algorithm operates as follows for a given finite difference interval h.

Step 1 : (Initialization) Choose $\varepsilon \in \mathbb{R}_+$, set: $k := 1$, and $(v_h^k(z,\beta) := 0,\ \forall \beta \in \mathcal{E},\ \forall z \in G_z^h$, guess an initial stationary policy $u^k \in G_u^h$.

Step 2 : (Policy Evaluation) Given the stationary policy $u^k \in G_u^h$, compute the corresponding cost function from the equation (10.10).

Step 3 : (Policy Improvement) Obtain a new stationary policy $u^{k+1} \in G_u^h$ by using the equation (10.12)

Step 4 : Test:

$$\begin{aligned}\bar{c} &:= \min_{\substack{\beta\in\mathcal{S}\\ z\in G_z^h}} \{v_h^k(z,\beta) - v_h^{k-1}(z,\beta)\},\\ \underline{c} &:= \max_{\substack{\beta\in\mathcal{S}\\ z\in G_z^h}} \{v_h^k(z,\beta) - v_h^{k-1}(z,\beta)\}\end{aligned}$$

with $c_{min} \triangleq \frac{\rho}{1-\rho}\bar{c}, \quad c_{max} \triangleq \frac{\rho}{1-\rho}\underline{c}$.

If $|c_{max} - c_{min}| \leq \varepsilon$, then stop at $u^* = u^k$; otherwise let $k = k+1$, and go to step 2.

The use of the policy iteration will give a solution to the approximating optimization problem. The results established by Kushner and Dupuis [32] concerning the convergence of the approximating solution to the real solution when h goes to zero can be adapted to our case.

Theorem 4 If there exists a constant C and a constant k_1 such that

$$0 \leq v_h(\beta, z) \leq C(1 + |z|^{k_1}), \qquad \forall \beta \in \mathcal{S},$$

then we obtain $\lim_{h\to 0} v_h(z,\beta) = v(z,\beta), \quad \forall \beta \in \mathcal{S}$.

Proof. The proof of this theorem can be adapted from the ones presented in Boukas *et al.* [13] or Kushner and Dupuis [32]. ∇∇∇

4. MANUFACTURING SYSTEM WITH DETERIORATING TERMS

Most products, when they are stocked somewhere, deteriorate, and their value reduces with time. For instance, electronic products may become obsolete as technology changes (personal computers are a good example), fashion tends to depreciate the value of clothing over time, and batteries die out with time, etc. The effect of time is even more critical for perishable goods such as foodstuff and cigarettes (see Wee [46], Xu and Wang, [47], Hamid [26], and the references therein). The subject has been extensively studied during the past ten years. For more information regarding the contributions, we refer the reader to Xu and Wang [47] and Hamid [26] and the references therein.

In this section, we study the continuous-time inventory-production control problem with deteriorating rate of the produced items. Note that, in most of the models reported in the literature so far, it has been considered that the deteriorating rate of the items is constant, the demand rate is also constant and deterministic, and the system capacity is unbounded and deterministic. But in reality, it is very hard to verify the assumption that the system capacity is deterministic and known. Here, we will consider the case in which the system capacity is described by a continuous-time stochastic Markov process with finite state space and the deteriorating rate and the demand rate are both constants.

The goal of this section is to determine the optimal production policy that reduces the cost of the facility under consideration. A numerical example is presented to show the usefulness of this model.

4.1 Model Description. Let us assume that we have a production firm that produces one item type. Let the demand rate of the produced item, d, be a constant and known. Suppose that the deterioration rate, γ, is constant and known. Let the state of the production system be described by a continuous-time Markov process, $\{\xi(t), t \geq 0\}$, with finite state space $\mathcal{S} = \{0, 1, 2, \cdots, N\}$ and generator $Q = (q_{\alpha\beta}), \alpha, \beta \in \mathcal{S}$, where $q_{\alpha\beta} \geq 0$ for all $\alpha \neq \beta$, $q_{\alpha\alpha} = -\sum_{\beta\neq\alpha} q_{\alpha\beta}$ for all $\alpha \in \mathcal{S}$. The equation $\xi(t) = \alpha \neq 0$ indicates that the system is in the operational state α and that it can produce items, and $\xi(t) = 0$ indicates that the production system is under repair and that it can not produce items. When the machine is in the operational mode i, its production rate is assumed to be subject to the constraint $0 \leq u(t) \leq \bar{u}_\alpha$, where $\bar{u}_\alpha$ is a positive constant. If we introduce the notation $\bar{u}_0 = 0$, then the production rate satisfies the following constraint:

$$0 \leq u(t) \leq \bar{u}_{\xi(t)}.$$

Under these assumptions, the differential equation that describes the evolution of the inventory of our facility is therefore given by

$$\dot{x}(t) = f(x(t), u(t), \xi(t)), x(0) = x_0, \xi(0) = \alpha, \tag{10.13}$$

where

$$f(x(t), u(t), \xi(t)) = -\gamma I_{\{x(t)\geq 0\}} x(t) + u(t) - d. \tag{10.14}$$

The optimization problem is to seek a control law that minimizes the following cost function:

$$J(x_0, \alpha, u(\cdot)) = \mathbb{E}\left[\int_0^\infty e^{-\rho t} g(x(t), \xi(t)) dt | x(0) = x_0, \xi(0) = \alpha\right] \tag{10.15}$$

where $g(x(t), \xi(t)) = [c^+ x^+(t) + c^- x^-(t)]$, c^+ ($c^+ \neq 0$) is the inventory holding cost, and c^- ($c^- \neq 0$) is the shortage cost.

In the sequel, for notational simplicity, $x(t), \xi(t), u(t)$ will also be written as x_t, ξ_t, u_t so that no confusion is caused.

4.2 Characterization of the Optimal Control. The goal of this section is to determine what would be the optimal production rate $u(t)$ that minimizes the cost function (10.15). Before determining this control, let us introduce some useful definitions.

Definition 1 A control $u(\cdot) = \{u(t) : t \geq 0\}$ is said to be *admissible* if (i) $u(\cdot)$ is adapted to the σ-algebra generated by $\alpha(\cdot)$, denoted by $\sigma\{\xi(s) : 0 \leq s \leq t\}$, and (ii) $u(t) \in [0, \bar{u}_{\xi(t)}]$ for all $t \geq 0$.

Let $\mathcal{A}$ denote the set of all admissible controls of our control problem.

Definition 2 A measurable function $u(x(t), \xi(t))$ is an admissible *feedback* control, or simply a feedback control, if (i) for any given initial x and α of the continuous state and the mode, the following equation has a unique solution $x(\cdot)$:

$$\dot{x}(t) = -\gamma I_{\{x(t)\geq 0\}} x(t) + u(x(t), \xi(t)) - d, \; x(0) = x, \xi(0) = \alpha$$

and (ii) $u(\cdot) = u(x(\cdot), \xi(\cdot)) \in \mathcal{A}$.

Let the value function $v(x(t), \xi(t))$ be defined by

$$v(x(t), \xi(t)) = \min_{u(t)} J(x(t), \xi(t), u(\cdot)). \tag{10.16}$$

Formally, we can write the Hamilton-Jacobi-Bellman (HJB) equation as follows:

$$\min_{u(t)\in\Gamma(\xi(t))} [(\mathcal{A}_u v)(x(t), \xi(t)) + g(x(t), \xi(t))] = 0, \tag{10.17}$$

where $\Gamma(\alpha) = [0, \bar{u}_\alpha]$ and $(\mathcal{A}_u v)(x(t), \xi(t))$ is defined as

$$\begin{aligned}(\mathcal{A}_u v)(x(t), \xi(t)) &= f(x(t), \xi(t)) \frac{\partial v}{\partial x}(x(t), \xi(t)) \\ &\quad + \sum_{\beta \in \mathcal{S}} q_{\xi(t)\beta} v(x(t), \beta). \end{aligned} \tag{10.18}$$

Let $w(\cdot, \cdot)$ be a real-valued continuous function defined on $\mathbb{R} \times \mathcal{S}$, and let H be a real-valued functional defined by

$$H(\alpha, x, r, u, w(x, \cdot)) = f(x, u, \alpha) \cdot r + g(x, \alpha) + \sum_{\beta \in \mathcal{S}} q_{\alpha\beta} v(x, \beta). \tag{10.19}$$

Thus, the HJB equation can be rewritten as

$$\rho v(x_t, \xi(t)) = \min_{u(t)\in\Gamma(\xi(t))} \{H(\xi(t), x(t), v_x(x(t), \xi(t), u(t), w(x(t), \cdot))\}.$$

Note that HJB equation (10.17) involves the first order partial derivatives of the value function when there is no guarantee that those partial derivatives would exist. In fact, one can always construct an example in which the value function does not have partial derivatives at certain points. In order to deal with a possible nondifferentiability of the value function, the concept of viscosity solution is usually used. For more information and a discussion of viscosity solution, the reader is referred to the books by Fleming and Soner [23] and the references therein. In what follows, we briefly describe the concept. For $w \in C(\mathbb{R} \times \mathcal{S})$, let

$$\Phi_w(x, \alpha, h, r) \triangleq \frac{w(x+h, \alpha) - w(x, \alpha) - hr}{|h|},$$

for $r, h \in \mathbb{R}$.

Define the superdifferential $D^+w(x, \alpha)$ and subdifferential $D^-w(x, \alpha)$ as follows:

$$D^+w(x, \alpha) = \left\{ r \in \mathbb{R} : \limsup_{h \to 0} \Phi_w(x, \alpha, r, h) \leq 0 \right\},$$

$$D^-w(x, \alpha) = \left\{ r \in \mathbb{R} : \liminf_{h \to 0} \Phi_w(x, \alpha, r, h) \geq 0 \right\}.$$

Definition 3 Suppose $w(x, \alpha)$ is continuous in x and has at most polynomial growth, i.e. there exist positive constants C, k_0, such that

$$|w(x, \alpha)| \leq C(1 + |x|^{k_0}).$$

Then

(a) $w(x, \alpha)$ is said to be a viscosity subsolution to (10.17), if for all $r \in D^+w(x, \alpha)$,

$$\rho w(x, \alpha) \leq \min_{u \in \Gamma(\alpha)} \left\{ H(x, \alpha, r, u, w(x, \alpha)) \right\}, \alpha \in \mathcal{S}.$$

(b) $w(x, \alpha)$ is said to be a viscosity supersolution to (10.17), if for all $r \in D^-w(x, \alpha)$,

$$\rho w(x, \alpha) \geq \min_{u \in \Gamma(\alpha)} \left\{ H(x, \alpha, r, u, w(x, \alpha)) \right\}, \alpha \in \mathcal{S}.$$

(c) $w(x, \alpha)$ is said to be a viscosity solution to (10.17), if (a) and (b) hold.

It is easy to see that if $w(x, \alpha)$ is differentiable at x_0, then both $D^+w(x, \alpha)$ and $D^-w(x, \alpha)$ are singleton and equal to $\{w_x(x_0, \alpha)\}$. Hence, the equality in the HJB equation (10.17) holds at points whenever the value function $w(x, \alpha)$ is differentiable.

Theorem 5 For any control $u(\cdot) \in \mathcal{A}$, the state trajectory of (10.13) has the following properties.

(i) Let x_t be the state trajectory with initial state x_0; thus, there exists a constant C such that

$$|x_t| \leq |x_0| + Ct. \tag{10.20}$$

(ii) Let x_t^1, x_t^2 be the state trajectories corresponding to $(x_1, u(\cdot))$ and $(x_2, u(\cdot))$ respectively; thus there exists a constant C such that $|x_t^1 - x_t^2| \leq C|x_1 - x_2|$.

Proof: Since the control is bounded, the proof of (i) is trivial.
For the proof of (ii) is trivial, the reader is referred to Boukas and Liu [9] for details. $\nabla\nabla\nabla$

Theorem 6 (i) For each $\xi(t) \in \mathcal{S}$, the value function, $v(x(t), \xi(t))$, is convex,

(ii) There exists a constant C, such that $v(x(t), \xi(t)) \leq C(1 + |x(t)|)$,

(iii) For each $\xi(t) \in \mathcal{S}$, the value function, $v(x(t), \xi(t))$, is Lipschitz in x.

Proof: To prove $v(x, \alpha)$ is convex in x, it suffices to show that $J(x, \alpha, u(\cdot))$ is jointly convex in (x, u), which follows from the convexity of the function $c^+x^+ + c^-x^-$. The details of the proof is similar to Lemma H.1 in Sethi and Zhang [38] and is omitted here.

The proof of (ii) is trivial.

Now we proceed to prove (iii). Let $x_1, x_2 \in \mathbb{R}$, and $u(\cdot) \in \mathcal{A}$. Suppose x_t^1, x_t^2 are the state trajectories corresponding to $(x_1, u(\cdot))$ and $(x_2, u(\cdot))$ respectively, and $\xi(0) = \alpha$. Then using (10.15), we obtain

$$\begin{aligned}
&|J(x_1, \alpha, u(\cdot)) - J(x_2, \alpha, u(\cdot))| \\
&\quad = \mathbb{E}\left[\int_0^\infty e^{-\rho t}\Big(g(x_t^1, \xi(t)) - g(x_t^2, \xi(t))\Big)dt\right] \\
&\quad \leq \mathbb{E}\left[\int_0^\infty e^{-\rho t}(c^+ + c^-)|x_t^1 - x_t^2|dt\right] \\
&\quad \leq C|x_1 - x_2|.
\end{aligned} \tag{10.21}$$

In (10.21), the last inequality is due to (ii) of Theorem 5, and $C = (c^+ + c^-)/\rho$. From (10.21), it follows that

$$\begin{aligned} |v(x_1,\alpha) - v(x_2,\alpha)| &\leq \sup_{u(\cdot)\in\mathcal{A}} |J(x_1,\alpha,u(\cdot)) - J(x_2,\alpha,u(\cdot))| \\ &\leq C|x_1 - x_2|. \end{aligned} \tag{10.22}$$

This completes the proof of Theorem 6. ∇∇∇

Note that the HJB equation (10.17) involves the first order partial derivatives of the value function, which may not exist at some points. Thus, the concept of viscosity solution to the HJB equation is used to characterize the optimal control.

Theorem 7 *The value function, $v(x(t),\xi(t))$, is the unique viscosity solution of the Hamilton-Jacobi-Bellman equation (10.17).*

Proof: The uniqueness follows from Theorem G.1 in Sethi and Zhang [38], and the proof of the second part follows directly from Lemma H.2 in Sethi and Zhang [38], which are omitted here. ∇∇∇

Theorem 8 Suppose there is a continuously differentiable function $\hat{v}(x(t), \xi(t))$ that satisfies the Hamilton-Jacobi-Bellman equation (10.17). If there exists $u^\star(\cdot) \in \mathcal{A}$, for which the corresponding $x^\star(t)$ satisfies (10.13) with $x^\star(0) = x$, and

$$\min_{u\in[0,\bar{u}_{\xi(t)}]} \Big[(\mathcal{A}_u v)(x^*(t),\xi(t))\Big] = (\mathcal{A}_{u^*} v)(x^*(t),\xi(t)) \tag{10.23}$$

almost everywhere in t with probability one, then $u^\star(.)$ is optimal, i.e.

$$\hat{v}(x(t),\xi(t)) = v(x(t),\xi(t)) = J(x(t),\xi(t),u^\star(.)).$$

Proof: Since $\hat{v}(x(t),\xi(t))$ is differentiable and satisfies the HJB equation (10.17), it is the viscosity solution of (10.17). Using Theorem 7, we obtain

$$\hat{v}(x(t),\xi(t)) = v(x(t),\xi(t)). \tag{10.24}$$

Using the Dynkin's formula (see Fleming and Soner [23]), we obtain for any $T > 0$,

$$\begin{aligned} v(x,\alpha) &= \mathbb{E}\Big[e^{-\rho T} v(x^*(T),\xi(t))\Big] + \mathbb{E}\Big[\int_0^T e^{-\rho t}\Big(\rho v(x^*(t),\xi(t)) \\ &\qquad -(\mathcal{A}_{u^*} v)(x_t^*,\xi_t)\Big)dt\Big] \\ &= \mathbb{E}\Big[e^{-\rho T} v(x^*(T),\xi(t))\Big] + \mathbb{E}\Big[\int_0^T e^{-\rho t} g(x^*(t),\xi(t))dt\Big]. \end{aligned} \tag{10.25}$$

Using Theorem 6 and letting $T \to \infty$ in (10.25), we obtain $v(x,\alpha) = J(x,\alpha,u^*(\cdot))$. This completes the proof of Theorem 8. $\nabla\nabla\nabla$

Let us consider the case of $\mathcal{S} = \{0,1\}$. Obviously, in this case, $\bar{u} = \bar{u}_1$. Let us assume that the value function is continuously differentiable with respect to the argument x. Using the expressions for the functions $f(\cdot)$, $g(\cdot)$ and the HJB equation given by (10.17), we obtain

$$\begin{aligned}
\rho v(x,1) &= \min_u \Big[\left[-\gamma I_{\{x\geq 0\}} x + u - d \right] v_x(x,1) + q_{11} v(x,1) \\
&\quad + q_{10} v(x,0) + c^+ x^+ + c^- x^- \Big], \qquad (10.26) \\
\rho v(x,0) &= \Big[\left[-\gamma I_{\{x\geq 0\}} x - d \right] v_x(x,0) + q_{01} v(x,1) \\
&\quad + q_{00} v(x,0) + c^+ x^+ + c^- x^- \Big]. \qquad (10.27)
\end{aligned}$$

Due to the structure of the optimality conditions, we can derive the optimal control law as follows:

$$u^\star(t) = \begin{cases} \bar{u}, & \text{if } v_x(x(t),1) < 0\ ; \\ \gamma I_{\{x(t)\geq 0\}} x(t) + d, & \text{if } v_x(x(t),1) = 0\ ; \\ 0, & \text{if } v_x(x(t),1) > 0. \end{cases} \qquad (10.28)$$

Moreover, if we let x^* be the minimal point of $v(x,1)$, i.e. $v_x(x^*,1) = 0$, then because of the convexity of $v(x,1)$ we obtain

$$u^\star(t) = \begin{cases} \bar{u}, & \text{if } x < x^*; \\ \gamma I_{\{x\geq 0\}} x^* + d, & \text{if } x = x^*; \\ 0, & \text{if } x > x^*. \end{cases} \qquad (10.29)$$

To solve the HJB equations, we can use the numerical method provided in Section 3.2. This technique, for this special problem, the following MDP problem:

$$\begin{aligned}
v(x,1) &= \min_u \left[c(x,1) + \frac{1}{1+\frac{\rho}{Q^1_{h_x}}} \left[\frac{|-\gamma I_{\{x\geq 0\}} x + \bar{u} - d|}{h_x Q^1_{h_x}} v(x+h_x,1) \right.\right. \\
&\quad \left.\left. + \frac{q_{10}}{Q^1_{h_x}} v(x,0) \right]\right], \\
v(x,0) &= c(x,0) + \frac{1}{1+\frac{\rho}{Q^0_{h_x}}} \left[\frac{|-\gamma I_{\{x\geq 0\}} x - d|}{h Q^0_{h_x}} v(x-h_x,0) \right. \\
&\quad \left. + \frac{q_{01}}{Q^0_{h_x}} v(x,1) \right],
\end{aligned}$$

Table 10.1 Data

c^+	c^-	q_{12}	q_{21}	γ	ρ	$\bar{u}_1$	d
1	5	0.05	0.15	0.1	0.1	0.24	0.16

where h_x is the discretization step for the x, $c(x, \alpha)$, and Q_h^1 and Q_h^0 are defined by

$$c(x,\alpha) = \frac{c^+x^+ + c^-x^-}{Q^\alpha_{h_x}\left[1+\frac{\rho}{Q^\alpha_{h_x}}\right]}, \quad \text{for all } \alpha \in \mathcal{S}, \tag{10.30}$$

$$Q^1_{h_x} = \frac{|\gamma I_{\{x(t)\geq 0\}}x - u + d|}{h} + |q_{11}|, \tag{10.31}$$

$$Q^0_{h_x} = \frac{\gamma I_{\{x(t)\geq 0\}}x + d}{h} + |q_{00}|. \tag{10.32}$$

4.3 Simulation Results. To illustrate the usefulness of the results of the previous subsections, we provide a numerical example here. The data of the system simulated are listed in Table 10.1. By using the policy iteration algorithm, a Matlab program is developed to find the solution of the optimal control. Let $x \in [-2, 5]$, and let $h_x = 0.02, h_u = 0.02$ be the grid steps of the continuous state variable x and the control variable u, respectively. The computation results are plotted in Figures 10.1. Figures 10.1 represents the optimal value functions $v(x, 1)$ and $v(x, 0)$ in modes 1 and 0 respectively, and the optimal feedback control for the production rate.

The computation results show that the control policy is still a hedging policy with hedging point $x^* = 0.4$, and the only difference in this case is that when $x = x^*$, the control is given by $u^* = \gamma x^* + d = 0.2$ instead of $u^* = d = 0.16$ as was established by Akella and Kumar [2]. If we let $\gamma = 0$, our results are the same as the ones obtained by Akella and Kumar.

5. PRODUCTION SYSTEM WITH QUALITY CONTROL

In the simplest case, the machine in a manufacturing system is often assumed to have two modes: working and down. However, according to

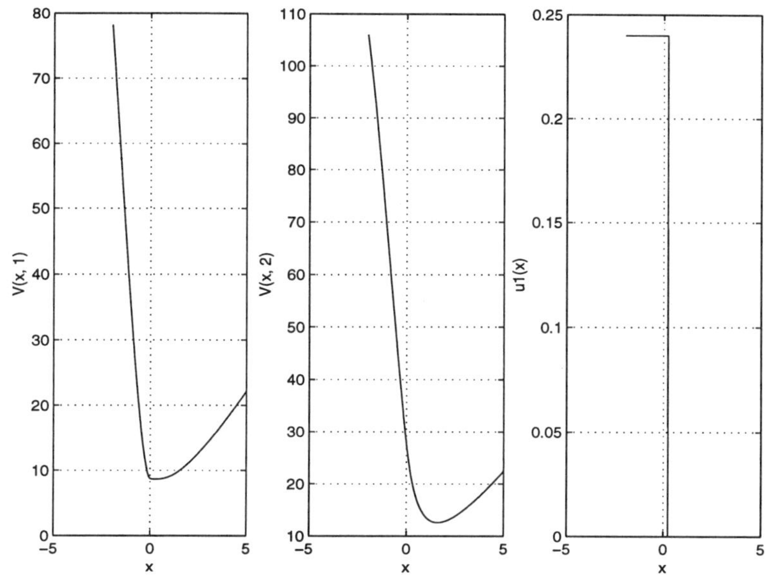

Figure 10.1 maintenance rates $w_p^1(x, 2)$, $w_p^2(x, 2)$.

the aging of the machine, the working mode can be divided further. In fact, as the machine ages, the production quality will become lower and lower, i.e. it will produce more and more bad products. To characterize this phenomena, Boukas and Liu [11] divided the machine's working state into three states: 'good', 'average' and 'bad' according to the aging of the machine. For each working state, we associate a rejected rate of bad products to quantify the production quality under this working mode. The quality control is to control the jump rates between the states to minimize the long run cost. This section will be devoted to introduce this model.

5.1 Model Description. Let us consider a manufacturing system consisting of one machine that produces one part type. Assume that the machine has multiple modes denoted by $\mathcal{S} = \{1, 2, 3, 4\}$. Mode 4 indicates that the machine is under repair. Mode 1 indicates that the machine is in good state, mode 2 indicates that the machine is in an average state and mode 3 indicates that the machine is in bad state. In mode 4, the machine does not produce parts. In modes 1, 2, and 3, the machine produces parts, but the rate of the rejected parts will depend on the state of the machine. The mode of the machine is assumed to switch between the four modes in a logical way, which is presumed to be governed by a continuous-time Markov process $\{\xi(t), t \geq 0\}$ taking value in $\mathcal{S}$ with generator matrix

$$Q = \begin{pmatrix} q_{11} & q_{12} & 0 & q_{14} \\ q_{21} & q_{22} & q_{23} & q_{24} \\ q_{31} & 0 & q_{33} & q_{34} \\ q_{41} & 0 & 0 & q_{44} \end{pmatrix}$$

where $q_{\alpha\beta} > 0, \alpha \neq \beta$, and $q_{\alpha\alpha} = -\sum_{\beta\neq\alpha} q_{\alpha\beta}, \forall\alpha$. Jump rates $q_{\alpha 1}$, $\alpha = 2, 3, 4$ are control variables; $q_{\alpha 1}, \alpha = 2, 3$ are preventive maintenance variables denoted by $w_p^1, w_p^2 \in [0, W_p]$; and q_{41} is the corrective maintenance variable denoted by $w_r \in [0, W_r]$, where W_p, W_r are positive constants. The generator of the controlled Markov process can be rewritten as

$$Q(w) = \begin{pmatrix} q_{11} & q_{12} & 0 & q_{14} \\ w_p^1 & q_{22}(w_p^1) & q_{23} & q_{24} \\ w_p^2 & 0 & q_{33}(w_p^2) & q_{34} \\ w_0 + w_r & 0 & 0 & q_{44}(w_r) \end{pmatrix}, \tag{10.33}$$

where $q_{22}(w_p^1) = -(w_p^1 + q_{23} + q_{24})$, $q_{33}(w_p^2) = -(w_p^2 + q_{34})$, and $q_{44}(w_r) = -(w_0 + w_r)$. Calculate $w = (w_p^1, w_p^2, w_r), q_\alpha(w) = -q_{\alpha\alpha}(w)$.

Remark 10.1 In the term $Q(w)$, w_0 is a positive constant, which is the smallest jump rate from repair state to good state. This transition rate is assumed to be free of charge. If a greater transition rate is required, i.e. shorter repair time is required, then a cost has to be paid.

Remark 10.2 The method of modeling preventive maintenance and corrective maintenance is different from that which is found in the literature and that implies immediate action. Here, we are controlling the jump rates from the corresponding state to the operational state, i.e. mode 1. By doing so, we are controlling the mean time for the technicians to arrive at the manufacturing system and start their repairs.

Let $x(t) \in \mathbb{R}$ denote the inventory when positive and backlogs when negative. The parts are assumed to deteriorate at a constant rate γ when they remain in the stock. The demand rate is assumed to be constant and denoted by d. When the inventory is negative and the machine is under repair, the demand will be canceled with a constant rate $1 - \theta$. Let $u(t)$ be the production rate satisfying constraint

$$0 \leq u(t) \leq \bar{u}_{\xi(t)},$$

where $\bar{u}_4 = 0$ and $\bar{u}_\alpha, \alpha \neq 0$ are positive constants. Let $\mathbf{u}(t) = (u(t), w(t))$ be the vector consisting of control variables. Using these assumptions, the inventory dynamic equation of this manufacturing system can be written as follows:

$$\dot{x}(t) = f(x(t), u(t), \xi(t)), x(0) = x_0 \text{ given}, \tag{10.34}$$

where

$$\begin{array}{rcl} f(x(t), u(t), \alpha) & = & -\gamma I_{\{x(t)\geq 0\}} x(t) + \mu(\alpha) u(t) - \theta(\alpha, x) d \\ \text{when } \xi(t) = \alpha, & & \end{array} \tag{10.35}$$

and

$$\theta(\alpha, x) = \begin{cases} \theta, & \text{if } \alpha = 4,\ x < 0 \\ 1, & \text{otherwise.} \end{cases}$$

Remark 10.3 In dynamics (10.34), the parameters γ and $1 - \mu(\xi(t))$ represent, respectively, the rates of the deteriorated and the rejected parts. The parameter $\theta(\xi(t), x(t))$ represents the rate by which we reduce the demand when the mode is $\xi(t)$ and the inventory is $x(t)$.

The optimization problem is to seek a control law which minimizes the following cost function:

$$J(x_0, \alpha_0, \mathbf{u}(\cdot)) = E\left[\int_0^\infty e^{-\rho t} h(x(t), \mathbf{u}(t), \xi(t)) | x(0) = x_0, \xi(0) = \alpha_0\right], \tag{10.36}$$

where

$$\begin{array}{r} h(x, u, w, \alpha) = c_u \mu(\alpha) u + c_p^1 w_p^1 I_{\{\alpha=2\}} + c_p^2 w_p^2 I_{\{\alpha=3\}} + c_r w_r I_{\{\alpha=4\}} \\ + c^+ x^+ + c^- x^-, \end{array} \tag{10.37}$$

with $c_u, c_p^1, c_p^2, c_r, c^+, c^-$ being positive constants.

Remark 10.4 In (10.37), the c^+, c^- are the inventory holding and backlog cost rates, where c_p^i, c_r are the maintenance cost rates, and $c_u \mu(\alpha) u$ is the cost resulting from the rejected bad parts. Just as in the previous section, we first establish some properties of the value function and characterize the optimal control; then we give an illustrative example by using the numerical technique provided in Section 3.2.

5.2 Characterization of the Optimal Control. Let us give some definitions.

Definition 4 A control $\{\mathbf{u}(\cdot), t \geq 0\}$ is said to be admissible if

(i) $\{\mathbf{u}(t), t \geq 0\}$ is adapted to the σ-algebra generated by $\{\xi(t), t \geq 0\}$, denoted by $\sigma\{\xi(s), 0 \leq s \leq t\}$.

(ii) $\mathbf{u}(t) \in U_{\xi(t)} \triangleq [0, \bar{u}_{\xi(t)}] \times [0, W_p] \times [0, W_p] \times [0, W_r]$.

(iii) $\mathbf{u}(t)$ is right-continuous and have left-hand limits.

Remark 10.5 In the above definition, the assumption that $\mathbf{u}(t)$ is right continuous and has left-hand limits is only used to prove that the value function is the viscosity solution of the Hamiltonian-Jacobi-Bellman equation.

Let $\mathcal{A}$ be the collection of all admissible control.

Definition 5 A measurable function $\{\mathbf{u}(x(t), \xi(t)), t \geq 0\} \in \mathcal{A}$, is called an admissible feedback control, or simply a feedback control if

(i) for any $x \in \mathbb{R}$ and $\alpha \in \mathcal{S}$, the following equation has a unique solution:

$$\begin{aligned} \dot{x}_t &= -\gamma I_{\{x_t>0\}} x_t + \mu(\xi(t)) u(x_t, \xi(t)) - \theta(\xi(t), x_t) d, x(0) \\ &= x, \xi(0) = \alpha, \end{aligned} \tag{10.38}$$

where $\xi(t)$ is a Markov process with generator $Q(w)$ as in (10.33) and

(ii) $\mathbf{u}(\cdot) \in \mathcal{A}$.

Let the function $v(x, \alpha)$ be defined as follows:

$$v(x, \alpha) = \min_{(u(\cdot), w(\cdot)) \in \mathcal{A}} J(x, \alpha, \mathbf{u}(\cdot)).$$

Theorem 9 For any admissible control $\mathbf{u}(\cdot) \in \mathcal{A}$ and $x_1, x_2 \in \mathbb{R}$, let x_t^1, x_t^2 be the state trajectories with initial point x_1, x_2, respectively. Thus, there exist a constant C_s such that

$$|x_t^1 - x_t^2| \leq C_s |x_1 - x_2|. \tag{10.39}$$

Proof: The proof can be adapted from Boukas and Liu [11].

Theorem 10 The optimal value function $v(x, \alpha)$ has the following properties.

(i) There exists constant C and k_0 such that

$$v(x, \alpha) \leq C(1 + |x|^{k_0}). \tag{10.40}$$

(ii) $v(x,\alpha)$ is locally Lipschitz in x, i.e. there exist constants C and k_0 such that for any $x_1, x_2 \in \mathbb{R}$,

$$|v(x_1,\alpha) - v(x_2,\alpha)| \le C(1 + |x_1|^{k_0} + |x_2|^{k_0})|x_1 - x_2|. \quad (10.41)$$

Proof: The proof of (i) is trivial.
Now we come to proof (ii). Let

$$v^0(x,\alpha) = \min_{\mathbf{u}(\cdot)\in\mathcal{A}} J^0(x,\alpha,\mathbf{u}(\cdot)), \quad (10.42)$$

where

$$J^0(x,\alpha,\mathbf{u}(\cdot)) = J(x,\alpha,\mathbf{u}(\cdot)), \quad (10.43)$$

when there is no jump in $\{\xi(t), t \ge 0\}$.
For $N \ge 1$, J^N and v^N are defined inductively as follows:

$$\begin{aligned} J^N(x,\alpha,\mathbf{u}(\cdot)) &= E_{\mathbf{u}}\left[\int_0^\tau e^{-\rho t} h(x_t, u_t, w_t, \xi(t))dt \right. \\ &\quad \left. + e^{-\rho\tau} v^{N-1}(x_\tau, \xi_\tau)\right], \end{aligned} \quad (10.44)$$

where $\{(x(t),\xi(t)), t \ge 0\}$ is the state trajectory corresponding to control $\mathbf{u}(\cdot)$, and τ is the first jump of $\{\xi(t), t \ge 0\}$. Let $q_\alpha(w) = -q_{\alpha\alpha}(w)$, thus, τ is exponentially distributed with failure rate function $q_\alpha(w_t)$. Obviously, there exists a constant $\varrho_0 > 0$ such that $q_\alpha(w_t) > \varrho_0$. So, for any control law $\mathbf{u}(\cdot)$,

$$e^{-\int_0^\infty q_\alpha(w_s)ds} = 0, \quad (10.45)$$

which implies that $P(\tau < \infty) = 1$. By direct computation, we obtain

$$\begin{aligned} J^N(x,\alpha,\mathbf{u}(\cdot)) = \int_0^\infty e^{-\rho t} e^{-\int_0^t q_\alpha(w_s)ds} \Big(& h(x_t, \mathbf{u}_t, \xi(t)) \\ & + \sum_{\beta\neq\alpha} v^{N-1}(x(t),\beta) q_{\alpha\beta}(w_t)\Big) dt. \end{aligned} \quad (10.46)$$

Define

$$v^N(x,\alpha) = \min_{(\mathbf{u}(\cdot)\in\mathcal{A}} J^N(x,\alpha,\mathbf{u}(\cdot)). \quad (10.47)$$

The rest of the proof is divided into the following three steps.

(a) $\{v^N(x,\alpha), \alpha \in \mathcal{S}, N \ge 0\}$ is uniformly locally Lipschitz in x, i.e. $v^N(x,\alpha), N \ge 0$ have the same Lipschitz coefficient.

(b) $v(x,\alpha) = \lim_{N\to\infty} v^N(x,\alpha)$,

(c) $v(x,\alpha)$ is also locally Lipschitz in x.

Now we proceed to prove (a).

It follows from (10.43) and (10.39) that

$$\begin{aligned}
&|J^0(x_1,\alpha,\mathbf{u}(\cdot)) - J^0(x_2,\alpha,\mathbf{u}(\cdot))| \\
&\leq \mathbb{E}\Big[\int_0^\infty e^{-\rho t}|c^+x_t^{1+} + c^-x_t^{1-} - c^+x_t^{2+} - c^-x_t^{2-}|dt\Big] \\
&\leq \mathbb{E}\Big[\int_0^\infty e^{-\rho t}(c^+ + c^-)|x_t^1 - x_t^2|dt\Big]. \qquad (10.48)
\end{aligned}$$

From (10.48) and (10.39), it follows that there exists constants C_0 and k_0 which are independent of the control law such that
$|J^0(x_1,\alpha,\mathbf{u}(\cdot)) - J^0(x_2,\alpha,\mathbf{u}(\cdot))|$

$$\leq \; C_0(1 + |x_1|^{k_0} + |x_2|^{k_0})|x_1 - x_2|. \qquad (10.49)$$

It follows that

$$\begin{aligned}
|v^0(x_1,\alpha) - v^0(x_2,\alpha)| &\leq \sup_{\mathbf{u}(\cdot)\in\mathcal{A}} |J^0(x_1,\alpha,\mathbf{u}(\cdot)) - J^0(x_2,\alpha,\mathbf{u}(\cdot))| \\
&\leq C_0(1 + |x_1|^{k_0} + |x_2|^{k_0}). \qquad (10.50)
\end{aligned}$$

This proves that $v^0(x,\alpha)$ is locally Lipschitz in x.

For $N \geq 0$, let us define

$$T_N = \sup_{x_1,x_2,\alpha} \frac{|v^N(x_1,\alpha) - v^N(x_2,\alpha)|}{(1 + |x_1|^{k_0} + |x_2|^{k_0})|x_1 - x_2|}. \qquad (10.51)$$

Then (10.50) indicates $T_0 \leq C_0$.

For $N \geq 1$ and a given control law $(u(\cdot), w(\cdot))$, using (10.46) and (10.51), we obtain

$$\begin{aligned}
&|J^N(x_1,\alpha,\mathbf{u}(\cdot)) - J^N(x_2,\alpha,\mathbf{u}(\cdot))| \\
&\leq \int_0^\infty e^{-\rho t} e^{-\int_0^t q_\alpha(w_s)ds}\Big(|h(x_t^1,\mathbf{u}_t,\xi(t)) - h(x_t^2,\mathbf{u}_t,\xi(t))| \\
&+ \sum_{\beta\neq\alpha}(v^{N-1}(x_t^1,\beta) - v^{N-1}(x_t^2,\beta))q_{\alpha\beta}(w_t)\Big)dt \\
&\leq \int_0^\infty e^{-\rho t} e^{-\int_0^t q_\alpha(w_s)ds}\Big((c^+ + c^-)|x_t^1 - x_t^2|\Big)dt \\
&+ \int_0^\infty e^{-\rho t} e^{-\int_0^t q_\alpha(w_s)ds} T_{N-1}(1 + |x_1|^{k_0} + |x_2|^{k_0})|x_1 - x_2| q_\alpha(w_t)dt \\
&\triangleq I + II. \qquad (10.52)
\end{aligned}$$

From (10.39), we obtain

$$I \leq C_0(1 + |x_1|^{k_0} + |x_2|^{k_0})|x_1 - x_2|. \tag{10.53}$$

$$\begin{aligned} II &\leq \int_0^\infty e^{-\rho t} e^{-\int_0^t q_\alpha(w_s)ds} T_{N-1} \\ &\quad \Big((1 + |x_1|^{k_0} + |x_2|^{k_0})|x_1 - x_2| q_\alpha(w_t) \Big) dt \\ &\leq T_{N-1}(1 + |x_1|^{k_0} + |x_2|^{k_0})|x_1 - x_2|) \\ &\quad \cdot \int_0^\infty e^{-\rho t} e^{-\int_0^t q_\alpha(w_s)ds} q_\alpha(w_t)dt. \end{aligned} \tag{10.54}$$

Let $M = \sup_{\alpha,w} q_\alpha(w)$. Thus,

$$\begin{aligned} \int_0^\infty e^{-\rho t} e^{-\int_0^t q_\alpha(w_s)ds} q_\alpha(w_t)dt &= 1 - \rho \int_0^\infty e^{-\int_0^t (\rho + q_\alpha(w_s))ds} dt \\ &\leq 1 - \int_0^\infty \rho e^{-\int_0^t (\rho+M)ds} dt = \frac{M}{\rho + M} \triangleq \lambda_0. \end{aligned} \tag{10.55}$$

By using (10.53)–(10.55), we obtain

$$\begin{aligned} &|J^N(x_1, \alpha, \mathbf{u}(\cdot)) - J^N(x_2, \alpha, \mathbf{u}(\cdot))| \\ &\qquad \leq (C_0 + \lambda_0 T_{N-1})(1 + |x_1|^{k_0} + |x_2|^{k_0})|x_1 - x_2|, \end{aligned} \tag{10.56}$$

from which it follows that

$$\begin{aligned} &|v^N(x_1, \alpha, \mathbf{u}(\cdot)) - v^N(x_2, \alpha, \mathbf{u}(\cdot))| \\ &\qquad \leq (C_0 + \lambda_0 T_{N-1})(1 + |x_1|^{k_0} + |x_2|^{k_0})|x_1 - x_2|. \end{aligned} \tag{10.57}$$

So, $v^N(x, i)$ is locally Lipschitz. To prove that $v^N(x, \alpha), N \geq 0$ have the same Lipschitz coefficient, it suffices to show that $\{T_N, N \geq 0\}$ is bounded. Indeed, from (10.57) we obtain

$$T_N \leq C_0 + \lambda_0 T_{N-1}, \tag{10.58}$$

which implies

$$T_N \leq C_0 + C_0 \sum_{t=0}^{N-1} \lambda_0^t \leq C_0(1 + \frac{1}{1 - \lambda_0}). \tag{10.59}$$

This completes the proof of (a).

We now prove (b). For $V \in C(\mathbb{R} \times \mathcal{S})$, let us define $\|V\| = sup_{x\in\mathbb{R},\alpha\in\mathcal{S}} V(x,\alpha)$ and an operator $\mathcal{T}$ on $C(\mathbb{R} \times \mathcal{S})$ as follows:

$$\mathcal{T}(V)(x,\alpha) = \min_{\mathbf{u}(\cdot)\in\mathcal{A}} \left\{ \int_0^\infty e^{-\rho t}(e^{-\int_0^t \lambda_\alpha(w_s)ds}) \right.$$
$$\left. \left[h(x_t,u_t,w_t,\alpha_t) + \sum_{\beta\neq\alpha} V(x_t,\beta)q_{\alpha\beta}(w_t)\right] dt \right\}. \tag{10.60}$$

Then it can be checked that $\mathcal{T}$ is contractive. Moreover, by using the definition of v^N, we obtain

$$\mathcal{T}(v^{N-1}) = v^N. \tag{10.61}$$

Thus, using the fixed point theorem, we obtain

$$\lim_{N\to\infty} v^N(x,\alpha) = v(x,\alpha). \tag{10.62}$$

Since $v^N(x,\alpha)$ is locally Lipschitz, the locally Lipschitz property of $v(x,\alpha)$ follows from (10.62). This completes the proof of Theorem 10. $\nabla\nabla\nabla$

Theorem 11 Under the assumptions of Theorem 10, the optimal value function $v(x,\alpha), \alpha \in \mathcal{S}$ is the unique viscosity solution to the following HJB equations:

$$\begin{aligned} \rho v(x,1) \quad = \quad & \min_{u\in[0,\bar{u}_1]} \Big\{ c^+x^+ + c^-x^- + c_u\mu(1)u \\ & +(-\gamma x^+ + \mu(1)u - d)\frac{dv(x,1)}{dx} + q_{12}(v(x,2)-v(x,1)) \\ & +q_{14}(v(x,4)-v(x,1))\Big\}, \end{aligned} \tag{10.63}$$

$$\begin{aligned} \rho v(x,2) \quad = \quad & \min_{u\in[0,\bar{u}_2], w\in[0,W_p]} \Big\{ c^+x^+ + c^-x^- + c_u\mu(2)u + c_p^1 w \\ & +(-\gamma x^+ + \mu(2)u - d)\frac{dv(x,2)}{dx} + w(v(x,1)-v(x,2)) \\ & +q_{23}(v(x,3)-v(x,2)) + q_{24}(v(x,4)-v(x,2))\Big\}, \end{aligned} \tag{10.64}$$

$$\rho v(x,3) \quad = \quad \min_{u\in[0,\bar{u}_3], w\in[0,W_p]} \Big\{ c^+x^+ + c^-x^- + c_u\mu(3)u + c_p^2 w$$

$$+(-\gamma x^+ + \mu(3)u - d)\frac{dv(x,3)}{dx} + w(v(x,1) - v(x,3))$$
$$+q_{34}(v(x,4) - v(x,3))\Big\}, \qquad (10.65)$$

and

$$\rho v(x,4) = \min_{w\in[0,W_r]} \Big\{ c^+x^+ + c^-x^- + c_r w + (-\gamma x^+ - \theta(x,4)d)\frac{dv(x,1)}{dx}$$
$$+(w_0 + w)(v(x,1) - v(x,4))\Big\}. \qquad (10.66)$$

Proof: The proof is similar to Theorem G.1 and Lemma H.2 in [38] and is omitted here. ∇∇∇

Theorem 12 (Verification Theorem) Suppose $v(x,\alpha) \in C^1(\mathbb{R}), \forall \alpha \in \mathcal{S}$, such that

$$|v(x,\alpha)| \le C(1 + |x|^{k_0}) \qquad (10.67)$$

and satisfies the HJB equation

$$\rho v(x,\alpha) = \min_{\mathbf{u}\in U_\alpha} \Big\{ h(x,u,w,\alpha) + f(x,u,\alpha)\frac{dv(x,\alpha)}{dx}$$
$$+\sum_\beta v(x,\beta)q_{\alpha\beta}(w)\Big\}. \qquad (10.68)$$

Then, the following assertions hold:

(i) $v(x,\alpha) = V(x,\alpha) \le J(x,\alpha,\mathbf{u}(\cdot))$ for any control $\mathbf{u}(\cdot) \in \mathcal{A}$.

(ii) Suppose there exists a control $\mathbf{u}^*$ and $x_t^*, \xi(t)^*$ which satisfies

$$\dot{x}_t^* = f(x_t^*, u_t^*, \xi(t)^*),\ x^*(0) = x, \qquad (10.69)$$

and

$$\min_{\mathbf{u}\in U_i} \Big\{ h(x_t^*, u, w, \xi(t)^*) + f(x_t^*, u, \xi(t)^*)\frac{dv(x_t^*, \xi(t)^*)}{dx}$$
$$+\sum_\beta v(x_t^*,\beta)q_{\alpha_t^*\beta}(w)\Big\}$$
$$= h(x_t^*, u_t^*, w_t^*, \xi(t)^*) + f(x_t^*, u_t^*, \xi(t)^*)\frac{dv(x_t^*, \xi(t)^*)}{dx}$$
$$+\sum_\beta v(x_t^*,\beta)q_{\xi(t)^*\beta}(w_t^*), \qquad (10.70)$$

a.e. in t with probability one, where $\{\alpha_t^*, t \geq 0\}$ is a Markov process with generator $Q(w_t^*)$ and $\{x_t^*, t \geq 0\}$ is the state trajectory with control $\mathbf{u}^*$. Then $\mathbf{u}^*(\cdot)$ is optimal, i.e.

$$v(x,\alpha) = V(x,\alpha) = J(x,\alpha,\mathbf{u}^*). \tag{10.71}$$

Proof: A differentiable function solving HJB equation (10.67) is, specifically, the viscosity solution to the HJB equation. So $v(x,\alpha) = V(x,\alpha)$ follows from the uniqueness of the viscosity solution of the HJB equation. According to the definition of the value function and Theorem 11, we get

$$V(x,\alpha) = v(x,\alpha) \leq J(x,\alpha,\mathbf{u}(\cdot)).$$

This proves (i).

Using Dynkin's formula, we obtain

$$\begin{aligned} v(x,\alpha) &= \mathbb{E}\Bigg[\int_0^T e^{-\rho t}\Bigg(\rho v(x_t^*, \xi(t)^*) - f(x_t^*, u_t^*, \xi(t)^*)\frac{dv(x_t^*, \xi(t)^*)}{dx} \\ &\quad - \sum_\beta v(x_t^*, \beta) q_{\xi(t)^*\beta}(w_t^*)\Bigg)dt + e^{-\rho T} v(x_T^*, \xi(t)^*)\Bigg], \end{aligned} \tag{10.72}$$

which combined with (10.70) yields

$$v(x,\alpha) = \mathbb{E}\Bigg[\int_0^T e^{-\rho t} h(x_t^*, u_t^*, w_t^*, \xi(t)^*)dt + e^{-\rho T} w(x_T^*, \xi(t)^*)\Bigg]. \tag{10.73}$$

Noting the polynomial growth condition (10.67) and letting $T \to \infty$ in (10.73) produces (ii). This completes the proof of Theorem 12. ∇∇∇

Based on Theorem 12, we get the optimal feedback control as follows.

Corollary 10.1 With the assumptions in Theorem 10, if the value function is differentiable, then the optimal control is as follows:

$$u(x,\alpha) = \begin{cases} \bar{u}_\alpha, & \frac{dv(x,\alpha)}{dx} < -c_u \mu(\alpha) \\ \frac{\gamma x + d}{\mu(\alpha)}, & \frac{dv(x,\alpha)}{dx} = -c_u \mu(\alpha) \quad \alpha \in \{1,2,3\} \\ 0, & \text{otherwise}, \end{cases} \tag{10.74}$$

$$w_p^1(x) = \begin{cases} W_p, & v(x,2) > c_p^1 + v(x,1) \text{ and } \alpha = 2 \\ 0, & \text{otherwise}, \end{cases} \tag{10.75}$$

$$w_p^2(x) = \begin{cases} W_p, & v(x,3) > c_p^2 + V(x,1) \text{ and } \alpha = 3 \\ 0, & \text{otherwise}, \end{cases} \tag{10.76}$$

Table 10.2 example data

c^+	c^-	c_u	c_p^1	c_p^2	c_r	ρ	θ	$\bar{u}_i$	d	W_p
1	10	1	3	6	10	0.1	0.8	0.2	0.18	0.2
W_r	w_0	γ	$\mu(1)$	$\mu(2)$	$\mu(3)$	q_{12}	q_{14}	q_{23}	q_{24}	q_{34}
1	0.2	0.1	0.98	0.82	0.73	0.05	0.01	0.1	0.05	0.1

$$w_r(x) = \begin{cases} W_r, & V(x,4) > v(x,1) + c_r \text{ and } \alpha = 4 \\ 0, & \text{otherwise .} \end{cases} \tag{10.77}$$

Remark 10.6 At the point x^* where $\left.\frac{dv(x,\alpha)}{dx}\right|_{x=x^*} = -c_u\mu(\alpha)$, the optimal production rate can take any value. The purpose of letting $u(x^*, \alpha) = \frac{\gamma x^* + d}{\mu(\alpha)}$ in (10.74) is to keep the optimal trajectory at the level x^* as long as possible.

5.3 Numerical Example. This subsection will provide a numerical example to support the theoretical results in this section. The data from the system are listed in Table 10.2. Using the computation technique described in previous section 3.2, a Matlab program is developed to find the optimal solution of the controls. The computation results are plotted in Figures 10.2 to Figures 10.4. Figure 10.2 represents the value functions with the machine in mode 1 to 4, respectively. Figure 10.3 represents the optimal production rates in modes 1 to 3. Figures 10.4 is the jump rates from modes 2, 3 and 4 to mode 1.

Based on the results of the computation, we can make the following comments:

- the value function, $v(x, 1)$, attains its minimum at zero, and $v(x, 2)$, $v(x, 3)$ and $v(x, 4)$ seem to have their minimum at positive points, meaning that the hedging point is 0 for $\alpha = 1$ and that the hedging points are positive for $\alpha = 2, 3, 4$.

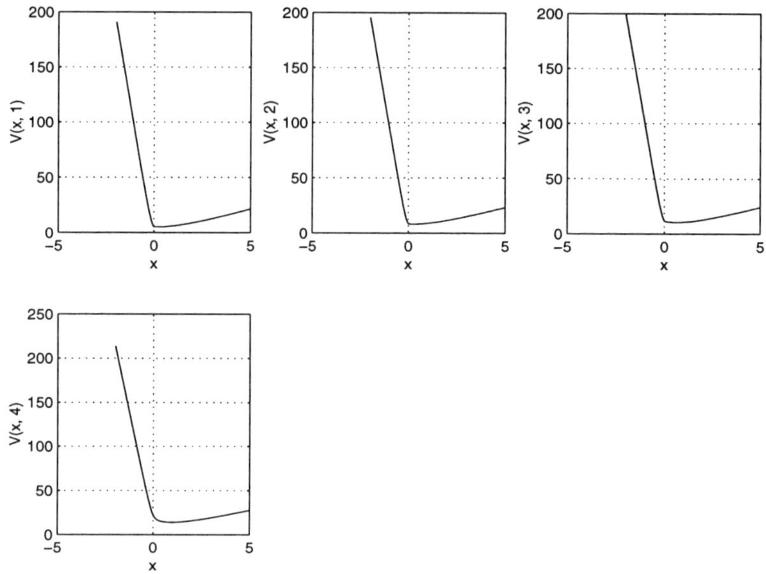

Figure 10.2 Value function $v(x, i)$ versus x.

- the production rates, $u(x, 1)$, $u(x, 2)$ and $u(x, 3)$, are almost similar in the three modes with the particularity that they take the value, $\frac{\gamma x^\star + d}{\mu(\alpha)}$ at the hedging level $x^\star$ in each mode. These controls are a hedging point-type policy which suggests that we can use the machine at its maximum capacity when the stock level is less than the hedging level $x^\star$, $\frac{\gamma x^\star + d}{\mu(i)}$ when the stock level is equal to $x^\star$, and 0 when it is greater than $x^\star$. The hedging point depends on the mode system.

- the preventive maintenance rates, $w_p^1(x, 1)$, $w_p^2(x, 2)$ and $w_p^1(x, 3)$, have the same form except that the switching values of the stock level are different from one mode to another and the switching value is higher for mode 3. In all the modes, 1, 2 and 3, the preventive maintenance rates are bang-bang controls.

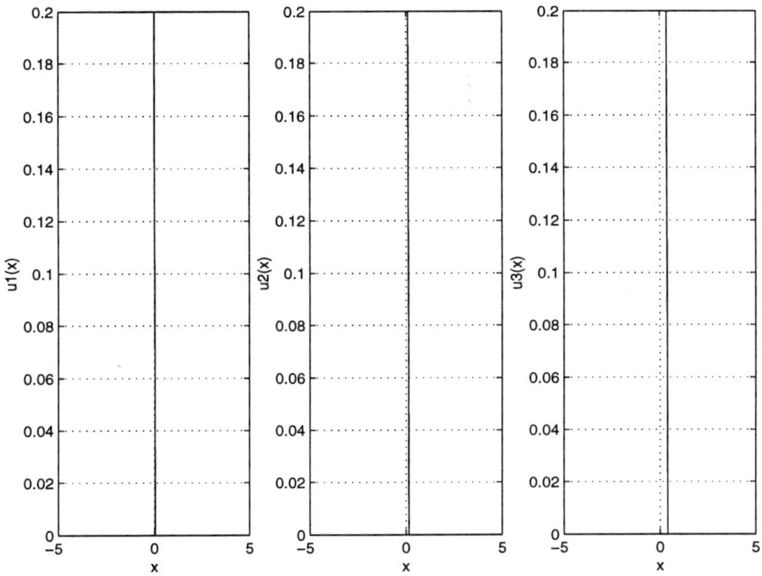

Figure 10.3 Production rate $u(x, i)$ versus x.

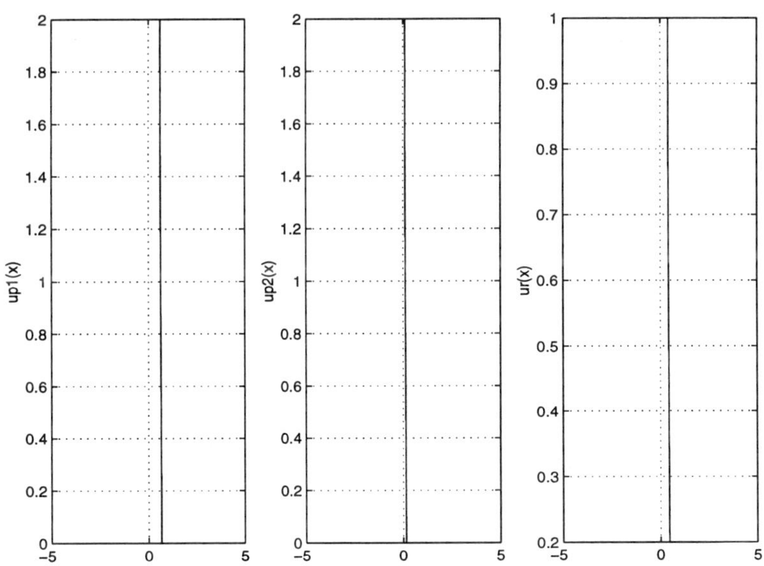

Figure 10.4 maintenance rates $w_p^1(x, 2)$, $w_p^2(x, 2)$.

- the corrective maintenance rate, $w_r(x,4)$, is a function of $v(x,4)$, $v(x,1)$, and the instantaneous cost c_r and it is also a bang-bang control.

6. PRODUCTION AND MARKETING CONTROL

It is critical for a manufacturing system to have a good market for its products. It is common knowledge that marketing actions, such as advertisement, pricing etc. can increase the demand rate. In the past two decades, a lot have been achieved and reported related to the problems including promotional and production decisions. We quote Abad [1], Sethi and Zhang [38], Sogomonian and Tang [41], Zhang *et al.* [52] and the references therein.

In this section, we will model the demand rate using a continuous-time Markov process with finite state space corresponding to the multiple demand levels and will consider the jump rate from one demand level to another as control variables. The optimization problem is to seek a short-term decision to maximize the profit of the system in the long run.

6.1 Model Description.

Consider a manufacturing firm that is facing stochastic demand for its production over time. For $t \geq 0$, let $x_t, u_t, d(t)$ be the inventory level, production rate, and demand rate, respectively. Thus, the inventory level has the following dynamics:

$$\dot{x}_t = u(t) - d(t), \; x(0) = x. \tag{10.78}$$

Suppose $u(t) \in [0, \bar{u}]$, with $\bar{u}$ being a positive constant denoting the production capacity of the system. The production of the firm has 2 demand levels, d_1, d_2, belonging to $\mathcal{D}$, i.e. $\mathcal{D} = \{d_1, d_2\}$. The demand rate $d(t)$ is a continuous-time Markov process, taking value in $\mathcal{D}$. The controlled demand process has a generator

$$Q(w(t)) = \begin{pmatrix} -w_1(t) & w_1(t) \\ W - w_2(t) & w_2(t) - W \end{pmatrix}, \tag{10.79}$$

where $w_1(t), w_2(t)$ are the control variables when the demand level is at level 1 and level 2, respectively. W is a constant denoting the jump rate from demand level 2 to level 1 if no marketing action is executed.

Let $h(x) = c^+x^+ + c^-x^-$, where c^+, c^-, denote, respectively the inventory and backlog cost rates. Let c_m denote the marketing cost rate and c_r be the net revenue rate. Consider the performance index

$$J(x, d, \mathbf{u}(\cdot)) = \mathbb{E}\left[\int_0^\infty e^{-\rho t}(h(x_t) + c_m w_t - c_r d(t))dt \right|$$

$$\left. x(0) = x, d(0) = d \right], \qquad (10.80)$$

where $\mathbf{u}(t) = (u(t), w(t))$. The optimization problem is to seek a control policy $(u(\cdot), w(\cdot))$ to minimize $J(x, d, u(\cdot), w(\cdot))$.

6.2 Characterization of the Optimal Policy.

It is not easy to get a closed-form solution for this optimal control problem. Just as in previous sections, we first characterize the optimal control and then give a numerical technique. To this end, let us introduce some definitions.

Definition 6 A control $\mathbf{u}(\cdot) = \{(u_t, w_t), t \geq 0\}$ is said to be admissible if

(i) $(u(\cdot), w(\cdot)) = \{(u_t, w_t), t \geq 0\}$ is adapted to the σ algebra generated by $\{d(t), t \geq 0\}$, denoted by $\sigma\{d(s), 0 \leq s \leq t\}$.

(ii) $0 \leq u_t \leq \bar{u}, w_t \in [0, W]$, where $\{d(t), t \geq 0\}$ is a Markov process with generator $Q(w(t))$.

Let $\mathcal{A}$ be the collection of all admissible control.

Definition 7 A measurable function $(u(x_t, d(t)), w(x_t, d(t)), t \geq 0$, is said to be an admissible feedback control or simply the feedback control if

(i) for any $x \in \mathbb{R}$ and $d \in \mathcal{D}$, the following equation has a unique solution:

$$\dot{x}(t) = u(x(t), d(t)) - d(t), x(0) = x, d(0) = d, \qquad (10.81)$$

where $d(t)$ is a Markov process with generator $Q(w)$ as shown in (10.79), and

(ii) $(u(\cdot), w(\cdot)) \in \mathcal{A}$.

Let us define the value functions by

$$v(x, d) = \min_{\mathbf{u}(\cdot) \in \mathcal{A}} J(x, d, \mathbf{u}(\cdot)),$$

which has the following properties

Theorem 13 The optimal value function $v(x, d)$ has the following properties,

(i) There exits a constant C such that

$$|v(x,d)| \leq C(1+|x|). \tag{10.82}$$

(ii) $v(x,d)$ is Lipschitz, i.e. there exists a constant C such that, for any $x_1, x_2 \in \mathbb{R}$,

$$|v(x_1,d) - v(x_2,d)| \leq C|x_1 - x_2|. \tag{10.83}$$

Proof: The proof can be adapted from that of Theorem 6.

Associated with this optimal control problem, we have the following Hamilton-Jacobi-Bellman (HJB) equations:

$$\begin{aligned} \rho v(x,d_1) &= \min_{u\in[0,U],w\in[0,W]} \Big\{ h(x) + c_m w - c_r d_1 + (u-d_1)\frac{dv(x,d_1)}{dx} \\ &\quad + w(v(x,d_2) - v(x,d_1)\Big\} \end{aligned} \tag{10.84}$$

and

$$\begin{aligned} \rho v(x,d_2) &= \min_{u\in[0,\bar{u}],w\in[0,W]} \Big\{ h(x) + c_m w - c_r d_2 (u-d_2)\frac{dv(x,d_2)}{dx} \\ &\quad + (W-w)(v(x,d_1) - v(x,d_2))\Big\}. \end{aligned} \tag{10.85}$$

The following is an equivalent definition of viscosity solution (see Sethi and Zhang [38]).

Definition 8 Suppose $v(x,d), d \in \mathcal{D}$ is continuous in x and there exist constants C and k_0 such that $|v(x,d)| \leq C(1+|x|^{k_0})$.

(a) $v(x,d)$ is called the viscosity subsolution of (10.83) and (10.85) if, for any $x_0 \in \mathbb{R}$,

$$\begin{aligned} -\rho v(x_0,d_1) + \min_{u\in[0,\bar{u}],w\in[0,W]} &[h(x_0) + c_m w - c_r d_1 + (u-d_1)\phi_x^1(x_0) \\ &+ w(v(x_0,d_2) - v(x_0,d_1))] \geq 0, \end{aligned} \tag{10.86}$$

$$\begin{aligned} \min_{u\in[0,\bar{u}],w\in[0,W]} &[-\rho v(x_0,d_2) + h(x_0) + c_m w - c_r d_2 + (u-d_2)\phi_x^2(x_0) \\ &+ (W-w)(v(x_0,d_1) - v(x_0,d_2))] \geq 0, \end{aligned} \tag{10.87}$$

whenever $\phi^i(x) \in C^1(\mathbb{R}), i = 1,2$ and $v(x,d_i) - \phi^i(x)$ attains local maximum at x_0.

(b) $v(x,d)$ is called the viscosity supersolution of (10.83) and (10.85) if, for any $x_0 \in \mathbb{R}$,

$$\min_{u\in[0,\bar{u}],w\in[0,W]} [-\rho v(x,d_1) + h(x_0) + c_m w - c_r d_1 + (u-d_1)\phi_x^1(x_0)$$
$$+w(v(x_0,d_2) - v(x_0,d_1))] \leq 0, \quad (10.88)$$

$$\min_{u\in[0,\bar{u}],w\in[0,W]} [-\rho V(x_0,d_2) + h(x_0) + c_m w - c_r d_2 + (u-d_2)\phi_x^2(x_0)$$
$$+(W-w)(v(x_0,d_1) - v(x_0,d_2))] \leq 0, \quad (10.89)$$

whenever $\phi_d(x) \in C^1(\mathbb{R}), d \in \mathcal{D}$ and $v(x,d_i) - \phi_2(x)$ attains local minimum at x_0,

(c) $v(x,d)$ is called the viscosity solution of HJB equations (10.83) and (10.85), if (a) and (b) hold.

Theorem 14 The optimal value function $v(x,d), d \in \mathcal{D}$ is the unique viscosity solution to the HJB equations (10.83) and (10.85).

Proof: We first prove that $v(x,d)$ is the viscosity subsolution of (10.83) and (10.85).

With no loss of generality, suppose $d(0) = d_2$ and $\{d(t), t \geq 0\}$ is right-continuous having left hand limits. Let $x_0 \in \mathbb{R}$ and $\phi(x) \in C^1(\mathbb{R})$ such that $v(x,d_2) - \phi(x)$ attains its local maximum at x_0, i.e. there exists a neighborhood $N(x_0)$ of x_0 such that

$$v(x,d_2) - \phi(x) \leq v(x_0,d_2) - \phi(x_0), \ \forall x \in N(x_0). \quad (10.90)$$

Note that the first jump time of $\{d(t), t \geq 0\}$ is exponentially distributed with parameter $W - w_t$.

For any $w \in [0,W], u \in [0,\bar{u}]$, choose a jump rate control law $w(t)$ such that $w(0) = w$, let τ be the first jump time of $\{d(t), t \geq 0\}$ corresponding to this control law, and let $u(t)$ be an admissible control satisfying $u(t) = u, t \in [0,\tau]$. Moreover, let $\theta \in (0,\tau]$ be such that $x_t \in N(x_0), \ t \in [0,\theta]$. Define

$$\psi_1(x,d) = \begin{cases} \phi(x) + v(x_0,d_2) - \phi(x_0), & d = d_2; \\ v(x,d_1), & d = d_1. \end{cases} \quad (10.91)$$

From (10.90) and (10.91), we obtain

$$\psi_1(x,d_2) \geq v(x,d_2), \ x \in N(x_0). \quad (10.92)$$

According to Dynkin's formula and the fact that $d(t) = d_2, t \in [0,\theta]$,

$$\mathbb{E}[e^{-\rho\theta}\psi_1(x_\theta, d_2) - v(x_0, d_2)] =$$
$$\mathbb{E}\bigg[\int_0^\theta e^{-\rho t}\Big(- \rho\psi_1(x_t, d_2) + (u - d_2)\phi_x(x_t)$$
$$+(W - w_t)(\psi_1(x_t, d_1) - \psi_1(x_t, d_2))\Big)dt\bigg]. \tag{10.93}$$

From (10.91), (10.92), and (10.92), it follows that

$$\begin{aligned}\mathbb{E}[e^{-\rho\theta}v(x_\theta, d_2) - v(x_0, d_2)] \;&\leq\; \mathbb{E}[e^{-\rho\theta}\psi_1(x_\theta, d_2) - v(x_0, d_2)]\\ &\leq\; \mathbb{E}\bigg[\int_0^\theta \Big(- \rho v(x_t, d_2) + (u - d)\phi_x(x_t)\\ &\quad +(W - w_t)(v(x_t, d_1)\\ &\quad -v(x_t, d_2))\Big)dt\bigg].\end{aligned} \tag{10.94}$$

Using the dynamic programming principle, we obtain

$$\begin{aligned}v(x_0, d_2) \;&\leq\; \mathbb{E}_u\bigg[\int_0^\theta e^{-\rho t}\Big(h(x_t) + c_m w_t - c_r d(t)\Big)dt\\ &\quad +e^{-\rho\theta}v(x_\theta, d_2)\bigg],\end{aligned} \tag{10.95}$$

which, combined with (10.94), yields

$$\begin{aligned}0 \;\leq\; \mathbb{E}_u\bigg[\int_0^\theta e^{-\rho t}\Big(&- \rho v(x_t, d_2) + h(x_t) + c_m w_t - c_r d(t)\\ &+(u - d)\phi_x(x_t)\\ &+(W - w_t)(v(x_t, d_1) - v(x_t, d_2))\Big)\bigg]dt.\end{aligned} \tag{10.96}$$

Noting the arbitrariness of $w \in [0, W], u \in [0, \bar{u}]$ and letting $\theta \to 0$ in (10.96) produces (10.87).

Likewise, one can prove that, for $\phi(x) \in C^1(\mathbb{R})$, if satisfying that $v(x, d_1) - \phi(x)$ attains its local minimum, then (10.86) holds for $v(x, d_1)$. This proves $v(x, d)$ is the subsolution of the HJB equations (10.83) and (10.85).

Next, we prove that $v(x, d)$ is also the viscosity supersolution. For this purpose, it suffices to show that (10.88) and (10.89) hold for $v(x, d)$. Without loss of generality, we suppose $v(x, d_2)$ does not satisfy (10.89), i.e. there exists a constant $\delta > 0$ such that, for any $u \in [0, \bar{u}], w \in [0, W]$:

$$\begin{aligned}-\rho v(x, d_2) + h(x) + c_m w - c_r d_2 + (u - d_2)\phi_x(x)&\\ +(W - w)(v(x, d_1) - v(x, d_2)) \geq \delta,&\end{aligned} \tag{10.97}$$

where $\phi(x) \in C^1(\mathbb{R})$ such that $v(x, d_2) - \phi(x)$ attains its local minimum at x_0, i.e.

$$v(x, d_2) \geq \phi(x) + v(x_0, d_2) - \phi(x_0),\ x \in N(x_0). \tag{10.98}$$

Assume $d(0) = d_2$, $u \in [0, \bar{u}], w \in [0, W]$. Let $(u(t), w(t))$ be an arbitrary control, where $w_0 = w, u(0) = u$ and τ is the first jump of $\{d(t), t \geq 0\}$, corresponding to this control law. Just as in the first part of the proof, we can $\theta \in (0, \tau]$ be such that $x_t \in N(x_0)$ when $t \in [0, \theta]$.

Define

$$\psi_2(x, d) = \begin{cases} v(x, d_1), & d = d_1 \\ \phi(x) + (v(x_0, d_2) - \phi(x_0)), & d = d_2. \end{cases} \tag{10.99}$$

Obviously,

$$\psi_2(x) \leq v(x, d_2). \tag{10.100}$$

By using (10.96), one obtains

$$\begin{aligned} J(x_0, d_2, u(\cdot), w(\cdot)) \geq\ & \mathbb{E}\bigg[\int_0^\theta e^{-\rho t}\Big(h(x_t) + c_m w_t - c_r d(t)\Big)dt \\ & + e^{-\rho\theta} v(x_\theta, d_2)\bigg] \\ \geq\ & \mathbb{E}\bigg[\int_0^\theta e^{-\rho t}\Big(\rho v(x_t, d_2) - (u_t - d)\phi_x(x_t) \\ & -(W - w)\Big(v(x_t, d_1) - v(x_t, d_2)\Big) + \delta\Big)dt \\ & + e^{-\rho\theta} v(x_\theta, d_2)\bigg]. \end{aligned} \tag{10.101}$$

Moreover, using Dynkin's formula

$$\begin{aligned} \mathbb{E}[e^{-\rho\theta}\psi_2(x_\theta, d_2) - \psi_2(x_0, d_2)] = \mathbb{E}\bigg[\int_0^\theta e^{-\rho t}\Big(-\rho\psi_2(x_t, d_2) \\ (u_t - d)\phi_x(x_t) + (W - w_t)(\psi_2(x_t, d_1) - \psi_2(x_t, d_2))\Big)dt\bigg]. \end{aligned} \tag{10.102}$$

From (10.99), (10.100) and (10.102), we obtain

$$\begin{aligned} & V(x_0, d_2) = \psi_2(x_0, d_2) \leq \mathbb{E}[e^{-\rho\theta} v(x_\theta, d_2)] + \\ & \mathbb{E}\bigg[\int_0^\theta e^{-\rho t}\Big(\rho v(x_t, d_2) - (u_t - d)\phi_x(x_t) \\ & -(W - w_t)(v(x_t, d_1) - v(x_t, d_2))\Big)dt\bigg], \end{aligned} \tag{10.103}$$

which, combined with (10.101), yields

$$J(x_0, d_2, u(\cdot), w(\cdot)) \geq v(x_0, d_2) + \mathbb{E}\left[\int_0^\theta e^{-\rho t}\delta dt\right], \tag{10.104}$$

from which it follows that

$$v(x_0, d_2) \geq v(x_0, d_2) + \delta\frac{1 - e^{-\rho\theta}}{\rho}, \tag{10.105}$$

which is a contradict. This proves that $v(x, d), d \in \mathcal{D}$ is the unique viscosity solution to the HJB equations (10.83) and (10.85).

The proof of the uniqueness of the solution to the HJB equations is similar to that of Theorem G.1 of [38]. $\nabla\nabla\nabla$

Theorem 15 (Verification Theorem) Suppose $V(x, d) \in C^1(\mathbb{R}), \forall d \in \mathcal{D}$, such that

$$|V(x, d)| \leq C(1 + |x|^{k_0}), \tag{10.106}$$

and satisfies the HJB equations (10.83) and (10.85). Then, the following assertions hold:

(i) $v(x, d_i) = V(x, d_i) \leq J(x, d_i, u(\cdot), w(\cdot))$ for any control $(u(\cdot), w(\cdot)) \in \mathcal{A}$.

(ii) Suppose there exists a control (u^*, w^*) and $x_t^*, d^*(t)$ which satisfy

$$\dot{x}_t^* = u^*(t) - d^*(t), x^*(0) = x, d^*(0) = d, \tag{10.107}$$

and

$$\begin{aligned}
&\min_{u_t \in [0, \bar{u}], w_t \in [0, W]} \left\{ h(x_t^*) + c_m w_t - c_r d_1 + (u_t - d_1)\frac{dv(x_t^*, d_1)}{dx} \right. \\
&\left. + w_t(v(x_t^*, d_2) - v(x_t^*, d_1)) \right\} \\
&= h(x_t^*) + c_m w_t^* - c_r d_1 + (u_t^* - d_1)\frac{dv(x_t^*, d_1)}{dx} \\
&+ w_t^*(v(x_t^*, d_2) - v(x_t^*, d_1)) \qquad (10.108) \\
&\min_{u_t \in [0, \bar{u}], w_t \in [0, W]} \left\{ h(x_t^*) + c_m w_t - c_r d_2 + (u_t - d_2)\frac{dv(x_t^*, d_2)}{dx} \right. \\
&\left. + w_t(v(x_t^*, d_1) - v(x_t^*, d_2)) \right\} \\
&= h(x_t^*) + c_m w_t^* - c_r d_2 + (u_t^* - d_2)\frac{dv(x_t^*, d_2)}{dx} \\
&+ w_t^*(v(x_t^*, d_1) - v(x_t^*, d_2)), \qquad (10.109)
\end{aligned}$$

a.e. in t with probability one, where $\{d_t^*, t \geq 0\}$ is a Markov process with generator $Q(w_t^*)$ and $\{x_t^*, t \geq 0\}$ is the state trajectory with control (u^*, w^*). Then, (u_t^*, w_t^*) is optimal, i.e.

$$v(x, d_i) = V(x, d_i) = J(x, d_i, u^*, w^*). \tag{10.110}$$

From Theorem 15, we obtain the following optimal condition.

Corollary 10.2 Suppose the value function $v(x, d_i)$ is differentiable in x, then the optimal control is as follows:

$$u(x, d_i) = \begin{cases} \bar{u}, & \text{if } \frac{dv(x,d_i)}{dx} < 0 \\ d_i, & \text{if } \frac{dv(x,d_i)}{dx} = 0 \\ 0, & \text{if } \frac{dv(x,d_i)}{dx} > 0 \end{cases} \tag{10.111}$$

$$w(x, d_i) = \begin{cases} 0, & c_m + v(x, d_2) > v(x, d_1) \\ W, & \text{otherwise.} \end{cases} \tag{10.112}$$

If, furthermore, $v(x, d_i), i = 1, 2$ are convex in x, then the optimal production is hedging point policy

$$u(x, d_i) = \begin{cases} \bar{u}, & x > x_i^* \\ d_i, & x = x_i^* \\ 0, & \text{otherwise,} \end{cases} \tag{10.113}$$

where x_i^* is the solution of $\left.\frac{dv(x,d_i)}{dx}\right|_{x=x_i^*} = 0.$

For the general case, it is impossible to find an analytical solution and the approximate solution can be obtained by the same numerical techniques as in the previous Sections. Note that, under the conditions of Corollary 10.2, the optimal marketing control is bang-bang, taking values only in $\{0, W\}$. In the special case where $w(t) \in \{0, W\}$ and $w_2(t) \equiv W$, the optimal production and marketing problem has been analytically solved by Zhang *et al.* [52] .

6.3 A Special Case. Consider a special case of the above marketing model. The demand is assumed to have two levels $\mathcal{D} = \{d_1, d_2\}$. Suppose the production capacity to be $\bar{u} = 1, c_m = 1$ and the marketing action to be $w(t) \in \{0, w_d\}$. The controlled demand process has the following generator:

$$Q(w(t)) = \begin{pmatrix} -kw(t) & kw(t) \\ 0 & 0 \end{pmatrix}, \tag{10.114}$$

where k is a given constant.

In order to give the optimal control law, we will introduce some notions. Let

$$C_0^+ = \frac{c^+(d_2-d_1)}{\rho^2} + \frac{c_r(d_2-d_1)}{\rho} - \frac{1}{k},$$
$$C_0^- = \frac{c^-(d_2-d_1)}{\rho^2} + \frac{c_r(d_2-d_1)}{\rho} - \frac{1}{k},$$

and $b = \frac{\rho+kw_d}{1-d_1}$, and let d^+, d^- be the unique solution of the following equations:

$$\begin{cases} \frac{c^+d_1}{\rho^2}e^{-\rho d^+/d_1} - \frac{c^+d_2}{\rho^2}e^{-\rho d^+/d_2} + C_0^+ = 0, \\ \frac{c^-(d_2-d_1)}{b(1-d_2)-\rho}\left(-e^{bx^-} + \frac{b(1-d_2)}{\rho}e^{\rho x^-/(1-d_2)}\right) - \rho C_0^- = 0, \quad \text{if } b \neq \frac{\rho}{1-d_2} \\ c^-(d_2-d_1)\left(e^{bz^-} - bd^-e^{bz^-}\right) - \rho^2 C_0^- = 0, \quad \text{if } b = \frac{\rho}{1-d_2}. \end{cases}$$

The following theorem is a result of these notions.

Theorem 16 In this case, the optimal production rate is

$$u^*(x,d) = \begin{cases} 1, & \text{if } x < 0 \\ d, & \text{if } x = 0 \\ 0, & \text{if } x > 0, \end{cases} \tag{10.115}$$

and the optimal marketing control is

$$w^*(x,d) = \begin{cases} 0, & \text{if } x < z^* \text{ and } d = d_1 \\ w_d, & \text{if } x \geq z^* \text{ and } d = d_1 \\ 0, & \text{otherwise,} \end{cases} \tag{10.116}$$

where

$$z^* = \begin{cases} -\infty, & \text{if } c_r(d_2-d_1) \geq \frac{\rho}{k} + \frac{c^-(d_2-d_1)}{\rho}, \\ d^-, & \text{if } \frac{\rho}{k} < c_r(d_2-d_1) < \frac{\rho}{k} + \frac{c^-(d_2-d_1)}{\rho}, \\ 0, & \text{if } c_r(d_2-d_1) = \frac{\rho}{k}, \\ d^+, & \frac{\rho}{k} - \frac{c^+(d_2-d_1)}{\rho} < c_r(d_2-d_1) < \frac{\rho}{k}, \\ \infty, & \text{if } c_r(d_2-d_1) \leq \frac{\rho}{k} - \frac{c^+(d_2-d_1)}{\rho}. \end{cases} \tag{10.117}$$

Proof: The proof can be found in [52].

Remark 10.7 We dealt with a simple manufacturing system consisting of one machine with many states and producing one part type. This choice has been made in order to clarify the proposed models, but this does not restrict our results. Our results will remain valid multiple machine, multiple part-type manufacturing systems. The optimality condition can be obtained using a similar equation, requiring the use of the proposed numerical technique in Section 3.2.

7. CONCLUDING REMARKS

This chapter has addressed the continuous flow model of a manufacturing system with deteriorating terms, quality and marketing control included. The system is modeled as a controlled piecewise-deterministic process, and stochastic dynamic programming is used. We have seen that all the stochastic optimal control problems don't have closed-form solution and the proposed numerical approximation technique can be used to find the suboptimal solution of the optimal control problem. To support the theoretical results many numerical examples were provided.

The following are some future research directions.

- In this chapter, the system mode, i.e. the state of the machine and the demand level are assumed to be perfectly available for feedback. In practice, for multiple modes system, the observation of the system modes may contain some disturbance. Further investigation is required to determine how to handle this case.
- The approximation technique used in this chapter often leads to the "curse of dimensionality", and, thus, this numerical technique is suitable only for a small system. The perturbation analysis put forward by Ho and his co-workers is a single simulation sample path-based stochastic approximation technique, which has been shown to be very successful in finding the optimal production rates for the continuous flow model (see Bremaud *et al.* [14], Caramanis and Liberopoulos [16] and Tu, Song and Lou [43] and references therein). To develop a similar numerical algorithm for the models considered in this chapter is of great importance.
- Evidently, marketing actions often have a time-delay effect. How to model this factor and study the control problem is still an open question.

Acknowledgment: This work is supported by the natural Sciences and Engineering Research Council of Canada under grants OGP0036444.

References

[1] Abad, P. L. Approach to Decentralized Marketing-Production Planning. *International Journal of Systems Sciences*, **13** (1982) 227-235.

[2] Akella, R., and Kumar, P. R. Optimal Control of Production Rate in a Failure Prone Manufacturing System. *IEEE Transactions on Automatic Control*, **AC-31**(2) (1986) 106-126.

[3] Algoet, P. H. Flow Balance Equations for the Steady-State Distribution of a Flexible Manufacturing System. *IEEE Transactions on Automatic Control*, **34**(8) (1988) 917-921.

[4] Boukas, E. K. Numerical Methods for HJB Equations of Optimization Problems for Piecewise Deterministic Systems. *Optimal Control Applications & Methods*, **16** (1995) 41-58.

[5] Boukas, E. K. Control of Systems with Controlled Jump Markov Disturbances. *Control Theory and Advanced Technology*, **9**(2) (1993) 577-595.

[6] Bielecki, T. and Kumar, P. R. Optimality of Zero-Inventory Policies for Unreliable Manufacturing Systems, *Operation Research*, **36**(4) (1988) 532-541.

[7] Boukas, E. K. and Haurie, A. Manufacturing Flow Control and Preventive Maintenance: A Stochastic Control Approach. *IEEE Transactions on Automatic Control*, **35**(9) (1990) 1024-1031.

[8] Boukas, E. K. and Kenne, J. P. Maintenance and Production Control of Manufacturing System with Setups. *Lectures in Applied Mathematics*, **33** (1997) 55-70.

[9] Boukas, E. K. and Liu, Z. K. Manufacturing Systems with Random Breakdowns and Deteriorating Items. *Automatica*, (forthcoming 2000).

[10] Boukas, E. K. and Liu, Z. K. Modeling and Numerical Methods in Manufacturing System. *Handbook of Statistics and Allied Probability*, **20** (1999).

[11] Boukas, E. K. and Liu, Z. K. Production and Maintenance Control for Manufacturing System. *Proceedings of 38th IEEE CDC*, Phoenix, Arizona, Dec, 1999.

[12] Boukas E. K., and Yang, H. Manufacturing Flow Control and Preventive Maintenance: A Stochastic Control Approach. *IEEE Transactions on Automatic Control*, **41**(6) (1996) 881-885.

[13] Boukas, E. K., Zhang, Q. and Yin, G. On robust design for a class of failure prone manufacturing systems. *Recent Advances in Control and optimization of Manufacturing Systems*, Yin and Zhang ed. Lecture Notes in control and information sciences 214, Springer-Verlag, London, 1996.

[14] Bremaud, P., Malhame, R. P., and Massoulie, L. A Manufacturing System with General Failure Process: Stability and IPA of Hedging Control Policies. *IEEE Transactions on Automatic Control*, **42**(2) (1997) 155-170.

[15] Buzacott, J. A. and Shanthikumar, J. G. *Stochastic Models of Manufacturing Systems.* Prentice-Hall, Englewood Cliffs, NJ, 1993.

[16] Caramanis, M. and Liberopoulos, G. Perturbation Analysis for the Design of Flexible Manufacturing System flow controllers. *Operations research,* **40** (1992) 1107-1125.

[17] Connolly, S., Dallery, Y., and Gershwin, S. B. A Real-Time Policy for Performing Setup Changes in a Manufacturing System. *Proceedings of the 31st IEEE Conference on Decision and Control,* Tucson, AZ, Dec, 1992.

[18] Dassios, A. and Embrechts, P. Martingales and Insurance Risk. *Commun. Statist. - Stochastic Models,* **5**(2) (1989) 181-217.

[19] Davis, M. H. A. *Markov Modeling and Optimization,* Chapman and Hall, New York, 1993.

[20] Davis, M. H. A. Control of Piecewise-Deterministic Processes via Discrete-Time Dynamic Programming. *Stochastic Differential Systems,* 140-150, Helmes K. and Christopeit, Kohlman, Springer-Verlag, 1986.

[21] Davis, M. H. A. Piecewise Deterministic Markov Processes: a General Class off Non-Diffusion stochastic models. *Journal of Royal Staistical Society,* **46**(3) (1984) 353-388.

[22] Davis, M. H. A., Dempster, M. A. H. Sethi, S. P. and Vermes, D. Optimal Capacity Expansion under Uncertainty. *Advance in Applied Probability,* **19** (1987) 156-176.

[23] Fleming, W. H., and Soner, H. M. *Controlled Markov Processes and Viscosity Solutions,* Springer-Verlag, New York, 1992.

[24] Gershwin, S. B. *Manufacturing Systems Engineering,* Prentice Hall, Englewood Cliffs, 1993.

[25] Gershwin, S. B., Caramanis, M, and Murray, P. Simulation Experience with a Hierarchical Scheduling Policy for a Simple Manufacturing System. *Proceedings of the 27th IEEE Conference on Decision and Control,* 1941-1849, Austin, TX, Dec, 1988.

[26] Hamid, B. Replenishment Schedule for Deteriorating Items with Time proportional Demand. *Journal of Operations Research Society,* **40** (1989) 75-81.

[27] Haurie, A. and L'Ecuyer, P. Approximation and Bounds in Discrete Event Dynamic Programming. *IEEE Transactions on Automatic Control,* **31**(3) (1986) 227-235.

[28] Hu, J. Q., and Caramanis, M. Dynamic Set-Up Scheduling of Flexible Manufacturing Systems: Design and Stability of Near Optimal

General Round Robin Policies. *Discrete Event Systems, IMA volumes in mathematics and applications series*, Kumar, P.R. and Varaiya, P. (eds.), Springer-Verlag, 1994.

[29] Hu J. Q., Vakili, P. and Yu, G. X. Optimality of Hedging Point Policies in the Production Control of Failure Prone Manufacturing Systems. *IEEE Transactions on Automatic Control*, **39**(9) (1994) 1875-1880.

[30] Hu, J. Q. and Xiang, D. Structural Properties of Optimal Flow Control for Failure Prone Production System. *IEEE Transactions on Automatic Control*, **39**(3) (1994) 640-642.

[31] Kimemia, J. and Gershwin, S. B. An Algorithm for Computer Control of a Flexible Manufacturing System. *IIE Transactions*, **15**(4) (1983) 353-362.

[32] Kushner, H. J. and Dupuis, P. G. *Numerical Methods for Stochastic Control Problems in Continuous Time*, Springer-Verlag, New York, 1992.

[33] Liberopoulos, G. and Caramanis, M. Production Control of Manufacturing System with Production Rate-Dependent Failure Rates. *IEEE Transactions on Automatic Control*, **39**(4) (1994) 889-895.

[34] Menaldi, J. L. Some Estimates for Finite Difference Approximations. *SIAM Journal on Control and Optimization*, **27** (1989) 579-607.

[35] Olsder, G. J. and Suri, R. Time Optimal of Parts-Routing in a Manufacturing System with Failure Prone Machines. *Proceedings of 19th IEEE. Conference on Decision and Control*, 722-727, Alburquerque, New Mexico, 1980.

[36] Rishel, R. Control of systems with jump Markov disturbances. *IEEE Transactions on Automatic Control* , **20** (1975) 241-244.

[37] Sethi, S. P., Taksar, M. I., and Zhang Q. Optimal Production Planning in a Stochastic Manufacturing System with Long-Run Run Average Cost. *Journal of Optimization Theory and Applications*, **92**(1) (1997) 161-188.

[38] Sethi, S. P., and Zhang, Q. *Hierarchical Decision Making in Stochastic Manufacturing Systems.* Birkhauser, Boston, 1994.

[39] Sharifnia, A. Production Control of a Manufacturing System with Multiple Machine States'. *IEEE Transactions on Automatic Control*, **33**(7) (1988) 620-625.

[40] Sharifnia, A., Caramanis, M., and Gershwin, S. B. Dynamic Setup Scheduling and Flow Control in Manufacturing Systems. *Discrete*

Event Dynamic Systems: Theory and Applications, **1** (1991) 149-175.

[41] Sogomonian, A. G. and Tang, C. S. A Modeling Framework for Coordinating Reduction and Production Decisions within a Firm. *Management Science,* **39** (1993) 191-203.

[42] Soner, H. M. Optimal Control with State-Space Constraint II. *SIAM Journal on Control and Optimization*, **24**(6) (1986) 1110-1122.

[43] Tu, F. S., Song, D. P. and Lou, S. X. C. Preventive Hedging Point Control Policy and its Realization. *Preprint of the XII World IFAC congress,* 13-16 Sydney, Australia, Vol. 5, July, 1993.

[44] Veatch, M. H. and Caramanis, M. C. Optimal Average Cost Manufacturing Flow Controllers: Convexity and Differentiability. *IEEE Transactions on Automatic Control*, **44**(4) (1999) 779-783.

[45] Viswanadham, N. and Narahari, Y. *Performance Modeling of Automated Manufacturing Systems*, Prentice Hall, Englewood Cliffs, 1992.

[46] Wee, H. Economic Production Lot Size Model for Deteriorating Items with Partial Back-Ordering. *Computer Industrial Engineering*, **24** (1993) 449-458.

[47] Xu, H., Wang, U. H. An Economic Ordering Policy Model for Deteriorating Items with Time Proportional Demand. *European Journal of Operational Research*, **46** (1990) 21-27.

[48] Yan, H. and Zhang, Q. A Numerical Method in Optimal Production and Setup Scheduling of Stochastic Manufacturing System. *IEEE Transactions on Automatic Control*, **42**(10) (1997) 1452-1555.

[49] Yin, G. and Zhang, Q. *Continuous-Time Markov Chains and Applications: A Singular Perturbation Approach*, Springer-Verlag, 1998.

[50] Yin, G., and Zhang, Q. *Recent Advances in Control and Optimization of Manufacturing System*, Lecture Notes in Control and Information Sciences 214, Springer-Verlag, London, 1996.

[51] Yu, X. Z. and Song, W. Z. Further Properties of Optimal Hedging Points in a Class of Manufacturing System. *IEEE Transactions on Automatic Control*, **44**(2) (1999) 379-382.

[52] Zhang, Q., Yin, G. and Boukas, E. K. (1998) Optimal Control of a Marketing-Production System, preprint.

Chapter 11

A SINGLE PERIOD INVENTORY MODEL TO ACCOUNT FOR DEMAND SURPRISES

Dogman Serel and Herbert Moskowitz
Krannert Graduate School of Management
Purdue University
West Lafayette, IN 47907-1310, USA

Abstract In stochastic inventory literature, the demand distribution is generally assumed to be known and stable. This assumption results in a suboptimal ordering decision and excessive expected cost if the demand distribution is affected by a surprise event after the ordering decision is made. The optimal ordering policy is found by taking into account uncertainties in the timing of the surprise and the demand distribution after the surprise. This paper is an extension of the traditional newsvendor problem to account for random demand surprises. A numerical example is presented to illustrate the reduced expected cost values resulting from using the proposed model.

Keywords: Single period inventory model, optimal ordering policy, surprise event, demand distribution, demand surprise, newsvendor problem.

1. INTRODUCTION

The classical newsvendor problem used in inventory literature entails finding the optimal amount of stocks to hold against uncertain demand in a future selling period. Given the forecast for demand distribution and the cost parameters, viz. purchase, holding, and shortage costs associated with the product, the buyer seeks the best ordering decision that minimizes his expected cost (Lee and Nahmias, 1993).

In this paper, we study a single period inventory problem that extends the standard newsvendor problem to the case characterized by random changes in demand distribution. We consider the impact of possible shifts in the demand distribution that may occur after the ordering decision is made. Thus demand distributions are considered which can be regarded as more uncertain than those assumed in the traditional newsvendor problem. A higher level of demand uncertainty increases the importance of identifying the optimal order quantity due to the associated increase in the buyer's expected cost.

Random disruptions in the supply of products are studied extensively in inventory literature (Chao, 1987; Parlar and Berkin, 1991). However, the impact of random demand changes on inventory management costs is not explored to the same extent. An argument may be made, for example, that products with sales potential that is difficult to estimate, such as fashion goods, are particularly prone to experience sudden changes in the demand distribution subsequent to the initial demand forecast. Song and Zipkin (1993) write that shifts in demand at random time points occur primarily for three types of products: products sensitive to basic economic variables, products subject to obsolescence, and new products entering the market. Brill and Chaouch (1995) suggest that labor disputes and natural disasters at the demand point may result in sharp changes in demand. Both papers use a continuous-time Markov chain to model the random transitions between different demand states over time. Song and Zipkin assume that in state i , the demand follows a Poisson process with rate μ_i. Brill and Chaouch assume that in state i, the demand rate μ_i is constant.

We take a different approach, focusing on a single-period problem with normally distributed demand in each demand state. Only one change in the demand distribution is allowed after the initial forecast. This assumption is not implausible during a relatively short sale period. The rationale for this type of assumption can also be found in Rahim *et al.* (2000). It can also be argued that a surprise, by definition, occurs rarely. The normal distribution has been commonly used in literature for representing demand uncertainty. The optimal policy in this model is based on closed–form and simple expressions; consequently, the sensitivity of the optimal solution to changes in problem parameters can be easily explored.

Other related research on inventory control under non stationary or imperfectly known demand includes multi-period models in which the demand distributions change from period to period (Karlin, 1960) and the Bayesian approaches for demand distributions with unknown parameters (Azoury, 1985). Recent literature contains two–period inventory models that seek to mitigate the impact of demand uncertainty by dynamically adjusting the second period order quantity based on observed demand in the first period (Eppen and Iyer, 1997; Fisher and Raman,

1996). These models assume that demands in two periods are probabilistically dependent. This line of research is applied to analyze supply contracts for seasonal goods such as fashion clothes.

2. MODEL ASSUMPTIONS

Assume the purchase cost per unit is c, the leftover stock at the end of the sale period costs h per unit, and the penalty cost for unsatisfied demand is p per unit, $p > c > -h$. The unit cost h represents the net cost of holding an item in stock minus salvage value; hence, it can be negative. The buyer has only one chance to order. Let L be the delivery lead time and T be the length of the period in which sale of the product occurs. Supply quantity is ordered at time zero, the order arrives, and the sale of the product begins at time L. Let the forecasted (at time zero) probability distribution function for demand per unit time be $F_0(.)$. Furthermore, let $F_0(.)$ be a normal distribution with mean μ_0, and variance σ_0^2 and demands in non overlapping intervals be independent, i.e., the demand generating process is a Brownian motion with drift μ_0 and volatility σ_0. At a random point in time after the ordering decision is made, the demand probability distribution characteristics suddenly change. The demand distribution following the shock (surprise) is also normally distributed, but with a different mean and/or variance.

The timing of a demand surprise can be restricted to occur during a prespecified time interval after the time of ordering. Two cases of interest are analyzed separately. First, it is considered that the shift in demand may occur any time after the ordering decision. Demand shifts triggered by unexpected changes in weather conditions illustrate a surprise of this kind. In the second case, the timing of surprise is restricted to the selling period. The second case is more likely when, for example, consumers' purchasing behaviors are influenced by changing fashion trends after the sale of the product begins.

For another example for the inventory problem with demand surprises, consider the connection between the uncertainty in the quality level of a process and the usage rate of process inputs. In a supply chain, from the viewpoint of a supplier operating in a make-to-stock fashion, the demand distribution in consideration is the demand for the items which are delivered by the supplier to be used as input in a manufacturer's process. Sudden changes in this demand distribution may occur due to unexpected short-term quality problems arising in the manufacturer's process. For example, an increase in the variance of the demand distribution can result from a temporary disturbance in the stability of the manufacturer's process. Similarly, unanticipated changes in the yield of the manufacturer's process can create wild fluctuations in the amount of inputs needed. It is essential to identify the true demand distribution by taking into account all possible uncertainties for determining the optimal stocking level for the supplier.

The time until the occurrence of the change in the demand distribution is assumed to be exponentially distributed with a mean of $1/\lambda$. Note that $\lambda = 0$ corresponds to the special case of no surprise, i.e., the classical newsvendor problem. Suppose there are n different probability distributions (states), one of which randomly becomes the distribution representing the after-shock demand. Let (μ_i, σ_i^2) be the demand parameters (mean and variance) in state $i, i = 1, ..., n$. Also let state 0 represent the original demand forecast at time zero. It is assumed that there may be, at most, one shift in demand during the entire planning horizon.

3. THE OPTIMAL ORDER QUANTITY

Case 1: Time of Demand Surprise is Unrestricted

The probability of remaining in state 0 during the supply lead time, α_0, is

$$\alpha_0 = 1 - \int_0^L \lambda e^{-\lambda t} dt = e^{-\lambda L} \tag{11.1}$$

In order to assign the probabilities of shifting to state i during the time interval between 0 and L, α_i, the probability distribution developed by Knappenberger and Grandage (1969) is used:

$$\alpha_i = \frac{n!\pi^i(1-\pi)^{n-i}(1-e^{-\lambda L})}{i!(n-i)![1-(1-\pi)^n]} \qquad 0 < \pi < 1, \qquad i = 1, ..., n \tag{11.2}$$

Notice that α_i also represents the probability that the demand state is i at time L, i.e., at the beginning of the sale period. Knappenberger and Grandage use equation (11.2) in the context of the economic design of $\overline{X}$ control charts. Because of its flexibility to describe a variety of ways in which shocks can occur, the probability distribution defined by equation (11.2) is also suitable to analyze the problem of random changes in demand in an inventory management framework. A variety of demand scenarios can be modeled by the appropriate choice of the distribution parameters λ, π, and n. The probability of a change in the initial demand distribution can be increased by increasing the value of λ. The parameter π influences the allocation of probabilities of movements to different demand states. The number of possible after-shock demand states and the demand distribution characteristics in those states will depend on the particular problem environment.

Duncan (1956) shows that, given the demand state is 0 at time L, the average fraction of time within the sale period before a change in the demand state occurs is given by

$$f \equiv \frac{1-(1+\lambda T)e^{-\lambda T}}{\lambda(1-e^{-\lambda T})} \tag{11.3}$$

Analogous to equation (11.2), given that demand state is 0 at time L, the probability of shifting to state i during the sale period, p_i, is

$$p_i = \frac{n!\pi^i(1-\pi)^{n-i}(1-e^{-\lambda T})}{i!(n-i)![1-(1-\pi)^n]} \qquad 0<\pi<1, \qquad i=1,....,n \tag{11.4}$$

Note also that

$$p_0 = e^{-\lambda T} \tag{11.5}$$

The expected percentage of time that the demand state is i during the sale period, γ_i, is given by (Saniga, 1977):

$$\gamma_0 = \alpha_0 p_0 + \alpha_0 f(1-p_0) \tag{11.6}$$

$$\gamma_i = \alpha_i + \alpha_0 p_i (1-f), \qquad i=1,...,n. \tag{11.7}$$

Finally, because of the assumption of Brownian motion for the demand generating process, demand for the whole period is normally distributed with mean μ_T and variance σ_T^2 given by

$$\mu_T = \sum_{i=0}^{n} \gamma_i \mu_i T \tag{11.8}$$

$$\sigma_T^2 = \sum_{i=0}^{n} \gamma_i \sigma_i^2 T \tag{11.9}$$

Thus, the problem is reduced to a traditional newsvendor problem with demand uncertainty described by a normal distribution with parameters defined by equations (11.8) and (11.9). Let $F_T(.)$ and $f_T(.)$ be the cumulative distribution function and probability density function for a normally distributed random variable with mean μ_T and variance σ_T^2. Without loss of generality, suppose the initial inventory on hand is zero. Thus, the buyer's expected inventory cost, EC, as a function of the order quantity, S, is

$$EC(S) = cS + h\int_0^S (S-y)f_T(y)dy + p\int_S^\infty (y-S)f_T(y)dy \quad (11.10)$$

Since $f_T(.)$ is the density of a normal random variable, we have (e.g., Hadley and Whitin, 1963, p. 446)

$$\int_S^\infty y f_T(y)dy = \mu_T[1-\Phi(z)] + \sigma_T\phi(z) \quad (11.11)$$

where $z = (S-\mu_T)/\sigma_T$, and $\phi(z)$ and $\Phi(z)$ are the density and cdf of the standard normal distribution. Making the standard assumption that the area under $f_T(y)$ to the left of $y = 0$ is negligible, i.e.,

$$\int_0^S y f_T(y)dy = \mu_T - \int_S^\infty y f_T(y)dy \quad (11.12)$$

and combining equations (11.11) and (11.12), equation (11.10) can be rewritten as

$$EC(S) = (c-p)S + p\mu_T + (p+h)[(y-\mu_T)\Phi(z) + \sigma_T\phi(z)] \quad (11.13)$$

Using the first-order condition, $EC(S)$ is minimized when the order quantity S is chosen according to the critical percentile point of the demand distribution; i.e.,

$$F_T(S^*) = (p-c)/(p+h) \quad (11.14)$$

S^* is the well-known order-up-to level in the classical newsvendor model (e.g., Lee and Nahmias, 1993). If the initial inventory on hand is x, the optimal ordering policy is as follows: if $x < S^*$, purchase $S^* - x$, otherwise, do not purchase.

Case 2: No Demand Surprise During the Lead Time

Now consider the situation in which a change in demand distribution is not expected to occur during the lead time. The values of γ_i in this case can be easily found by substituting $\alpha_0 = 1$ and $\alpha_i = 0, \quad i = 1, ..., n$ in equations (11.6) and (11.7). The optimal order quantity is determined using these modified γ_i values.

4. NUMERICAL EXAMPLE

For example, suppose $x = 0$, $L = \lambda = 1$, $T = 2$, $n = 2$, $\pi = 0.2$, $c = 5$, $h = -2$, and $p = 10$. The parameters for demand states are as follow: $\mu_0 = 100$, $\sigma_0^2 = 400$, $\mu_1 = 180$, $\sigma_1^2 = 900$, $\mu_2 = 60$, and $\sigma_2^2 = 200$. First, Case 1 is considered. Using equations (11.1) through (11.3), the values obtained are $\alpha_0 = 0.368$, $\alpha_1 = 0.562$, $\alpha_2 = 0.07$, and $f = 0.687$. The probabilities of demand distribution movements to each state during the sale period are found from equations (11.4) and (11.5): $p_0 = 0.135$, $p_1 = 0.768$, and $p_2 = 0.096$. Equation (11.4) where $n = 2$ implies that $p_1 > p_2$ for $\pi < 0.667$, and $p_1 < p_2$ for $\pi > 0.667$ (Saniga, 1977). Hence, by changing the value of π, the relative probabilities of moving to new demand states can be manipulated. Equations (11.6) and (11.7) give $\gamma_0 = 0.268$, $\gamma_1 = 0.65$, and $\gamma_2 = 0.081$. Finally, substitutions into equations (11.8) and (11.9) yield

$\mu_T = 297.56$, $\sigma_T^2 = 1417.88$.

Using equation (11.14), the optimal order quantity is:

$S^* = 309.56$.

The buyer's minimum expected cost is computed from equation (11.13):

$EC(S^*) = 1602.02$.

The resulting expected cost if the order quantity is chosen based on the demand forecast at time zero can also be computed. Using the distribution with parameters $(T\mu_0, T\sigma_0^2)$ in equation (11.14), the order quantity $S = 209.01$. Substituting this value into equation (11.13) results in $EC(S = 209.01) = 1931.48$. Thus, the buyer's expected cost increases by 20.6 % if the potential demand shifts are not taken into account in the ordering decision.

Similarly, considering Case 2, in the sample problem, $\gamma_0 = 0.73$, $\gamma_1 = 0.24$, and $\gamma_2 = 0.03$ if no surprise is expected during the lead time. Then we obtain the corresponding $\mu_T = 236.1$, $\sigma_T^2 = 1028.6$, and $S^* = 246.3$. The resulting minimum expected cost, $EC(S^*) = 1277.7$. The expected cost based on a single demand forecast at time zero (i.e., for $S = 209.01$) is 1344.3, 5.2 % higher than the minimum expected cost. Since 5.2 % $<$ 20.6 %, as can be expected, the buyer's expected cost savings from taking into account the possible demand surprise are lower when there is "less" uncertainty in the timing of surprise.

5. SENSITIVITY ANALYSIS

In this section, the buyer's savings in expected inventory costs under different values of model parameters are computed. To be concise, our attention is limited to Case 1, i.e., no restriction in the timing of surprise. The main purpose of this section is to present some examples of the types of questions that can be considered to explore the robustness of the model solution to input parameters for the cases discussed below.

First, using the example in Section 4, the changes in percentage savings as λ and T change are investigated. Recall that $1/\lambda$ is the expected time until the demand surprise. Table 11.1 lists the numerical results for three selected π values. The entries in the %Savings column are $[EC(S) - EC(S^*)]/EC(S^*)$, where S is the optimal order quantity based on the single initial demand forecast. As λ increases, the probability of a surprise during the planning horizon increases. Interestingly, %Savings seem quite robust in response to changes in λ as λ increases. This means that an accurate estimation of λ is not very critical as long as the probability of a surprise is sufficiently high.

In the second part of the analysis, the effects of the post-surprise demand characteristics are explored using $\mu_0 = \mu_2 = 100$, $\sigma_0^2 = \sigma_2^2 = 400$ and allowing μ_1 and σ_1^2 to vary. Table 11.2 shows that, as can be expected, the savings in the expected cost are greater when demand distributions in the after-shock states are more "dissimilar" to the initial demand distribution; in other words, the ordering decision based on the initial demand forecast is likely to yield higher costs when that forecast is charecterised by more error. Along the same lines, it can also be observed that the impact of π on the expected cost is higher when the value of π implies a high probability of movement to a state in which the values of the demand distribution parameters are far from those of the initial distribution. It also appears that the expected cost is more sensitive to changes in the mean than changes in the variance of the demand distribution.

6. SUMMARY AND EXTENSIONS

In this paper, we discuss how random changes in the demand distribution can be analyzed in the framework of a single period inventory model. The likelihood of a change in the demand distribution increases as the length of the lead time and sale period increase. If there is significant uncertainty surrounding the demand generating process, the standard newsvendor analysis based on a single demand forecast will not minimize the buyer's expected cost.

Table 11.1 Sensitivity to λ and T $\mu_0 = 100$, $\mu_1 = 180$, $\mu_2 = 60$, $\sigma_0^2 = 400$, $\sigma_1^2 = 900$, $\sigma_2^2 = 200$, $L = 1$.

a. $\pi = 0.2$

λ	T	μ_T	σ_T^2	$EC(S^*)$	$\%Savings$
1	1	151.2	724	837.4	17.5
1	5	714.8	3360.6	3750	22.1
1	10	1421.5	6669.7	7355.4	23.9
5	1	166.6	821.7	919.8	23.4
10	1	166.7	822.2	920.3	23.5

b. $\pi = 0.6$

λ	T	μ_T	σ_T^2	$EC(S^*)$	$\%Savings$
1	1	121.9	553.5	681	5.2
1	5	592.1	2644.5	3116.4	7.9
1	10	1180.7	5264.6	6123.4	9.6
5	1	128.5	599.7	717	7.8
10	1	128.6	600	717.2	7.8

c. $\pi = 0.8$

λ	T	μ_T	σ_T^2	$EC(S^*)$	$\%Savings$
1	1	100	425.6	562.6	0.0005
1	5	500	2107.4	2639.3	0.0002
1	10	1000	4210.8	5196.8	0.0001
5	1	100	433.8	563.1	0.0009
10	1	100	433.3	563.1	0.0009

Table 11.2 Effect of Demand Distribution Parameters $\mu_0 = \mu_2 = 100$, $\sigma_0^2 = \sigma_2^2 = 400$, $\lambda = L = 1$, $T = 2$

π	σ_1^2/σ_0^2	μ_1/μ_0	$EC(S^*)$	$\%Savings$
0.1	0.5	0.5	722.8	22.9
		1	1069.4	0.02
		2	1762.5	32.8
		3	2455.7	51.8
	1	0.5	739.2	20.2
		1	1085.8	0
		2	1779	31.6
		3	2472.2	50.8
	4	0.5	804	11.7
		1	1150.6	0.1
		2	1843.8	27
		3	2537	46.9
0.9	0.5	0.5	1016.4	0.9
		1	1082.9	0.0004
		2	1215.9	3.2
		3	1349	10.6
	1	0.5	1019.3	0.8
		1	1085.8	0
		2	1218.8	3.2
		3	1351.9	10.4
	4	0.5	1035	0.6
		1	1101.5	0.01
		2	1234.5	3
		3	1367.6	9.6

The issue of demand distributions with random shifts may arise in different contexts. When the flow of materials in a multi-echelon system is considered, it may be the case that an input item is used in a manufacturing process with an unpredictable quality level. In particular, when a new production technology or inexperienced workforce are involved, it may be harder to keep the quality of the process under control. Thus, stocking decisions at the upstream stages are influenced by the possibility of quality related disruptions in the downstream processes. The quality level of the manufacturing process also affects the process output quality, which may lead to shifts in the quantity of the finished product demanded by consumers during the selling period.

A selling period with random duration is a special case of the proposed model. For products that can quickly become obsolete, the demand may disappear before the end of the planned sale period. In this model, this situation can be represented by two demand states with the demand in the second state being zero.

It is assumed that the demand distribution changes instantaneously. It might be interesting to extend this model to the case of gradual change in the mean demand following the surprise event. This situation may arise if, for example, the mean demand moves from the pre-surprise level to the stable post-surprise level by following a linear trend over a time interval.

The buyer's inventory problem can also be formulated as a profit maximization problem. This version would be more appropriate if the selling price of the product in the consumer market is related to the demand state. It is possible to modify the proposed model to incorporate the impact of different unit profits in different demand states on the buyer's expected profit. The time until the occurrence of the change in the demand distribution is assumed to be exponentially distributed. In a future study, this assumption could be relaxed (for example, see Rahim and Costa (2000)).

References

[1] Azoury, K. Bayes Solution to Dynamic Inventory Models under Unknown Demand Distribution. *Management Science*, **31** (1985) 1150-1160.

[2] Brill, P.H. and Chaouch, B. A. An EOQ Model with Random Variations in Demand. *Management Science*, **41**(1995) 927-936.

[3] Chao, H. Inventory Policy in the Presence of Market Disruptions. *Operations Research*, **35**(1987) 274-281.

[4] Duncan, A. J. The Economic Design of $\overline{X}$ Charts Used to Maintain Current Control of A Process. *Journal of the American Statistical Association*, **51** (1956) 228-242.

[5] Eppen, G. D. and Iyer, A. V. Backup Agreements in Fashion Buying- The Value of Upstream Flexibility. *Management Science*, **43**(1997) 1469-1484.

[6] Fisher, M. and Raman, A. Reducing the Cost of Demand Uncertainty Through Accurate Response to Early Sales. *Operations Research*, **44** (1996) 87-99.

[7] Hadley, G. and Whitin, T. *Analysis of Inventory Systems*, Prentice-Hall, Englewood Cliffs, NJ, 1963.

[8] Karlin, S. Dynamic Inventory Policy with Varying Stochastic Demands. *Management Science* **6** (1960) 231-258.

[9] Knappenberger, H. A. and Grandage, A. H. E. Minimum Cost Quality Control Tests. *IIE Transactions* **1**(1969) 24-32.

[10] Lee, H. L. and Nahmias, S. Single-Product, Single-Location Models. *Handbooks in OR & MS*, **4**, Graves et al. (Eds), North-Holland, New York, 1993.

[11] Parlar, M. and Berkin, D. Future Supply Uncertainty in EOQ Models. *Naval Research Logistics* **38**(1991) 107-121.

[12] Rahim, M. A. and Costa, A. F. B. Joint Economic Design of $\overline{X}$ and R Charts under Weibull Shock Models. *International Journal of Production Research*, **38** No. 1 (2000) 2871-2889.

[13] Rahim, M. A., Kabadi, S. N. and Banerjee, P. K. A Single Period Inventory Model where Deterioration Begins at a Random Point in Time. *International Journal of Systems Science*, **31** No. 1 (2000) 131-136.

[14] Saniga, E. M. Joint Economically Optimal Design of $\overline{X}$ and R Control Charts. *Management Science* **24** (1977) 420-431.

[15] Song, J. and Zipkin, P. Inventory Control in a Fluctuating Demand Environment. *Operations Research* **41** (1993) 351-370.

IV

QUALITY & MAINTENANCE

Chapter 12

IMPROVING AND MAINTAINING PROCESS PROFITABILITY

Elart von Collani and
Institut fur Angewandte Mathematik und Statistik,
Sanderring 2, D-97070 Wurzburg,
Germany

Gudrun Kiesmüller
Faculty of Technology Management
Eindhoven University of Technology,
P.O. Box 513, Paviljoen F14
NL-5600 MB Eindhoven, The Netherlands

Abstract The profitability of a production process may suffer from disturbances and, under certain circumstances, from inappropriate process adjustment, which may lead to either inferior quality of the output or to process failure. In order to reach and keep profitability on a high level, various actions such as adjustments, sampling, inspections, different level of repairs, replacements, and renewals are undertaken according to process monitoring policies and maintenance policies, respectively. Each of the two policies consists of rules about which action should be undertaken in which situation. Generally, the policies are designed and implemented independently; although, their efficiency is mutually dependent. Moreover, the policies are selected either heuristically or according to criteria which reflect only implicitly the overall aim, process profitability. In this paper, an integrated approach is developed which enables an objective-oriented joint design of monitoring and maintenance policies.

Keywords: Production process, process control, homogeneous poisson process, profitability.

1. THE PRODUCTION PROCESS

This paper deals with a production process operated in time for making profit. The larger the profit, the better the process; therefore, the quality of the process is determined by its profitability.

A retrospective evaluation of a production process with respect to profitability could be performed by comparing the overall expenditures with the overall returns obtained by selling the output. If, however, the process is evaluated not in retrospective but in advance in order to adjust and maintain it in a way resulting in as high a profitability as possible, then, the expenditures and the returns have to be predicted in order to select the most profitable policies. Since there is uncertainty with respect to almost everything related to a production process, the above outlined task of prediction and selection can be solved only by developing an appropriate stochastical model.

Process profitability is affected by various aspects, for instance

1. process design,
2. process maintenance,
3. process control including process monitoring,
4. staff training,
5. marketing policy,
6. warranty policy, etc.

Process profitability is investigated here only as function of process maintenance, but the results obtained may also have implications with respect to the other aspects. Process maintenance is defined as the totality of actions performed on given production facilities during the process history aiming at reaching and maintaining profitability.

A mathematical model is a necessary condition for investigating any phenomenon quantitatively. The phenomenon of interest here are the processes of production and maintenance. Hence, the model must include the relevant parts of the process itself and the actions to be performed within the maintenance policy, where relevance refers to the aim, i.e. improving and maintaining process profitability.

There have been only a few attempts made to integrate process monitoring and process maintenance so as to increase their joint efficiency. Two early papers on this topic are those of Rahim and Banerjee [20] and of Frahm [11]. Recently, Kiesmüller ([14], [15] and [16]) proposed an integrated model for a normal production process.

1.1 Process Time. The process runs in time; therefore, in order to model its relevant parts, a suitable process time unit has to be selected. It is assumed here that the output consists of discrete items and that the time for producing one item is constant, implying that it is not affected by quality deterioration and therefore cannot be used for an indicator of process disturbances. In this case, the time for producing one item is a natural process time unit.

With respect to time, three different types are to be distinguished.

- *Production time*, which is characterized by the fact that in each time unit the process produces one item
- *shut-down time*, which is characterized by the fact that although the process is supposed to produce, it had to be closed due to failure
- *Off-time* which is characterized by the fact that it is not available for production because of other reasons

It isn't necessary to take into account any time which is not available for production, and therefore off-time may simply be omitted, i.e. not counted. Hence, process time consists of the available time represented by the production time and the shut-down time, each of them measured in time units, where the time unit equals the constant time of producing one item.

1.1.1 Production Time. Production causes expenses, but during production time it yields an output. The output is utilized in order to compensate for the production expenses and to make a profit. Hence, the output is the vehicle of profitability. Modeling the output may be restricted to those features which affect the profit derived from it. An item produced is called better than another item if it yields a higher profit. Thus, profit is the appropriate quality characteristic for the items produced. It is assumed that the profit derived from one item produced is completely determined by a single item feature denoted by X according to the following simple function:

$$g(X) = \begin{cases} g^{+} & \text{for} \quad X \in [L, U] \\ g^{-} & \text{for} \quad X \notin [L, U] \end{cases} \tag{12.1}$$

where L is called the lower specification limit and U the upper specification limit. If $X \in [L, U]$, then the corresponding item is called *conforming*, and if $X \notin [L, U]$ then it is called *non-conforming.*

The profit or quality function for items (12.1) is extremely simple, but nevertheless, there are many cases in practice in which it can be used as a sufficiently good model.

The economic parameter g^+ is the profit derived from a conforming item, and g^- is the profit from a non-conforming item. Of course, $g^+ > 0$, $g^+ > g^-$ and g^- might be negative. Note that g^+ and g^- are defined as profit, i.e. revenue minus cost of manufacturing, where the cost of manufacturing consists of the direct costs like the cost of material and the indirect proportional costs incurred by salaries and production facilities.

There is output during production time only, hence, the item feature X determines the quality of one unit of production time. Let X_i denote the quality characteristic of the ith item produced. Then, the economically relevant part of the output process is completely described by the sequence

$$\left\{X_i\right\}_{i=1,2,\cdots}. \tag{12.2}$$

Assuming an uninterrupted production of $m_2 - m_1$ items from time m_1 to m_2, the profit derived during this time is given by means of the process quality function Q:

$$Q(m_2) - Q(m_1) = \sum_{i=1}^{m_2} g(X_{m_1+i}). \tag{12.3}$$

As soon as a process failure occurs, the situation, with respect to process quality, changes completely as the process stops producing output.

1.1.2 Shut-Down Time. The second type of process time starts with a process shut-down due to a failure, after which the process does not produce items, while the expenditures for salaries, production facilities, etc. continue. Consider a shut-down time from t_1 to t_2. The profit derived during this time is negative and modeled by

$$Q(t_2) - Q(t_1) = -(t_2 - t_1)c_f. \tag{12.4}$$

The economic parameter c_f includes all costs incurred during production, including salaries of workers and employees.

1.2 Process Quality. It is assumed that, at the start of the process at time $t = 0$, the process is in the production state. After some random time, say T_f, a process failure occurs, and the process is shut-down. According to (12.3) and (12.4) the profit derived from the process until time t is given by

$$Q(t) = \begin{cases} \sum_{i=1}^{[t]} g(X_i) & \text{for} \quad t < T_f \\ \sum_{i=1}^{[T_f]} g(X_i) - (t - T_f)c_f & \text{for} \quad t \geq T_f, \end{cases} \tag{12.5}$$

where $[a]$ denotes the largest integer not exceeding a.

Relation (12.5) implies that process quality up to time t is determined by the output X_i, obtained during the production time, and by the time of a process failure T_f which determines the length of the shut-down time $t - T_f$.

Dealing with stochastical quantities, we conclude that process quality is (except for the inherent uncertainty) completely determined by

- the joint distribution function of the output sequence, and
- the distribution function of the failure time.

Thus, a model for the joint distribution of $\{X_1, X_2, \cdots, X_{T_f}\}$ and for the distribution of T_f is needed.

Let the marginal distribution functions of $\{X_1, X_2, \cdots, X_{T_f}\}$ belong to the distribution family $\mathcal{F}_\Theta$, depending on the value $\theta \in \Theta$ of a process parameter. For given failure time t_f and given sequence $(\theta_1, \theta_2, \cdots, \theta_{t_f})$, the marginal distribution function of X_i is denoted by $F_{X_i}^{(\theta_i)}(x)$. For the joint distribution, we assume the following:

$$F_{X_1,\cdots,X_{[t_f]}}^{(\theta_1,\cdots,\theta_{[t_f]})}(x_1, \cdots, x_{[t_f]}) = \prod_{i=1}^{[t_f]} F_{X_i}^{(\theta_i)}(x_i). \tag{12.6}$$

The distribution function of the failure time T_f is an element of $\mathcal{G}_\Theta$ and it is denoted by

$$G_{T_f}^{(\theta)}(t). \tag{12.7}$$

Given $\mathcal{F}_\Theta$ and $\mathcal{G}_\Theta$, the development of output quality as well as failure behaviour is essentially determined by the development of the value θ

of the process parameter, which is described by the parameter process $\theta(t)$.

1.3 The Parameter process. It is assumed that the parameter process $\theta(t)$ may adopt values in a convex and compact parameter space Θ. Moreover, it is assumed that there are no changes of the parameter process during the shut-down time.

Any realization $\theta_i \in \Theta$ determines the actual item quality and the actual failure behaviour of the process at time t. The actual item quality is given by the distribution function $F_{X_i}^{(\theta_i)}(x)$ and the actual failure behaviour by $G_{T_f}^{(\theta_i)}(t)$ for $t \geq i$.

Assuming that the process can be adjusted to any value $\theta \in \Theta$ leads to the problem of determining a starting value $\theta_0 \in \Theta$ for the parameter process.

Once the starting value θ_0 is selected, the parameter process $\theta(t)$ develops in time until the occurrence of a process failure. After a process failure, the process remains in the shut-down state, i.e. it cannot repair itself.

The failure model, which is introduced in the next section, allows for the occurrence of many disturbances until process failure. It is assumed that each disturbance has an additive impact on the process parameter and, thus, affects item quality and failure behaviour.

We assume that the first disturbance causes a shift in the process parameter of size $|\delta_1| = \delta$ with sign $sign(\delta_1)$. It is assumed that δ is numerically known and that $sign(\delta_1) = +$ with probability w and $sign(\delta_1) = -$ with probability $1 - w$. For the k-th disturbance with $k > 1$, we assume:

$$|\delta_k| = |\delta_1| = \delta \tag{12.8}$$

$$sign(\delta_k) = sign(\delta_1). \tag{12.9}$$

Let $\theta(0) = \theta_0$ be the starting value of the parameter process and $N(t)$ be the number of disturbances until time t. Thus, from (12.8) and (12.9), we obtain the value of the process parameter at time $t < T_f$ as a function of the number of disturbances:

$$\theta(t) = \theta_0 + \sum_{k=1}^{N(t)} \delta_i = \theta_0 + sign(\delta_1)N(t)\delta. \tag{12.10}$$

The model (12.10) considered here takes into account only so-called random shocks. Any additional information about the occurrence mechanism of disturbances should, of course, be taken into account by modifying the model accordingly. For instance, if the parameter process is subject to a linear drift, then this should be reflected by the model.

The next assumption refers to the occurrence of disturbances. It is assumed that the time between two consecutive disturbances is exponentially distributed with parameter λ, implying that $N(t)$ is a homogeneous Poisson process with intensity λ:

$$P(N(t) = i) = \frac{(\lambda t)^i}{i!} e^{-\lambda t}. \tag{12.11}$$

With each disturbance, the value of the process parameter moves further away from the starting value θ_0.

1.4 The Failure Model. For any realistic model, the parameter process must necessarily be bounded. Thus, we assume the existence of a threshold value $\theta_f > 0$. As soon as the absolute value of the parameter process is equal or exceeds the threshold value, a process failure occurs and the process enters the shut-down state.

Let θ_0 be the starting value. A process failure will occur, when one of the following two events occurs:

$$\delta_1 > 0 \quad \text{and} \quad N(t) \geq \frac{\theta_f - \theta_0}{\delta_1} = N_r^{(+)} \tag{12.12}$$

$$\delta_1 < 0 \quad \text{and} \quad N(t) \geq \frac{\theta_f + \theta_0}{-\delta_1} = N_r^{(-)}. \tag{12.13}$$

implying that at the start $t = 0$ of the process we have

$$G_{T_f}^{(\theta_0)}(t) = \mathbf{P}\Big(N(t) > N_r^{(+)} \,\Big|\, \delta_1 > 0\Big)w + \mathbf{P}\Big(N(t) > N_r^{(-)} \,\Big|\, \delta_1 < 0\Big)(1-w). \tag{12.14}$$

At any other time point t_1, we obtain for $t > t_1$ with $\theta_1 = \theta(t_1)$ and $N_1 = N(t_1)$:

$$G_{T_f}^{(\theta_1)}(t) \;=\; \mathbf{P}\Big(N(t) > N_1 + \frac{\theta_f - \theta_1}{\delta_1} \,\Big|\, \delta > 0\Big)w +$$

$$\mathbf{P}\Big(N(t) > N_1 - \frac{\theta_f + \theta_1}{\delta_1} \,\Big|\, \delta_1 < 0\Big)(1-w). \quad (12.15)$$

Clearly, (12.14) and (12.15) imply that any disturbance increases the probability of a process failure reflecting process aging.

Using (12.12) and (12.13), the process quality function is obtained by

$$Q(t) = \begin{cases} \sum_{i=1}^{[t]} g(X_i) & \text{if} \quad N(t) = 0 \\ \sum_{i=1}^{[t]} g(X_i) & \text{if} \quad \delta_1 > 0 \ \&\ N(t) \leq N_r^{(+)} \\ \sum_{i=1}^{[T_f]} g(X_i) - (t - T_f)c_f & \text{if} \quad \delta_1 > 0 \ \&\ N(t) > N_r^{(+)} \\ \sum_{i=1}^{[t]} g(X_i) & \text{if} \quad \delta_1 < 0 \ \&\ N(t) \leq N_r^{(-)} \\ \sum_{i=1}^{[T_f]} g(X_i) - (t - T_f)c_f & \text{if} \quad \delta_1 < 0 \ \&\ N(t) > N_r^{(-)}. \end{cases} \quad (12.16)$$

1.5 Constant Parameter Process. For example, assume that the process parameter, once adjusted to a certain starting value $\theta_0 \in \Theta$, remains unchanged, i.e. $\delta_1 = 0$. Thus,

$$\theta(t) = \theta_0 \quad \text{for any } t > 0. \quad (12.17)$$

In this case, output quality does not change in time, and there is no process failure, i.e. the process remains in the production state all the time. The output is described by a sequence of independent identically distributed random variables. In this simple and unrealistic case, the process quality function $Q(t)$ adopts the simple form:

$$Q(t) = \sum_{i=1}^{[t]} g(X_i). \quad (12.18)$$

where the sequence $\{X_1, X_2, \cdots\}$ consists of independent and identically distributed random variables.

Process profitability does not refer to one single time unit of operation, but to the whole operation period. The quality of a unit of production time which is given by the quality of the item produced is evaluated by the net profit obtained from selling the item. Thus, it makes sense to select the average net profit per item as an appropriate criterion for process profitability. Moreover, it is assumed that the period of operation is unknown and lengthy; thus, the average net profit per item in the long run is used for measuring process profitability.

The average net profit per item in the long run is given by the limit of the average net profit per item up to time t for t tending to infinity:

$$\lim_{t\to\infty} \frac{\sum_{i=1}^{[t]} g(X_i)}{[t]} = E\Big[Q(1)\Big] = (g^+ - g^-) \int_L^U dF_X^{(\theta_0)}(x) + g^-, \qquad (12.19)$$

where $F_X^{(\theta_0)}(x) = F_{X_i}^{(\theta_0)}(x) \quad \text{for } i = 1, 2, \cdots.$

From (12.19), we ascertain that maximizing the long-run profit per item is equivalent to maximizing the conformance probability with respect to the process parameter θ. Thus, the optimum starting value θ_0^* of the process parameter is defined by

$$\theta_0^* = \arg\max_{\theta\in\Theta} \int_L^U dF_X^{(\theta)}(x). \qquad (12.20)$$

Relation (12.20) implies that in the case of a constant parameter process, the problem of optimizing process profitability, given by the average net profit per item, in the long run is reduced to the determination of a target value, which maximizes the conformance probability.

1.6 Unstable Production Process. Real production processes are, however, never stable in the above sense, implying that (12.19) is not true. Due to disturbances with positive shift size δ, the parameter process is subject to changes. Moreover, if the process parameter exceeds the critical value θ_f, a process failure occurs, and the process enters the shut-down state.

In general, the average net profit per item is obtained in the long run by means of the quality function given by (12.16) and by $\lim_{t\to\infty} F_{T_f}^{(\theta_0)}(t) = 1$:

$$\lim_{t \to \infty} \frac{Q(t)}{T_f} = -\infty, \tag{12.21}$$

Therefore, the average net profit per item tends to $-\infty$, reflecting the well-known experience that in the long run any process which is not maintained appropriately will produce losses.

Thus, to assure profitability of a production process over time, changes in the process parameter, which necessarily occur, should be removed early enough before they may grow and cause considerable losses. Moreover, any process failure should be omitted, and if, despite all efforts, a failure occurs, then the failure should be removed immediately. Actually, the cost of monitoring and maintaining production processes constitutes a considerable part of the operational costs of industrial production processes. Hence, the design of monitoring and maintaining policies aimed at ensuring process profitability is important.

2. THE CONTROL MODEL

When operating on target, the actual process parameter is equal to the target value θ_0, and process profitability is at maximum. Operating off target and, particularly, any process failure causes a decrease in profitability.

According to the assumption that the process cannot repair itself, the process can be brought back on target only by means of a corrective action. But any corrective action costs money, and its effect lasts only for a certain time; thus, an inappropriate corrective action may indeed be counter-productive in the sense that it does not increase, but decreases profitability. This happens if the cost for the corrective action is larger than its benefit. Even if the benefit of a corrective action is larger than its cost, it may be inappropriate. This is the case where there is another corrective action that leads to a larger increase of process profitability.

Therefore, the main problem is to determine which corrective action in a given situation (off target or failure) is best with respect to process profitability. Before this problem can be solved, the corrective actions must be properly introduced.

2.1 Process Corrections.

A corrective action may be performed if the process operates off target or if a process failure has occurred. In either case, the corrective action may be described by its effect on the value θ of the parameter process. Two cases may occur:

- When performing the corrective action, the actual value θ of the process parameter is known.
- When performing the corrective action, the actual value θ of the process parameter is unknown.

In the former case, any meaningful corrective action decreases the deviation of the actual value θ from the target value θ_0.

In the latter case, however, a corrective action might lead to an increase of the deviation of the actual value θ from the target value θ_0. This case has been examined in [9] and [8]. Consider the case where a false alarm is released, and a corrective action is performed while the process actually operates on target. If the corrective action consists of something like an adjustment usually performed by automatic process controllers, then the corrective action may deteriorates process quality. This effect is well-known in industry and is called the over-adjustment problem. There is a large body of literature dealing with this effect. For a recent reference, [3].

Here, we deal with the case where the actual value of the process parameter is determined before a corrective action is performed. In this case, any meaningful corrective action reduces the deviation of the actual parameter value from target. The effect of the corrective action is modeled in the following way:

$$R^{(\ell)}(\theta) = \theta_0 + \gamma(\theta - \theta_0) \quad \text{with } 0 \leq \gamma \leq 1. \tag{12.22}$$

where θ is the parameter value before the corrective action, and $R^{(\ell)}(\theta)$ is the parameter value after the corrective action.

The upper index $\ell \in \{1,2\}$ indicates whether or not the corrective action removes a process failure.

$$\ell = 1 \quad : \quad \text{corrective action before a process failure} \tag{12.23}$$
$$\ell = 2 \quad : \quad \text{corrective action after a process failure} \tag{12.24}$$

Distinguishing corrective actions before and after failure is necessary because removing a failure incurs additional costs.

For $\gamma = 0$, the corrective action is called renewal, and for $\gamma = 1$, it is called minimal repair.

Minimal repairs do not make sense in our model because of the failure model. A failure can be removed only by reducing the deviation of the process parameter from target, which necessarily leads to an improved failure behaviour. In this paper, it is assumed throughout that $\gamma = 0$, i.e. any corrective action restores the starting conditions.

Next, the corrective actions must be described economically. Any corrective action constitutes additional expenditures, but it also improves the output quality for some time and thus contributes to the profit. The effect on the output quality will be taken into account later. For the present, only the direct costs of corrective actions are modeled.

Any corrective action is preceded by an instruction to undertake it. Such an instruction is called an alarm. After an alarm is released, there are three possibilities with respect to the subsequent action:

- The search for the disturbances reveals a false alarm. The search is called inspection, and the resulting average costs are denoted by e^* and called inspection cost.
- The corrective action before a process failure reveals a large enough number of disturbances, i.e. a large deviation of the process parameter from target. Such an action is called correction, and the average costs accruing from removing the deviation are denoted by r_1^* and called correction cost.
- If the alarm is released by a process failure, the subsequent corrective action is called renewal, and the average costs are denoted by r_2^* and called renewal cost.

Often, the objection to economic models is that the determination of the actual values of the relevant cost parameters is difficult. In fact, if the economic environment is unknown and cannot be assessed, then it makes no sense to try to adapt a policy economically. However, the question which arises is whether it makes sense to have a production process operate in the case where the values of relevant cost parameters cannot be specified. As far as we understand, it should be possible to determine the actual values of the relevant parameters as soon as these parameters are identified, which can be done by means of the economic model. Moreover, many sensitivity studies performed for economic models have shown that the optimal economic design of policies is not prone to estimation errors with respect to the cost parameters. Therefore, one may conclude that taking into acount even preliminary information about the economic aspects of a production process is better than to neglect them.

To make the model economically meaningful, it is assumed that

$$0 < e^* < r_1^* < r_2^* \tag{12.25}$$

is true.

Any corrective action is preceded by a decision. The corresponding decision function is denoted by η:

$$\eta(\theta) = \begin{cases} 0 & : \text{ no action} \\ 1 & : \text{ process correction} \\ 2 & : \text{ process renewal.} \end{cases} \tag{12.26}$$

Because the model for corrective actions is simple, any process failure, i.e. $\theta = \theta_f$, implies $\eta(\theta_f) = 2$. However, in the case $\theta(t) < \theta_f$, a corrective action might be beneficial or not. In this context, two questions arise:

1. when to make a decision, and
2. how to make the decision.

Besides the decision function η and the decision time points $(t_k^{(d)})_{k=1,2,\cdots}$, the target value θ_0 determines essentially the success of any corrective action.

Therefore, the questions when and how to perform corrective actions are answered by selecting

- the target value θ_0,
- the decision time points $(t_k^{(d)})_{k=1,2,\cdots}$, and
- the decision function $\eta(\theta)$.

The three elements $\left(\theta_0, (t_k^{(d)})_{k=1,2,\cdots}, \eta(\theta)\right)$ are called maintenance policy.

3. PROCESS MONITORING

Unnecessarily ordered corrective actions resulting in an inspection should be avoided, because they diminish profitability. Therefore, an alarm should be released only in the case of a process failure, assumed to be immediately observed, or if there is sufficient reason to suspect a large enough number of process disturbances.

Generally, the occurrence of disturbances cannot be observed directly; therefore, other features are used to infer the actual number of disturbances.

Reasons sufficient for expecting a large enough number of disturbances follow.

- A long uninterrupted production time, suggesting the occurrence of a number of disturbances and thus a deviation from target and an early process failure.
- A decrease of the output quality, indicating the presence of disturbances.

In the first alternative, production time is an indicator of the occurrence of process disturbances. This approach is often used for determining maintenance policies (see, for instance [1] or [2]). The second approach, using output quality as an indicator, is used in statistical process control for determining so-called process monitoring policies (for details see [10]).

In the first alternative, production time has to be monitored, which is possible without further expenses. Selecting the second alternative means monitoring the output, which causes additional costs. However, the second alternative is often more effective in revealing disturbances and preventing failures.

Obviously, the question of whether the one or the other alternative contributes more to profitability can be answered only if the economic environment is taken into account. Therefore, the expenses for output monitoring have to be specified.

It is assumed here that the costs incurred by taking and evaluating a sample are proportional to the sample size, with proportionality factor denoted by a^* and called sampling cost:

- $a^*n =$ sampling cost.

As to process profitability, any shut-down time expenses must be taken into account these are contained in the economic parameters r_1^* if the process is shut down for a process correction, and r_2^* if a process renewal is performed after a process failure. Thus, besides the number of corrections and renewals determining overall costs, the output sequence, which refers to production time, must be considered.

The task is either to monitor production time or to monitor output quality.

Monitoring output quality is defined as observing the item quality characteristic X by means of samples taken from the output sequence

$\{X_i\}_{i=1,2,\cdots}$. Let the ℓth sample consist of n consecutively produced items with the characteristics $\vec{X}_\ell = (X_{t\ell-n+1}, \cdots, X_{t_\ell})$. The assumption that taking and evaluating the sample requires only negligible time implies that the ℓth decision before process failure, whether or not to release an alarm, is made at the time point t_ℓ, on the basis of $\vec{X}_\ell$, by means of a decision function denoted by ϕ:

$$\phi(\vec{X}_\ell) = \begin{cases} 1 & \text{alarm} \\ 0 & \text{no alarm.} \end{cases} \qquad (12.27)$$

The problem inherent to process monitoring is to select appropriately

- the monitoring time points $(t_\ell^{(m)})_{\ell=1,2,\cdots}$,
- the sample size n, and
- the decision function $\phi(\vec{X})$ for releasing alarms.

An appropriate selection is a selection that results in increased profitability. The three elements $\left((t_\ell^{(m)})_{\ell=1,2,\cdots}, n, \phi(\vec{X})\right)$ are called monitoring policy.

4. PROCESS PROFITABILITY

As already mentioned, the average net profit per item in the long run serves here as a measure for process profitability.

In order to specify the process profit function $Q(t)$ for a process that is monitored and maintained, some further notations are necessary.

$A(t)$	:	number of monitoring time points
$A_F(t)$	:	number of false alarms
$R_1(t)$	:	number of corrections
$R_2(t)$	:	number of renewals
$N(t)$	:	number of produced items

Shut-down times are taken into economic account by the cost for corrections and renewals. Therefore, the time t in the above functions may refer to production time, and, as a result of the time unit selected, the number of produced items is essentially the same as the length of production time:

$$Q(t) = \sum_{i=1}^{N(t)} g(X_i) - A(t)a^*n - A_F(t)e^* - R_1(t)r_1^* - R_2(t)r_2^*. \quad (12.28)$$

The average net profit per item until time t is given by the ratio

$$\Pi(t) = \frac{Q(t)}{N(t)}. \tag{12.29}$$

For evaluating process profitability, the whole time period, from process start to its ultimate shutdown, has to be taken into account. Generally, this period is long and the number of items produced is large. Therefore, it makes sense to define the long run average net profit per item produced as the limit of (12.29) for $t \to \infty$:

$$\Pi = \lim_{t \to \infty} \Pi(t). \tag{12.30}$$

4.1 The Long Run Average Net Profit Per Item.

Dividing the process time into the time periods between two successive corrective actions (corrections or renewals) gives stochastically equivalent process parts called renewal cycles. Let the following quantities refer to one renewal cycle.

A	:	number of monitoring points
A_F	:	number of false alarms
R_1	:	number of corrections
R_2	:	number of renewals
Q	:	net profit
N	:	number of produced items

Thus, we obtain:

$$\Pi = \frac{E[Q]}{E[N]} = \frac{E\left[\sum_{i=1}^{N} g(X_i)\right] - E[A]a^*n - E[A_F]e^* - E[R_1]r_1^* - E[R_2]r_2^*}{E[N]}. \tag{12.31}$$

The long run average net profit per item given by (12.31) constitutes an appropriate criterion for comparing and purposefully improving production processes.

The aim of this paper is to determine a monitoring and maintenance policy which contributes best to process profitability. Hence, we will look at Π as a function of the monitoring and maintenance policies.

5. MONITORING AND MAINTENANCE POLICY

Any monitoring and maintenance policy implemented for controlling process profitability consists of the monitoring time points. After each monitoring time point, a decision about an alarm is made according to the decision function ϕ. In the case of an alarm, an inspection is made which is completed by a decision on whether or not to perform a correction according to the decision function η. Thus, the decision time points are given by the monitoring time points completed by an alarm and the time points of process failure.

Thus, a monitoring and maintenance policy is completely given by

$$\left(\{t_\ell^{(m)}\}_{\ell=1,2,\cdots}, n, \phi; \theta_0, \eta\right). \tag{12.32}$$

From the viewpoint of industrial implementation, monitoring policies with constant periods between monitoring time points are of special importance. Therefore, we restrict the present investigations to the case where the monitoring time points are given by a fixed monitoring interval denoted by h.

In essence, the sample size determines the reliability of the decision function with respect to making decision errors. The decision function $\phi(\vec{X})$ for releasing alarms depends on the joint distribution function $F_{\vec{X}_i}^{(\theta)}(\vec{x})$ and the type of the process parameter θ.

The starting value θ_0, which also serves as a target value for the parameter process, is a very critical quantity since small changes in the value can greatly influence profitability. The decision function $\eta(\theta)$ distinguishes between those values of θ for which a process correction would decrease profitability, in traditional terminology called the in-control state, and those values for which a process correction increases profitability, usually called the out-of-control state.

Let θ_0 be the starting value. Thus, for the parameter process $\theta(t)$ considered here, the decision function $\eta(\theta)$ is completely determined by two real numbers denoted by θ_{c^-} and θ_{c^+}, where $\theta_{c^-} < \theta_0 < \theta_{c^+}$ and

$$\eta(\theta) = \begin{cases} 1 & : \quad \theta \leq \theta_{c^-} \text{ or } \theta \geq \theta_{c^+} \\ 0 & : \quad \theta_{c^-} < \theta < \theta_{c^+}. \end{cases} \tag{12.33}$$

Therefore, the problem is to determine a monitoring and maintenance policy given by

$$\left(h^*, n^*, \phi^*; \theta_0^*, \theta_{c-}^*, \theta_{c+}^*\right), \tag{12.34}$$

such that

$$\Pi\left(h^*, n^*, \phi^*; \theta_0^*, \theta_{c-}^*, \theta_{c+}^*\right) \geq \Pi\left(h, n, \phi; \theta_0, \theta_{c-}, \theta_{c+}\right) \tag{12.35}$$

for any other admissible monitoring and maintenance policy.

There are two tasks to accomplish.

- Modeling: An explicit expression for the average net profit per item in the long run as a function of monitoring and maintenance policy must be determined.
- Optimization: A monitoring and maintenance policy has to be determined maximizing the average net profit per item in the long run.

How to solve the two problems is illustrated by using the example of a normally distributed quality characteristic X.

6. NORMALLY DISTRIBUTED QUALITY CHARACTERISTIC

6.1 The Process. In order to illustrate the approach developed so far, a normally distributed quality characteristic X is used. There are two distribution parameters, namely, the expectation and variance, with actual values denoted by μ and σ^2, respectively. For simplicity it is assumed that the variance is constant and that the process parameter of interest is the process mean given by the expectation of the quality characteristic X. Thus, we assume that the marginal distribution of the quality characteristic X_i for fixed value $\theta = \mu$ of the process mean is given by

$$F_{X_i}^{(\mu)}(x) = \Phi\left(\frac{x-\mu}{\sigma}\right), \tag{12.36}$$

where $\Phi(x)$ denotes the distribution function of the standardized normal distribution $\mathcal{N}(0,1)$.

The actual value σ^2 of the process variance is known and constant. Therefore, without loss of generality, we set $\sigma^2 = 1$. For $0 < t_1 < \ldots < t_n$ and under the condition $\mu(t_1) = \mu_1, \ldots, \mu(t_n) = \mu_n$, the joint distribution of the random variables $X_{t_1}, \ldots, X_{t_n}$ is assumed to be

$$F_{X_{t_1},\ldots,X_{t_n}}^{(\mu_1,\cdots,\mu_n)}\left(x_1,\cdots,x_n\right) = \prod_{i=1}^{n} F_{X_{t_i}}^{(\mu_i)}\left(x_i\right) = \prod_{i=1}^{n} \Phi\left(x_i - \mu_i\right), \tag{12.37}$$

Following the usual practice, the value of the process mean that maximizes the probability for producing a conforming item is used as target value and can be set, without loss of generality, equal to zero, i.e. $\mu_0 = 0$, which is equivalent to $U = -L$. It should be remarked, however, that the above simple selection of the target value may not be optimal with respect to profitability. It is not optimal, if for instance $w = P\Big(sign(\delta_1) = +\Big) \neq 0.5$.

Thus, at process start $t = 0$, the process parameter is adjusted to $\mu_0 = 0$, implying that

$$N_r^{(+)} = N_r^{(-)} = N_r. \tag{12.38}$$

At random times $T_k^{(d)}$, $k = 1, 2, \cdots$, disturbances occur and shift the process quality indicator away from the target value $\mu_0 = 0$.

If the deviation from the target value is small, then, from an economic point of view, it is not beneficial to intervene in the process. On the other hand, removing a large deviation improves process profitability. If the deviation gets too large, then a process failure occurs and has to be removed immediately to avoid large losses. Thus, three process states with respect to the degree of profitability or equivalent to the necessary action for maintaining process profitability can be defined.

- *State I*: The process is working at a sufficiently high quality level given by $|\mu| < \mu_c$, such that a process correction is not justified economically.

- *State II*: The process is working at a sufficiently low quality level given by $\mu_c \leq |\mu| < \mu_r$, such that a process correction increases process profitability.

- *State III*: The process has stopped operating (= process failure) because the value of the process parameter is equal or exceeds the threshold value μ_r.

By means of the number of disturbances, an equivalent definition of the process state is obtained.

- *State I*: The number of disturbances is small, i.e. $N < N_c$ and does not justify economically any process correction.

- *State II*: The number of disturbances is large, i.e. $N_c \leq N < N_r$, such that a process correction increases process profitability.

- *State III*: The process has stopped operating (= process failure) because the number of disturbances is equal or exceeds the threshold value N_r.

The values μ_c or N_c are called correction boundary, and μ_r or N_r are called renewal boundary. The correction boundary N_c is defined by the minimum number of shifts for which a process correction becomes profitable. It is one of the design parameters. In contrast, the renewal boundary N_r is assumed to be given because of technical reasons.

Because of the simple correction model, only three states have to be distinguished according to the actions: no action, adjustmant to target, or renewal. Any further correction method (for instance adjustment to a value $\theta_1 \neq \theta_0$) would lead to the definition of another process state. In other words, each correction mode M defines a process state $\mathcal{S}_M$ in the following sense: If the process is operating in state $\mathcal{S}_M$, then correction mode M contributes best to process profitability.

6.2 The Monitoring and Maintenance Policy. Any failure is recognized immediately; therefore, no special monitoring actions for detecting a failure are necessary. However, operating in State II cannot be observed directly. In order to detect State II, a periodic sampling policy with monitoring interval h is used. At the monitoring time points $\ell \cdot h, \ell \in \mathbb{N}$ referring to the production time, samples of size n given by $\vec{X}_\ell = (X_{\ell h-n+1}, \cdots, X_{\ell h})$ are drawn from the process output, and the actual value μ of the process parameter is estimated by

$$\overline{X}_\ell = \begin{cases} \frac{1}{n} \sum_{i=0}^{n-1} X_{\ell h-i} & \text{if} \quad n > 0 \\ 0 & \text{if} \quad n = 0. \end{cases} \tag{12.39}$$

A decision about an alarm is made by means of the following decision function $\phi(T)$:

$$\phi(\overline{X}) = \begin{cases} \text{alarm} & \text{if} \quad \mid \overline{X} \mid \geq c \\ \text{no alarm} & \text{if} \quad \mid \overline{X} \mid < c \end{cases} \tag{12.40}$$

where the positive real number c is called alarm limit.

The decision function is completely determined by the alarm limit c, and, hence, a monitoring policy is given by (h, n, c).

In the case of $n = c = 0$, the decision function (12.40) gives an alarm in each monitoring time point ℓh. Instead of output quality, this type of monitoring policy takes simple time as an indicator for the occurrence of disturbances. It is known [15] as inspection policy because, at each monitoring time point, at least an inspection is undertaken (if the inspection reveals State II, an additional process correction is done).

Any alarm is followed by an inspection of the process with a subsequent decision by means of η. Let N_ℓ be the actual number of disturbances, then

$$\eta(\overline{X}_\ell) = \begin{cases} 0 & : \quad 0 \le N_\ell < N_c \\ 1 & : \quad N_c \le N_\ell < N_r \\ 2 & : \quad N \ge N_r. \end{cases} \tag{12.41}$$

The decision $\eta(\overline{X}_\ell) = 2$ does not depend on $\overline{X}_\ell$, but is caused by a process failure that is observed with certainty. For the two other decisions, however, either State II is detected, or a false alarm is realized. In the former case, a process correction restores the target value; in the latter case, no further actions are undertaken. A process renewal is undertaken immediately after a process failure.

Hence, any monitoring and maintenance policy is given by

$$\Big(h, n, c; \mu_0, N_c\Big). \tag{12.42}$$

A meaningful selection of the monitoring policy depends heavily on the correction boundary N_c, which, together with the starting value θ_0 determines the maintenance policy.

Thus, the problem is to determine the design of the monitoring and maintenance policy $(h, n, c; \mu_0, N_c)$, where the set of admissible policies is given by

$$\begin{aligned} \mathcal{M} \quad = \quad & \Big\{(h, n, c; \mu_0, N_c) \mid h \in \mathbb{N} \cup \{\infty\}, n \le h, n \in \mathbb{N}, c \in \mathbb{R}, \\ & \qquad \mu_0 \in \Theta, N_c \in \mathbb{N}, N_c \le N_r\Big\}. \end{aligned} \tag{12.43}$$

6.3 Expectations. The average long run profit per item given by (12.31) contains a number of expectations which have to be determined for enabling a comparison of different policies and for finding an optimal policy. Assume that the monitoring and maintenance policy $(h, n, c; \mu_0, N_c)$ is given.

6.3.1 Expected Number of False Alarms. Any false alarm is equivalent to a wrong decision concerning the actual value μ of the parameter process. The decision is based on the random variable $\overline{X}$. The probability of a wrong decision depends on the actual number of disturbances having occurred before the monitoring time point ℓh. Let

$$\overline{X}_\ell^{(j)} = \overline{X}_\ell \Big| N(\ell h) = j \tag{12.44}$$

denote the random variable $\overline{X}_\ell$, under the condition that exactly j disturbances have occurred prior to $t = \ell h$.

The results hold for $sign(\delta_1) = +$ as well as for $sign(\delta_1) = -$, therefore, only the former case is considered in the remainder. The arithmetic mean $\overline{X}_\ell^{(j)}$ is normally distributed with mean $j\delta$ and variance n^{-1}; thus, the probabilities for wrong decisions with respect to the process state are as follows:

- For $j \in \{0, 1, \ldots, N_c - 1\}$, i.e. in State I, the probability of a false alarm is given by

$$\alpha_j := \Phi\Big((-c + j\delta)\sqrt{n}\Big) + \Phi\Big((-c - j\delta)\sqrt{n}\Big). \tag{12.45}$$

- For $j \in \{N_c, N_c + 1, \cdots, N_r\}$, i.e. in State II, the probability of not releasing an alarm is given by

$$\beta_j := \Phi\Big((c - j\delta)\sqrt{n}\Big) - \Phi\Big((-c - j\delta)\sqrt{n}\Big). \tag{12.46}$$

By straightforward calculation (see [16]), the expected number of false alarms is obtained:

$$E[A_F] = \frac{1}{\lambda} \sum_{i=1}^{N_c} \alpha_{i-1} \sum_{l=1}^{\infty} \lambda \frac{(\ell\lambda h)^{i-1}}{(i-1)!} e^{-\ell\lambda h}. \tag{12.47}$$

6.3.2 A Markov Chain Approach. For determining the expectations $E[R_1]$ and $E[R_2]$, the renewal cycle is modeled by a Markov chain.

Let the random variable D describe the number of disturbances, and let $\mathbf{1}_a$ be an indicator variable for a maintenance action:

$$\mathbf{1}_a := \begin{cases} 0 & \text{no maintenance action} \\ 1 & \text{maintenance action (corrections and renewals).} \end{cases} \tag{12.48}$$

Any cycle ends either if State II is detected, or if at least N_r disturbances occur leading to a process failure. Hence, the number of disturbances within a renewal cycle may be one of the values $0, 1, \cdots, N_r$ (where N_r stands for $N \geq N_r$), which, in essence, represent the states of the considered Markov chain.

State No.	**Definition**
i	if $D = i$ $(0 \leq i \leq N_r - 1)$ and $\mathbf{1}_a = 0$
i_r	if $D \in \{N_c, \ldots, N_r, \cdots\}$ and $\mathbf{1} = 1$

Let Y_ℓ denote the state immediately after the ℓth monitoring point. Thus, the sequence $\{Y_\ell\}_{\ell \in \mathbb{N}_0}$ is a Markov chain with state probabilities

$$v_{\ell,i} = P(Y_\ell = i). \tag{12.49}$$

The vector

$$\vec{v}_\ell^T = (v_{\ell,0}, v_{\ell,1}, \cdots, v_{\ell,N_r}) \quad \ell \in \mathbb{N}_0 \tag{12.50}$$

defines the state distribution immediately after the ℓth decision point. The starting distribution is given by

$$\vec{v}_0^T = (1, 0, \cdots, 0). \tag{12.51}$$

For determining the transition probabilities of the Markov chain $Y_0, Y_1, \cdots$ denoted by $\varrho_{i,j}$, five different cases are distinguished (for details see [16]).

1) $0 \leq i \leq N_c - 1; i \leq j \leq N_c - 1$

   For a state transition within State I, the following transition probabilities are obtained:

$$\begin{aligned} \varrho_{i,j} &= P(Y_{\ell+1} = j | Y_\ell = i) \\ &= P\Big(\mu((\ell+1)h) = j\delta | \mu(\ell h) = i\delta\Big) \\ &= \frac{\frac{(\lambda h)^{j-i}}{(j-i)!}}{\sum\limits_{m=i}^{N_r} \frac{(\lambda h)^{m-i}}{(m-i)!}}. \end{aligned} \tag{12.52}$$

2) $0 \leq i \leq N_r - 1; N_c \leq j \leq N_r - 1; i \leq j$

For a transition into State II without alarm, we obtain the following:

$$\begin{aligned} \varrho_{i,j} &= P(Y_{\ell+1} = j | Y_\ell = i) \\ &= P\Big(\mu((l+1)h) = j\delta, |\overline{X}_{\ell+1}| < c \Big| Y_\ell = i\Big) \\ &= \beta(j) \cdot \frac{\frac{(\lambda h)^{j-i}}{(j-i)!}}{\sum_{m=i}^{N_r} \frac{(\lambda h)^{m-i}}{(m-i)!}}. \end{aligned} \tag{12.53}$$

3) $0 \leq i \leq N_r - 1; j = i_r$

The absorbing state i_r is reached from state $i < N_r$ either by a failure or by detecting State II:

$$\begin{aligned} \varrho_{i,i_r} &= P(Y_{\ell+1} = i_r | Y_\ell = i) \\ &= P\Big(\mu((\ell+1)h) = \mu_r \Big| Y_\ell = i\Big) + \\ &\quad P\Big(\mu((\ell+1)h) \in \{N_c\delta, \cdots, (N_r - 1)\delta\}, |\overline{X}_{\ell+1}| \geq c \Big| Y_\ell = i\Big) \\ &= \frac{\frac{(\lambda h)^{N_r-i}}{(N_r-i)!}}{\sum_{m=i}^{N_r} \frac{(\lambda h)^{m-i}}{(m-i)!}} + \\ &\quad \frac{1}{\sum_{m=i}^{N_r} \frac{(\lambda h)^{m-i}}{(m-i)!}} \sum_{m=max\{i,N_c\}}^{N_r-1} (1 - \beta(m)) \cdot \frac{(\lambda h)^{m-i}}{(m-i)!}. \end{aligned} \tag{12.54}$$

4) $i = i_r; j = i_r$

The state i_r is an absorbing state, hence

$$\varrho_{i_r,i_r} = 1. \tag{12.55}$$

5) $0 \leq i \leq N_r; 0 \leq j \leq N_r; j < i$

The number of disturbances within a cycle cannot decrease, thus

$$\varrho_{i,j} = 0. \tag{12.56}$$

Once the transition matrix $\left(\varrho_{i,j}\right)_{i,j=1,\cdots,i_r}$ has been computed, the state probabilities $\upsilon_{\ell;j}$ may be obtained by the matrix product:

$$\vec{\upsilon}_\ell^T = \vec{\upsilon}_0^T \left(\varrho_{i,j}\right)^\ell. \tag{12.57}$$

6.3.3 The Expected Number of Corrections and Renewals. By means of the state probabilities $\upsilon_{\ell;j}$, the expected numbers of corrections $E[R_1]$ and of renewals $E[R_2]$ are given in [16]:

$$E[R_1] = \sum_{\ell=0}^{\infty} \left\{ \sum_{m=0}^{N_r-1} \left(\frac{1}{\sum_{k=m}^{N_r} \frac{(\lambda h)^{k-m}}{(k-m)!}} \sum_{i=max\{N_c,m\}}^{N_r-1} \left(1-\beta(i)\right) \frac{(\lambda h)^{i-m}}{(i-m)!} \right) \upsilon_{\ell;m} \right\}. \tag{12.58}$$

Because $R_1 + R_2 = 1$ with certainty, we obtain the following:

$$E[R_2] = 1 - E[R_1]. \tag{12.59}$$

6.3.4 Expected Number of Decision Time Points. Again, by straightforward calculation, the expected number of decision points within a renewal cycle is determined in [16]:

$$\begin{aligned} E[A] &= \sum_{\ell=1}^{\infty} \ell \left\{ \sum_{m=0}^{N_r-1} \frac{1}{\sum_{k=m}^{N_r} \frac{(\lambda h)^{k-m}}{(k-m)!}} \left(\sum_{i=max\{N_c,m\}}^{N_r-1} \left(1-\beta(i)\right) \frac{(\lambda h)^{i-m}}{(i-m)!} \upsilon_{\ell-1;m} \right.\right. \\ &+ \left.\left. \frac{(\lambda h)^{N_r-m}}{(N_r-m)!} \upsilon_{\ell;m} \right) \right\}. \end{aligned} \tag{12.60}$$

6.3.5 Expected Number of Items Produced. The length of the renewal cycle refers to the number of time units of production time, and hence may be looked upon as the number of produced items:

$$\begin{aligned} E[N] &= E[N \mid R_2 = 1] \cdot P(R_2 = 1) \\ &+ E[N \mid R_1 = 1] \cdot P(R_1 = 1), \end{aligned} \tag{12.61}$$

where

$$E[N, R_1 = 1] = \sum_{\ell=0}^{\infty} \left\{ (\ell+1)h \sum_{m=0}^{N_r-1} \left(\frac{1}{\sum\limits_{k=m}^{N_r} \frac{(\lambda h)^{k-m}}{(k-m)!}} \sum_{i=max\{m,N_c\}}^{N_r-1} \Big(1-\beta(i)\Big) \cdot \frac{(\lambda h)^{i-m}}{(i-m)!} \right) v_{\ell;m} \right\}, \tag{12.62}$$

and

$$E[\tau | R_2 = 1] = \sum_{\ell=0}^{\infty} \int\limits_{\ell h}^{(\ell+1)h} \Big(1 - P(N \leq t \mid R_2 = 1)\Big)\, dt, \tag{12.63}$$

where

$$\begin{aligned} P(N \leq t, R_1 = 1) &= \sum_{\ell=0}^{\lfloor \frac{t}{h} \rfloor - 1} \sum_{m=0}^{N_r-1} \frac{\frac{(\lambda h)^{N_r-m}}{(N_r-m)!}}{\sum\limits_{k=m}^{N_r} \frac{(\lambda h)^{k-m}}{(k-m)!}} \cdot v_{\ell;m} \\ &+ \sum_{m=0}^{N_r-1} \frac{\frac{(\lambda(t-\lfloor \frac{t}{h} \rfloor h))^{N_r-m}}{(N_r-m)!}}{\sum\limits_{k=m}^{N_r} \frac{(\lambda(t-\lfloor \frac{t}{h} \rfloor h))^{k-m}}{(k-m)!}} \cdot v_{\lfloor \frac{t}{h} \rfloor;m}, \end{aligned} \tag{12.64}$$

and where $\lfloor x \rfloor = \max\{n \mid n \leq x, n \in \mathbb{N}_0\}$.

6.4 Expected Profit. We consider the process operating in the Markov state i, $i = 0, 1, \cdots, N_r - 1$, and define the following quantities:

$N^{(i)}$: number of items produced in state i,

$X_{i,1}, \cdots, X_{i,N^{(i)}}$: sequence of quality characteristics in state i,

$p^{(i)}$: non conformance probability in state i,

$g^{(i)}$: expected profit per item produced in state i.

Therefore,

$$p^{(i)} = \Phi(-U + i\delta) + \Phi(-U - i\delta), \tag{12.65}$$

and

$$g^{(i)} = g^- p^{(i)} + g^+(1 - p^{(i)}) = g^+ - p^{(i)}(g^+ - g^-), \tag{12.66}$$

The profit incurred by the output is given by

$$G = \sum_{i=0}^{N_r-1} \sum_{m=1}^{N^{(i)}} g(X_{i,m}), \tag{12.67}$$

with the expectation

$$E[G] = \frac{1}{\lambda} \sum_{i=0}^{N_c-1} g^{(i)} + \sum_{i=N_c}^{N_r-1} g^{(i)} E[N^{(i)}]. \tag{12.68}$$

The second sum on the right hand side of (12.68) is small compared to the first sum. Moreover, the derivation of an analytical expression for it is difficult; therefore, following a proposal in [16], the second sum is approximated by a simpler expression:

$$\sum_{i=N_c}^{N_r-1} g^{(i)} E[N^{(i)}] \approx \frac{1}{\lambda} \sum_{i=N_c}^{\tilde{N}-1} g^{(i)} + \Big(E[N] - \frac{\tilde{N}}{\lambda}\Big) g^{(\tilde{N})}, \tag{12.69}$$

where $\tilde{N} \in \mathbb{N}$ with $N_c \leq \tilde{N} \leq N_r - 1$ is defined by

$$\frac{\tilde{N}}{\lambda} < E[N] \leq \frac{\tilde{N}+1}{\lambda}. \tag{12.70}$$

With the expectations derived so far, in principle, it is possible to calculate approximately the average long run net profit per item, and to search for an optimal monitoring and maintenance policy. Further simplifying approximations are proposed and utilized in [16].

7. NUMERICAL EXAMPLE

By means of the expectations $E[G]$, $E[R_2$, $E[A_F]$, $E[A]$, and $E[N]$ derived above and by further approximations described in [16], a monitoring and maintenance policy for a production process may be determined, which is (approximately) optimal with respect to the average net profit per item in the long run and which is given by

$$\hat{\Pi}(h,n,c;N_c) := \frac{E[G] - r_1^* - (r_2^* - r_1^*)E[R_2] - e^*E[A_F] - a^*nE[A]}{E[N]}, \tag{12.71}$$

Optimization (except for the target value, which was set to be $\mu_0 = 0$) was performed by a numerical optimization algorithm, and the result was verified by means of simulation.

Consider the following set of input parameters.

- specification limits: $U = -L = 2.576$
- shift size of a disturbance: $\delta = 1$
- renewal boundary: $N_r = 5$
- mean time between disturbances: $\frac{1}{\lambda} = 100$
- profit per conforming item: $g^{+*} = 1$
- profit per non-conforming item: $g^{-*} = -0.2$
- sampling cost: $a^* = 0.01$
- correction cost: $r_0^* = 10$
- renewal cost: $r_1^* = 1000$

These input parameters describe the relevant aspects of a production process, for which the following approximately optimal monitoring and maintenance policy is obtained.

- $\left(h = 9, n = 2, c = 2.43; \mu_0 = 0, N_c = 1\right)$ (12.72)

Accordingly, for every $h = 9$ items produced, a sample of size $n = 2$ is drawn from the production line. If $|\overline{X}|$ is less than $c = 2.43$, no alarm is released. Otherwise, an alarm is given, and a process inspection is performed. If the inspection reveals the occurrence of $N_c = 1$ or more disturbances, a process correction is undertaken, and the process starts anew. If five disturbances occur without correction, then the process stops operating. The process failure is immediately followed by a process renewal.

Using the monitoring and maintenance policy (12.72) yields the following average net profit per item in the long run:

$$\Pi\Big(9, 2, 2.43; 0, 1\Big) = 0.9052, \tag{12.73}$$

which is only 10% less than the yield of an ideal stable process producing conforming items with certainty.

8. CONCLUSIONS

Process monitoring and maintenance are keys for high profitability of production processes. However, even though the efficiency of one depends heavily on the other, there are almost no integrated models for determining them in a way that ensures the highest efficiency. The approach introduced in this paper illustrates the difficulties and indicates possible solutions to arrive at integrated and economically optimal montoring and maintenance policies. The example of normally distributed quality characteristics shows that computational difficulties increase even in the case of simplifying assumptions.

However, any monitoring and maintenance policy adopted is generally used for a lengthy period of a production process until fundamental changes in the technical or economic conditions of the production process require an appropriate adaption of the monitoring and maintenance policy.

In other words, even time-consuming optimization methods, for instance, optimization by means of simulation, can and should be used in conjunction with a realistic model for increasing the efficiency of the monitoring and maintenance policy in order to improve and maintain high process profitability.

Acknowledgement: The research was supported financially by Deutsche Forschungsgemeinsachft (DFG), grant no. Co 136/7–2.

References

[1] Barlow ,R.E. and Proschan F. *Statistische Theorie der Zuverlässigkeit*, Harri Deutsch, Frankfurt/Main, 1978.

[2] Beichelt, F. and Franken P. *Zuverlässigkeit und Instandhaltung*, Hanser, München, 1984.

[3] Box and Luce *Statistical Control by Monitoring and Feedback Adjustment*, Wiley, New York, 1997.

[4] Collani, E.v. A Simple Procedure to Determine the Economic Design of an $\overline{X}$-Control Chart. *Journal of Quality Technology*, **18** No. 3 (1986) 145-151.

[5] Collani, E.v. *The Economic Design of Control Charts*, Teubner, Stuttgart, 1989.

[6] Collani, E.v. Determination of the Economic Design of Control Charts Simplified. *Eds. Al-Sultan, K.S. and Rahim, M.A.: Optimization in Quality Control*, Kluwer Academic Publishers, Boston 1997.

[7] Collani, E.v. Control of Production Processes Subject to Random Shocks. *Annals of Operation Research*, **91** (1999) 289-404.

[8] Collani,E.v., Saniga, E.M. and Weigand, C. Economic Adjustment Designs for $\bar{x}$ Control Charts. *IIE Transactions*, **26** No.10 (1994) 37-43.

[9] Collani, E.v. and Weigand, C. Economic Machine Adjustment in the Case of Product Screening. *Statistical Papers*, **33** (1992) 171-184.

[10] Dumitrescu, M. *Mathematical Basis of Process Monitoring*, ASU Technical Papers Series, 08-6. Quality and Reliability Engineering Group, Arizona State University, 1998.

[11] Frahm, P. Combining Economic Design of $\overline{X}$-Charts and Block Replacement Under Weibull Shock Models. *Economic Quality Control*, **10** No. 2 (1995) 77-115.

[12] Ho, C. and Case, K.E. Economic Design of Control Charts: A Literature Review for 1981–1991. *Journal of Quality Technology*, **26** No. 1 (1994) 39-53.

[13] Kiesmüller, G. Controlling a Process with Three Different States. *Advances in Stochastic Models for Reliability, Quality and Saftey.* Birkhäuser Verlag, Boston, 1998.

[14] Kiesmüller, G. Most Profitable process control - An Integrated Approach. *Economic Quality Control*, **14** (1999) 5-18.

[15] Kiesmüller, G. Most Profitable Inspection Policies. *Economic Quality Control*, **14** (1999) 65-83.

[16] Kiesmüller, G. Most Profitable Sampling Policies. *Economic Quality Control* **14** (1999) 105-122.

[17] Lorenzen, T.J. and Vance, L. C. Th Economic Design of Control Charts: A Unified Approach. *Technometrics* **28** (1986) 3-10.

[18] Montgomery, D.C. The Economic Design of Control Charts: A Review and Literature Survey. *Journal of Quality Technology*, **12** (1980) 75-87.

[19] Rahim, M.A. Economic Design of $\overline{X}$ Control Charts Assuming Weibull In-Control Times. *Journal of Quality Technology*, **225** (1993) 296-305.

[20] Rahim, M.A. and Banerjee P.K. A Generalized Model for the Economic Design of $\bar{x}$-Control Charts for Production systems with increasing Failure Rate and Early Replacement. *Naval Research Logistics*, **40** (1993) 787-809.

Chapter 13

A SIMPLE MODEL FOR THE EFFECT OF MANUFACTURING PROCESS QUALITY ON PRODUCT RELIABILITY

Michael Tortorella
Bell Laboratories
Lucent Technologies
Holmdel, NJ 07733

Abstract When a product is designed, a certain "ideal" reliability inheres in it. It is called "ideal" because it is the reliability that would be achieved if all post-design product realization processes, such as manufacturing and installation, were perfect. We introduce a framework for studying the quantitative relationship between reliability and the quality of post-design product realization processes. We illustrate these results with an example from a product realization process that adheres to "Six Sigma" goals. The notion of latent defect is used to develop a quantitative model for the damage that could be done by post-design product realization processes to the inherent reliability of a product design. The concepts introduced generalize to reliability quantities Taguchi's notion of continuous quality loss as the output of a process deviates from its target.

Keywords: Process quality, product reliability, latent defect, quality loss, Taguchi's loss function, six sigma goals.

1. INTRODUCTION

When a product is designed, a certain "ideal" reliability inheres in the design, whether through deliberate attention to reliability planning or through neglect. This reliability is called "ideal" because it is the reliability that would be achieved in service if all post-design ("downstream") product realization processes, such as steps in manufacturing or installation, were perfect. It is possible that the ideal reliability might not be very good if, for example, poor or no design-for-reliability practices were followed in the design. Thus, "ideal" refers not to how good is the reliability of the real product (*i.e.*, after manufacture, installation, etc.) but rather to whatever that reliability is intended to be after the design is complete. We introduce a framework for studying the quantitative relationship between reliability, *i.e.*, failures in service, and the quality of downstream product realization processes. The notion of latent defect is used to introduce a quantitative lifetime description associated with the damage that could be done by post-design product realization processes to the inherent reliability of a product design.

We illustrate these results with an example of a product realization process that adheres to "Six Sigma" goals. We find that, even if all downstream processes meet the Six Sigma goals, any latent defects introduced when a downstream process output falls outside its specification limits must still have very long lifetimes (*i.e.*, the time it takes to activate the defect) in order that the contribution of downstream process quality to the number of field failures of the product be kept to some small proportion, say five to ten percent, of the overall field failures. The concepts introduced generalize to reliability quantities Taguchi's notion of continuous quality loss as the output of a process deviates from its target.

Many organizations have adopted the Six Sigma program, first reduced to practice by Motorola [1], as a means of improving product quality and reliability. Six Sigma programs aim at reducing variability (by addressing process quality issues) and widening the product's specification limits (by addressing robust design issues) so that at least twelve (process output) standard deviations fit within the specification limits and the process center is no more than 1.5 standard deviations away from the center of the specification window. The connection between Six Sigma (hereafter abbreviated to 6σ) and product quality, as measured by parts-per-million (PPM) defective, has been made clear in many applications. However, that 6σ improves product reliability is an implicit claim, at best. What is needed to establish this claim in particular cases is a quantitative model to relate 6σ process management to product reliability in quantitative terms. We aim to enable investigation of a variety of questions of which "Suppose all manufacturing processes meet 6σ goals. What then is the reliability of the product?" is a prototype. The model is flexible in the sense that even with little specific information about individual processes, results can be obtained that provide at least some rough-and-ready guidance about the quantities of interest. If more detailed information about particular processes is used, then the results will be correspondingly more precise.

The paper is organized as follows. After this Introduction, there is a section describing the construction of the quantitative model we suggest to aid investigation of these questions. Following that, we describe a very simple (little information input) version of the model to obtain some insight into the following question: "Suppose all manufacturing processes for a particular product meet the 6σ goals. How long does the time before the defect introduced by that process manifests itself have to be in order that the probability that the product fails in its first year of operation because of such a defect be less than some stated value, say 10^{-5} (or so that these defects comprise no more than 10% of all failures)?" We close with some conclusions and related questions.

2. HOW DOES MANUFACTURING PROCESS QUALITY AFFECT PRODUCT RELIABILITY?

While it may seem obvious that better manufacturing processes should lead to greater product reliability, it is not immediately apparent how to express this quantitatively. We have found that a fruitful way to do this is by thinking about the possible long-term effects that manufacturing processes may have on the product. Thus, we focus on the aspect of a manufacturing process that is related to its propensity to introduce defects into the product during execution of the process. For purposes of this discussion, we sort the defects introduced by manufacturing processes into two types: those that are immediately apparent and those that are latent. The immediately apparent defects cause quality problems like in-circuit or functional test failures, system burn-in failures, or others that manifest themselves before the product gets into the customer's hands. They may also cause "dead on arrivals" that are visible to the customer if they are not caught first by an installer. In examining the connection between manufacturing process quality and product reliability, we need focus only on the latent defects, since those are the defects that cause failures *after* the initial product installation.

To clarify this approach, we have found it useful to consider an analogy between product reliability and the sharpness of a photograph, which we will now discuss at length. When a product is designed (including specification of the components to be used in the product), a certain degree of reliability inheres in the design, based on the components used and their interrelationships, the stresses on them, the architecture and operation of the product, and so on. This inherent reliability can never be realized in practice *precisely because of the imperfect nature of manufacturing and other "downstream" product realization processes* that we are trying to capture in this model. Similarly, when one takes a photograph, whether it is in focus is determined by the relative positions of the subject, the lens, and the film. Once the photograph is taken, these cannot change (leave aside for purposes of this analogy the manipulations made possible by digital photography). In this analogy, the manufacturing process for the product is like the process of developing the photograph. If the negative is out of focus, nothing you do when developing the photograph can bring it into focus. Conversely, if the negative is in focus, then the process of developing it, if imperfect, could render the print fuzzy. Similarly, once a product is designed with a certain inherent or ideal reliability, nothing you do in manufacturing can make the product more reliable than this. However, a manufacturing process that introduces latent defects will cause the realized reliability of the actual product to be worse than its inherent or ideal reliability. This analogy helps us understand the assertion that downstream product realization processes cannot but decrease the reliability inherent in a product's design.

Using the above considerations as our starting point, we focus on the manufacturing processes that impinge upon the product. Assume that the upper and lower specification limits for a process are a^U and a^L, respectively, that the center of the process's specification window is $a^0 = (a^U + a^L)/2$ (which we also assume to be the target of the process output), and that the process output is a random variable A that is normally distributed with mean a^* and variance σ^2. With this notation, we express that this process meets the 6σ goals as described above if there is an m, $4.5 \leq m \leq 7.5$, for which

$$a^L \leq a^* - m\sigma = a^* + (12 - m)\sigma \leq a^U. \tag{13.1}$$

If $m = 6$, then the process is centered, and the expected fraction of process outputs falling outside the specification window (outputs that we shall henceforth refer to as "defects") is approximately 9.87×10^{-10}, or about one part per billion (PPB). The expected fraction defective is largest when the process center is as far away from the center of the specification window as possible. The maximum distance allowed by the 6σ methodology is 1.5 standard deviations, corresponding to $m = 4.5$ (left of center) or $m = 7.5$ (right of center) in (13.1). In this case of maximum deviation, the expected fraction defective is approximately 3.3977×10^{-6}, or about 3.4 parts per million (PPM).

To introduce reliability, let us focus now on those processes that introduce latent defects. The model is based on the fundamental idea that a defect introduced by a process "fires" at a random time $X(a)$ when the output of the process is a, and that when the defect fires, the product fails. We will use $F(x, a)$ to denote the conditional distribution of the firing time (or lifetime, or time to failure) of the defect given that the output of the process is a:

$$F(x, a) = P\{X \leq x | A = a\}. \tag{13.2}$$

For now, the only assumption we will make about F is that $F(x, a) \leq F(x, a^{'})$ whenever $|a - a^0| \leq |a^{'} - a^0|$. That is, the time before the defect fires does not decrease as the output of the process moves closer to its target. This expresses the common-sense idea that if a process output is further away from the process center than another process output, any defect introduced by the output further away will fire sooner than any defect introduced by the output further in. This simple assumption can accommodate many different shapes of "loss functions" that may be used to describe in more detail how the defects are affected by the process location. We may view this as a concept of loss function that is a direct analogue for latent defects of Taguchi's loss functions used in the study of process quality.

Suppose the total number of relevant manufacturing and other downstream product realization processes is M and that the mean and standard deviation of the output of process j are a_j^* and σ_j^2, respectively, $j = 1, ..., M$. Let $\mathbf{a}^* = (a_1^*, ..., a_M^*)$ be the vector of process output means and $\sigma = (\sigma_1, ..., \sigma_M)$ be the vector of process output standard deviations. In a given product lot, these M processes contribute altogether, say, N opportunities for a latent defect to be introduced. Usually, $N > M$. For instance, in the manufacture of a lot of 200 circuit packs or printed wiring boards containing 10,000 solder joints on each board, one process (wave soldering) has 2×10^6 opportunities to introduce a defect into the lot. The time at which an item chosen at random from that lot fails is thus

$$Z_N(\mathbf{a}^*, \sigma) = min\{Y_1, ..., Y_N\}, \tag{13.3}$$

where, with φ denoting the standard normal density,

$$\Psi_i(y) := P\{Y_i \leq y\} = \int_{-\infty}^{\infty} F_i(y, a)\varphi\left(\frac{a - a_i^*}{\sigma_i}\right) da, i = 1, ..., N \tag{13.4}$$

is the distribution of the firing time Y_i of (latent) defect i. If the defect firing times are stochastically independent, then we may write

$$P\{Z_N > z\} = \prod_{i=1}^{N} P\{Y_i > z\} = \prod_{i=1}^{N} (1 - \Psi_i(z)). \qquad (13.5)$$

We should note here that, while stochastic independence may be reasonable for the lifetimes of defects arising from different processes, it seems less reasonable for those arising from the same process. This seems especially relevant when the outputs of a process are themselves correlated. In such cases, more descriptive models for the minimum in (13.3) are required.

Our goal is to say something about the distribution of Z_N. When defect lifetimes are stochastically independent, using the Maclaurin series for $log(1-x)$ we can show (see Appendix) that

$$\exp\left\{-\frac{1}{2}\sum_{i=1}^{N}\left[\int_{-\infty}^{\infty} F_i(z,a)\varphi\left(\frac{a-a_i^*}{\sigma_i}\right)\right]^2\right\} \le$$

$$\frac{P\{Z_N(\mathbf{a}^*,\sigma) > z\}}{\exp\left\{-\frac{1}{2}\sum_{i=1}^{N}\int_{-\infty}^{\infty} F_i(z,a)\varphi\left(\frac{a-a_i^*}{\sigma_i}\right)\right\}} \le 1. \qquad (13.6)$$

Equation (13.6) shows that an approximate complementary distribution for Z_N (namely, the numerator in the central term of (13.6)) can be obtained with a (multiplicative) error estimate given by the two inequalities in (13.6). These inequalities should be tight when the downstream processes meet the 6σ goals because in that case the total mass of the normal distribution of process outputs outside the specification interval does not exceed 3.4×10^{-6}, as in (13.1), above.

3. AN ILLUSTRATION

In this section, we give some insight into the numbers typical of such situations by addressing the question, "When all manufacturing processes meet the 6σ goals, how long do defect lifetimes have to be in order that the probability that the product fails in the first year of its operation because of a manufacturing defect be not larger than 10^{-5}?" To do this, we will have to make very simple assumptions about the conditional distributions F_i because this analysis is not geared to any particular product but is intended rather to give a broad illustration.

The first assumption is as follows: If a process output is within the specification window, then the firing time, or lifetime, of any defect introduced by that process is taken to be infinite (really, any number larger than the intended service life for the product would do as well). That is, if a process output is within specification limits, then *no* defect is introduced. Further, if a process output falls outside the specification limits, then a defect is introduced whose firing time distribution does not depend on the particular value of the process output (that is, labeling the defect i, $F_i(x,a)$ is independent of a whenever a is outside the specification limits). Finally, we further assume that each process just barely meets the 6σ goal. This occurs when $m = 4.5$ or $m = 7.5$ in (13.1), and corresponds to an expected fraction $p = 3.4 \times 10^{-6}$ of process outputs falling outside the specification limits. Choosing $z = 1$ year and denoting F_i (13.1) by f_i, (13.5) and (13.6) lead to

$$\exp\left[-\frac{1}{2}p^2\sum_{i=1}^{N} f_i^2\right] \leq \frac{P\{Z_N > 1\}}{\exp\left[-p\sum_{i=1}^{N} f_i\right]} \leq 1. \tag{13.7}$$

Taking all f_i to be the same (clearly an unrealistic assumption but one that we nonetheless use to make progress in the absence of more specific information about the various defect types), we use (13.7) to tabulate, for several values of N, the value of the defect firing time distribution at one year needed for the product probability of failure at one year not to exceed 10^{-5} (see Table 13.1 , column 2). Use of (13.7) leads to bounds on this probability, but in this case the bounds are the same to five decimal places so we display in Table 13.1 only the single probability estimate. The third column of Table 13.1 lists the hazard rate in FITs (one FIT is one failure per 10^{-9} unit-hours) of the defect lifetime distribution, *assuming that it is exponential, i.e., has a constant hazard rate*, and the fourth column lists the mean time to defect firing (MTTF), again under the assumption that the defect firing time distribution is exponential.

One of the observations we are led to from these results is that the assumption that defect lifetimes are independent and identically distributed (*iid*) is probably not reasonable. Using the example of the lot of 200 circuit packs manufactured, there are in that lot 2,000,000 defect opportunities from the wave soldering process. Under the *iid* and exponential distribution assumptions, the mean life of each solder joint needs to be more than 617,000 years in order to meet the reliability requirement stated. Equivalently, the probability that each solder joint survives 25 years is one to at least 9 decimal places. Few would believe that wave solder processes are quite so capable. So what must be happening is that most of the time the wave solder process is essentially perfect (in the sense that it produces nothing but joints that with probability 1 last beyond the end of the user-defined service life of the product), but that it occasionally goes out of control and begins introducing latent defects

Table 13.1 Example

No. of Opportunities	$f = f_1 = F(1)$	**FITs**	**MTTF (hours)**
500	0.00588	670	1,492,537
1000	0.00294	335	2,985,074
2500	0.00118	134	7,462,686
5000	0.00059	67	14,925,373
10000	0.00029	33	30,303,030
2000000	0.00000162	0.18	5,410,000,000

(e.g., cold solder joints), some of which may fire before the user-defined service life is over. It would be interesting to extend the model described here to incorporate non-stationary behavior such as this.

For the record, the "loss function" used in this illustration is very similar to a pre-Taguchi process quality loss function: any output within the specification window is acceptable, and any output outside the specification window is defective. Here is another example that is closer to Taguchi's idea that the loss increases continuously as the process output moves further from its center. We take the conditional distributions (13.2) to be

$$F_i(x, a) = 1 - \exp\left[-\lambda_i(|a - a_i^0| + 1)x\right]. \qquad (13.8)$$

In this case, the firing time distribution is exponential with a "base" hazard rate λ_i when the process output is centered. As the process output moves further away from its center, the hazard rate increases (or, the expected time to defect firing decreases). The unconditional distribution (13.4) for the firing time of defect i is then

$$P\{Y_i \le y\} = 1 - \sqrt{\frac{2}{\pi}} \exp\left(-\lambda_i x + \frac{(\lambda_i \sigma x)^2}{2}\right) \Phi(-\lambda_i \sigma x), \qquad (13.9)$$

where Φ is the standard normal integral. Further development of this example into a complete illustration would require knowledge of the particular λ_i for each process.

4. CONCLUSIONS AND RELATED QUESTIONS

We have attempted to make a quantitative connection between downstream product realization process quality and product reliability, thereby providing a foundation for analysis of data collected to support the common engineering understanding of how better manufacturing quality can lead to better product reliability. We have used the notion of latent defect introduced during post-design processes to make a quantitative connection between product reliability and downstream process quality. The general results introduced here can be used as a foundation for more detailed studies for specific product types. Several important directions for extending the model to accommodate more realistic features of manufacturing are indicated. Overall, however, the conclusion we are led to is that even if the inherent reliability of the design is high, the quality of the manufacturing and other downstream product realization processes-even those meeting 6σ goals-must be extremely high in order not to compromise the inherent reliability of the design by introducing latent defects that will cause the product to fail in the customer's hands.

References

[1] Harry, M. J. *The Nature of Six-Sigma Quality*, Rolling Meadows, IL: Motorola University Press, 1988.

APPENDIX

In this Appendix, we prove the inequality in equation (13.6).

Begin with equation (13.5).

$$\begin{aligned} P\{Z_N > z\} &= \prod_{i=1}^{N}(1-\Psi_i(z)) = \exp\left[\sum_{i=1}^{N}\ln(1-\Psi_i(z))\right] \\ &= \left[\exp\left(-\sum_{i=1}^{N}\Psi_i(z)\right)\right]\left[\exp\left(\sum_{i=1}^{N}[\Psi_i(z)\right.\right. \\ &\quad \left.\left. + \ln(1-\Psi_i(z))]\right)\right] \end{aligned}$$

By the Maclaurin series for $ln(1-x)$, we have

$$x + \ln(1-x) = -\frac{x^2}{2} - \frac{x^3}{3} - \cdots, 0 \le x < 1,$$

which leads to the inequality

$$-\frac{\Psi_i(z)^2}{2} \leq \Psi_i(z) + \ln(1 - \Psi_i(z)) \leq 0.$$

From this it follows that

$$\exp\left[-\left(\frac{1}{2}\sum_{i=1}^{N}\Psi_i(z)^2\right)\right] \leq \exp\left(\sum_{i=1}^{N}[\Psi_i(z) + \ln(1 - \Psi_i(z))]\right) \leq 1,$$

and (13.6) follows.

V

WARRANTY, MANUFACTURING & QUALITY

Chapter 14

WARRANTY AND MANUFACTURING

K.F. Lyons and D.N.P. Murthy
Department of Mechanical Engineering,
The University of Queensland
Brisbane Q 4072, Australia

Abstract Warranty is becoming an increasingly important element in the marketing of a new product and in the commercial success of a manufacturing business. Offering warranty results in additional costs incurred by the manufacturer and these costs, which depend on product reliability and warranty terms, are determined by management decisions made prior to the release of the product. This implies that warranty can be viewed as a link to integrate the different stages of manufacturing - design, engineering, production, marketing, and post sale service - in an effective manner. This chapter develops a warranty management system to assist in decision making at the different stages and discusses the elements of the system and the management of information.

Keywords: Warranty, Design, Engineering, Manufacturing, Marketing, Post Sale Servicing, Warranty Management System, Models, Data Management.

1. INTRODUCTION

In the decision to purchase a product, buyers typically compare characteristics of similar models of competing brands. When competing brands are nearly identical, it is very difficult, in many instances, to choose a particular product solely on the basis of product-related characteristics such as product price, special features, perceived product quality and reliability, financing offered by the manufacturer and so on. In such situations, post-sale factors – warranty, parts availability and cost, service, maintenance, and so forth – take on added importance in product choice (see Lele and Karmarkar (1983), Ives and Vitale (1988), and

Ritchken, *et al.* (1989)). Of these factors, the warranty is one that is known (or at least potentially known) to the buyer at the time of purchase.

Warranties play an important role in providing product assurance to customers. Each new generation of product is more complex than the earlier generation it replaces, and often customers are uncertain about product performance. Different types of warranties are offered depending on the product and buyer.

Protection of consumer interests has become an important issue in all industrialised societies. In the USA, the Magnuson-Moss Warranty Act was legislated in the early sixties, and the EU recently passed new legislation that requires all products sold in the EU to have a two-year warranty.

The notion of post-sale support is becoming an important feature of any product sale. In this context, warranty (and extended warranty) is an element of post-sale support, and manufacturing businesses need to view it as part of the post-sale service strategy.

Warranty has a significant impact on the total costs for a manufacturing business. A warranty of any type, since it involves an additional service associated with a product, has potential costs beyond those associated with the design, manufacture, and sale of the product. These costs, in fact, are unpredictable future costs, which typically range from 2% to as much as 15% of net sales (McGuire (1980)).

Warranty has a significant impact on commercial aspects such as product promotion, sales, total costs, profits, etc. Technical issues during the design and production stages (such as reliability development, quality control, testing and so on) have an impact on warranty costs. This implies that warranty can be viewed as a link to integrate the different stages of manufacturing (design, engineering, production, marketing and post-sale support) in an effective manner. Warranty decisions need to be linked to decision making at other stages in manufacturing. Murthy and Blischke (2000) discuss warranty strategy as part of the overall business strategy for a new product and how it needs to be linked to other functional strategies in the different stages of manufacturing.

In this chapter, we develop a warranty management system to assist in making warranty-related decisions that are effectively linked to management making at other stages of manufacturing. This is achieved by using a framework comprised of four modules: (i) Design and Engineering, (ii) Production, (iii) Marketing and (iv) Post-sale Servicing. Each module involves several models to assist in decision making. The model building involves data exchange between modules. The chapter discusses the models and data flow issues.

The outline of the chapter is as follows. In Section 2, we provide a brief overview of product warranty. Section 3 deals with an integrated framework that shows the links with warranty and other elements of manufacturing. Section 4 is a discussion of the structure for the warranty management system and the four modules. A brief discussion of the models and some data issues are presented in Section 5. We conclude with a brief summary and suggest some topics for future research in Section 6.

2. PRODUCT WARRANTY: AN OVERVIEW

2.1 Concept and Role. A warranty is a manufacturer's assurance to a buyer that a product or service is or shall be as represented. It may be considered to be a contractual agreement between the buyer and manufacturer entered into upon the sale of the product or service. A warranty may be implicit or it may be explicitly stated. In broad terms, the purpose of a warranty is to establish liability among the two parties (manufacturer and buyer) in the event that an item fails. An item is said to fail when it is unable to perform satisfactorily its intended function when properly used. The contract specifies both the performance that is to be expected and the redress available to the buyer if a failure occurs.

The normal warranty is an integral part of the product sale and is factored into the sale price. In recent years, manufacturers and third parties (dealers, financial institutions in the credit business) have been offering extended warranties. These can be viewed as maintenance service contracts that the buyer can choose by paying an extra amount.

2.2 Product Categories. Products can be divided into two categories: (i) new and (ii) used. New products can be divided into the following three categories:

1. Consumer durables, such as household appliances, cars and personal computers bought by individual households as a single item, encompass the first category.

2. Industrial and commercial products bought by businesses for the provision of services, such as equipment used in a hospital to provide medical care, aircraft used by airlines, or products such as computers or copy machines, comprise the second category. These are bought either individually (for example, a single x-ray machine bought by a hospital) or as a batch, or lot, of L ($L > 1$) items (for example, batteries bought by a car manufacturer, a fleet of trucks bought by a car rental agency). We need to differentiate "standard" off-the-shelf products from "custom-built" products built

to buyer's specifications. We shall call the latter category "specialised" products.

3. Government acquisitions, such as a new fleet of tanks or jet fighters, make up the third category of new products. These are usually "custom built" and are products involving new and evolving technologies. As such, they are characterized by a high degree of uncertainty in the product development process. (Note: A government is also a large buyer of "standard" industrial and commercial products.) Used products are, in general, sold individually and can be either consumer durables, industrial, or commercial products.

2.3 Warranty Policies. Both new and used products are sold with many different types of warranties. Blischke and Murthy (1992) developed a taxonomy for new product warranty and Murthy and Chattophadhyay (1999) developed a similar taxonomy for used products. Armistead and Clark (1991) developed a framework for formulating a post-sale support strategy. O'Leary (1994) provided a classification method for warranties and a cost model for tracking warranty costs.

3. WARRANTY AS AN INTEGRATIVE ELEMENT IN MANUFACTURING

Warranty is affected by technical and commercial issues. The technical issues concern the performance of the product and relate to design, engineering, and manufacturing. The commercial issues relate to the price and warranty terms and are concerned with marketing and post-sale support.

3.1 Technical Issues. The main technical issues relating to warranty and product development are the measurement of product performance (reliability) and performance improvement (reliability growth). Reliability affects warranty costs and is determined by management decisions early in the development process. During the pre-design stage, product features such as performance specifications (reliability), unit cost, warranty terms, and warranty costs (as % of sale price) are defined.

Measurement and modelling of reliability allows the estimation of the number of failures during the warranty period, and hence the estimation of the expected warranty costs.

The manufacturing stage affects the conformity of the product to the design. As the design is developed, specifications may be altered based on manufacturing factors such as manufacturability, modularity, and redundancy. The main technical issues relating to warranty and manufacture are process variability, process control, testing, and inspection.

3.2 Commercial Issues. The sale price and the warranty terms affect the level of sales. Increasing sale price typically decreases the demand for the product. Better warranty terms imply that the product is more reliable and can lead to greater sales.

Better warranty terms, such as longer warranty length, lead to more claims and greater warranty costs. To be profitable, the product's sale price must generate enough revenue to more than offset all of the product's costs; therefore, warranty costs should be considered when setting the sale price. This consideration requires an estimate of the warranty costs prior to product launch.

3.3 Interactions. The technical and commercial warranty-related issues (including product reliability and improvement, sale price, demand, warranty terms and costs, and profits) interact with one another, as shown in Figure 14.1.

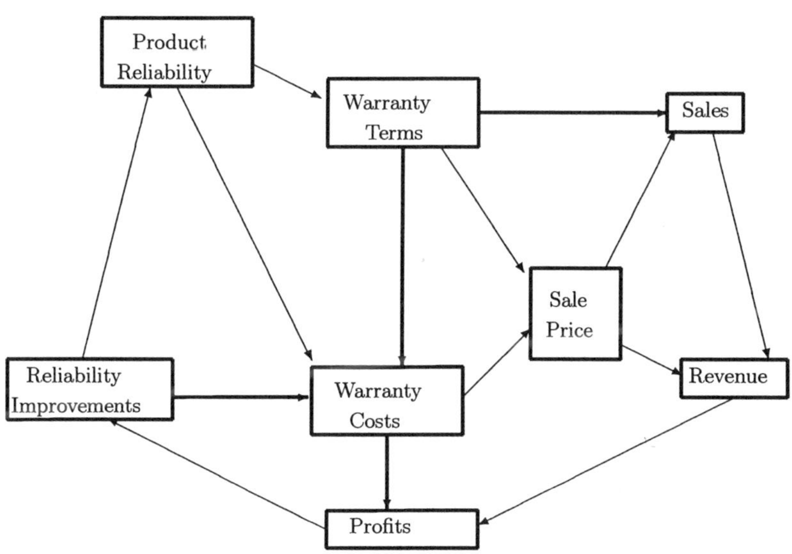

Figure 14.1 Interactions between warranty related issues (from Murthy and Blischke (1994)).

Blischke and Murthy (2000) deal with a strategic warranty management approach that takes into account these interactions.

4. WARRANTY MANAGEMENT SYSTEM

The purpose of a warranty management system is to evaluate trade-offs between competing objectives or variables by using models and to provide information to management for the development of functional and operational strategies. Data that is relevant to a warranty is extracted from the business operations, organised, analysed, and used to assist managers in decision making. Examples of these objectives or variables include technical variables such as reliability, quality, testing, etc. and commercial variables such as sale price, warranty type, warranty length, warranty costs, and manufacturing (unit) cost.

The decision maker can use this information to propose warranty terms and then evaluate the costs or to specify the cost objectives and then determine the optimal warranty terms. The warranty management system uses models to estimate the value of the objective (warranty cost) from the decision variables and presents this to the decision maker. This may be in the form of expected results of proposed actions (e.g. increasing reliability will reduce expected warranty costs by a certain amount) or optimal values for decision variables (e.g. to reduce expected warranty costs by X, increase reliability to Y). These models are grouped into four modules, Design & Engineering (D&E), Production, Marketing, and Post-Sale Servicing, that interact as shown in Figure 14.2.

The purpose of the D&E module is to measure, improve, and control product design, performance, and developmental costs. The purpose of the Production module is to measure, improve, and control manufacturing quality and costs. The purpose of the Marketing module is to manage market information, risks, and costs. The purpose of the Post-Sale Servicing module is to measure, improve, and control product field performance, warranty service, and warranty costs.

While warranty involves many decisions, the most important to the financial success of a product is the total warranty cost, the total amount expended on warranty claims throughout the product line's entire life. Total warranty cost is dependent on information and decisions from each of the four modules as indicated in Figure 14.3. The accuracy of the estimation of this cost depends on the quality, quantity, and timeliness of data gathered from operational processes.

The total warranty cost is dependent on the warranty cost for each item and the number of items sold (demand) over the lifetime of the product line.

The warranty cost per item is in turn affected by the number of claims made per item during the warranty period and the cost incurred for each claim. The warranty terms influence the number of claims that are made

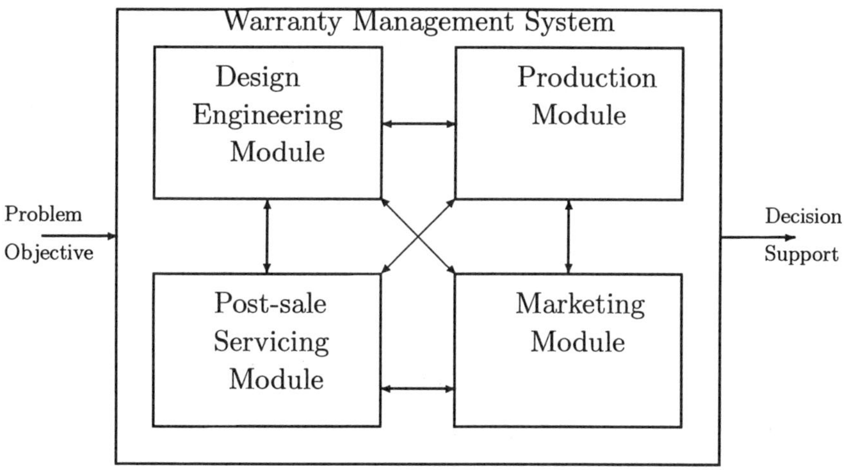

Figure 14.2 Warranty Management System Modules

and determine the cost per claim. Not all claims may be valid, but even invalid claims result in administration costs for the manufacturer.

The number of claims made during the warranty period is influenced by the performance characteristics of the manufactured item (as measured by the expected number of failures), the claim execution patterns of the customers, and the warranty terms (specifically the type and length). The number of failures occurring within the warranty period is estimated based on consideration of the design, the manufacturing process quality, the conditions of use, and the quality of the repair. The design and manufacturing process determines the inherent reliability of the items produced, and the usage patterns affect how well the items actually perform. The repair quality affects how the products perform after repairs have been made.

The following subsections discuss each module of the warranty management system and the flow of activities in each module.

4.1 Design and Engineering Module. The purpose of the D&E Module is to measure, improve, and control product performance and development costs by identifying areas where improvements can be made, making changes, and repeating the process. Estimating

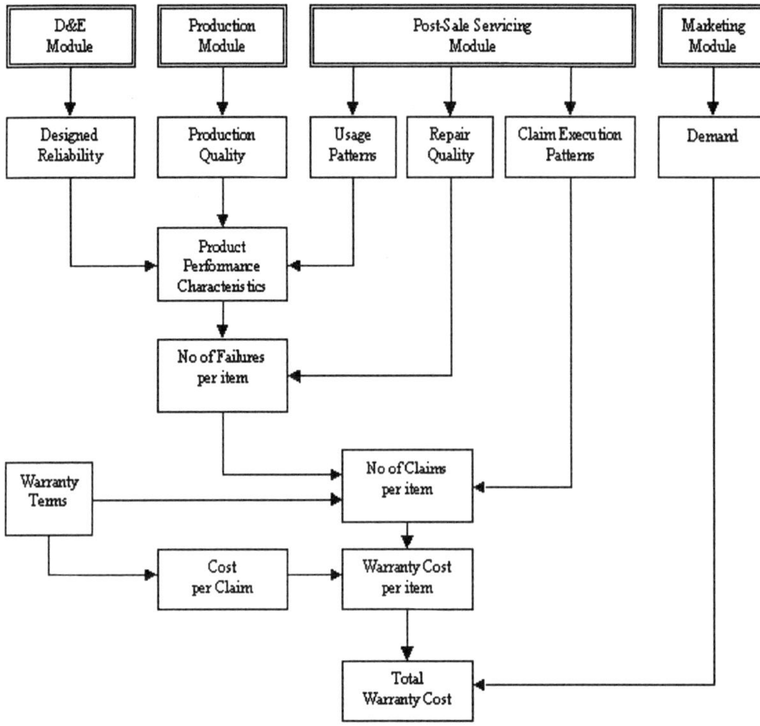

Figure 14.3 Characterisation of the Total Warranty Cost

the reliability throughout the process allows the estimation of indicators such as the expected number of failures. This has implications for warranty since it is this indicator that affects the warranty cost.

Forecasting and budgeting for product development involves estimating the number of development cycles (and the associated costs) required to reach a reliability target or estimating the reliability that will be achieved at the end of a specified number of cycles.

Figure 14.4 illustrates some of the possible inputs and outputs to the D&E Module, and some of the types of models that make up the module. D&E models include failure/reliability models, development time (reliability growth) models, and development cost models. Inputs to the module include the product design and the reliability target. Using these models and inputs, a variety of outputs can be obtained including the expected time to failure, system reliability, the expected time to achieve target reliability, and the expected cost to achieve reliability target.

D&E involves a series of activities, Product Design, Design Analysis, Research and Development, Prototype Construction and Prototype

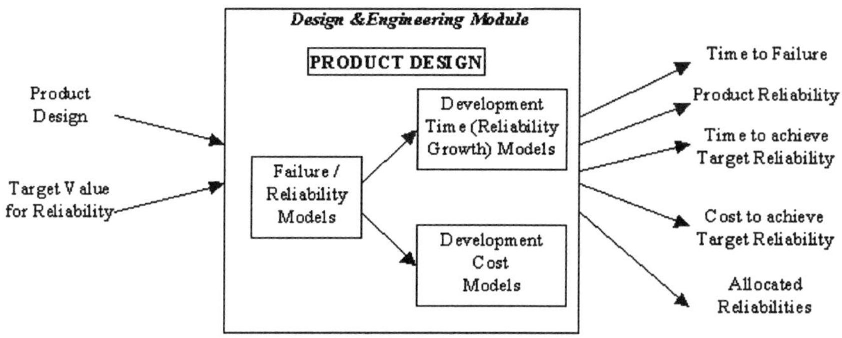

Figure 14.4 D&E Module

Testing, to produce a feasible design for a product that meets specifications, as shown in Figure 14.5.

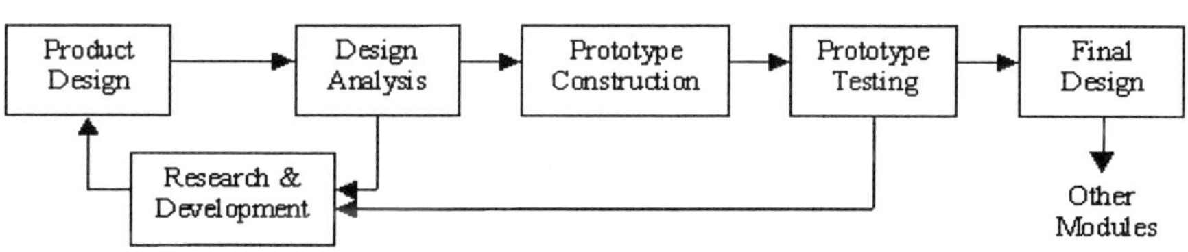

Figure 14.5 Flow of Activities in the D&E Module

At the product design stage, functional specifications, manufacturing process considerations, aesthetic guidelines, and repair considerations are used to develop one or more design options. Product planning methods such as Quality Function Deployment (QFD) provide a structured approach to weighting and ranking customer needs and benefits, and technical characteristics (Cohen (1995)). From a modeling perspective, input variables may include targets for the development cost, development duration, reliability, and repair cost.

Design analyses are performed to identify failure modes for improvement, to estimate reliability, and to select the best choice from a range of potential designs. To identify failure modes, fault tree analysis (FTA), failure mode and effect analysis (FMEA), failure mode effects and criticality analysis (FMECA), and event tree analysis can be performed (Fussel (1976), IEEE Std.352, MIL-STD882).

The failure modes identified should correspond to field failure modes for which the cause is repeatable because a repeatable cause can be accurately tested for, identified, and eliminated.

The design is used to develop reliability block diagrams and 'structure functions'; from these, system reliability can be estimated. Reliability block diagrams are networks that describe the function of a system by reducing its design to combinations of series and parallel connections. A structure function is a mathematical expression that leaves out components that do not have a direct effect on the product's ability to function (Hoyland and Rausand (1994)). A binary variable is used to indicate whether a component is functioning or not. Incorporating component reliability estimates into the structure function results in an estimate of the entire system's reliability. The structure function can be used as a decision tool to allocate reliability to critical components to increase system reliability (see Blischke and Murthy (2000)).

Design analyses can also be performed to estimate the development time required to reach target reliability, to estimate the reliability at the end of a specified development time, or to estimate the total development cost for each of those cases.

Information is fed back to the design stage to improve the design and manufacturing process, and the cycle repeats. Once a design is selected and analysed with satisfactory results, a prototype is constructed and tested to yield data on the types of failure mode that occur, their causes, and (to a limited extent) their frequency. Estimates of the reliability are made. As the product is improved through successive development cycles, analysis becomes more difficult since less test data is available for analysis because tests take more time (better products fail less often).

At the conclusion of the development process, a final design is produced for manufacturing.

4.2 Production Module. The purpose of the production module is to measure, improve, and control manufacturing quality and costs. Models are used to better understand and plan for the effect that production has on product performance and warranty costs. Variability in the manufacturing process affects the conformance of the final product with the design reliability. Warranty servicing costs depend on the number of items that do not conform to the design specifications. These non-conforming or defective items result in higher warranty costs than conforming or non-defective items.

Figure 14.6 illustrates some of the inputs and outputs to the Production Module and some of the types of models that make up the module. These models include production cost models that are used to understand and estimate the cost of producing items, production quality models that describe the reliability of outgoing items, and plant maintenance models that are used to monitor and optimise maintenance schedules and the quality of the production facility. Inputs include target values for process quality, product design, process design, and production strategies. Outputs include outgoing reliability, production costs and times, and process quality.

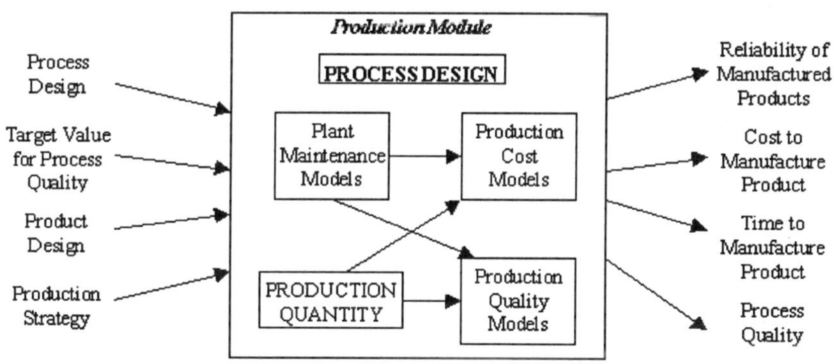

Figure 14.6 Production Module

The production module has three main functions, as shown in Figure 14.7: to produce products, to inspect, test, and control product conformity to specifications, and to estimate the performance characteristics of the outgoing items.

The first function involves designing the production process and producing units. The production equipment must be set up and calibrated. At the end of each production run, the process is checked to determine whether it is in control or out of control, and reset if necessary.

The second function of the production module is to inspect and test items to better control their conformity. Process control tools such as

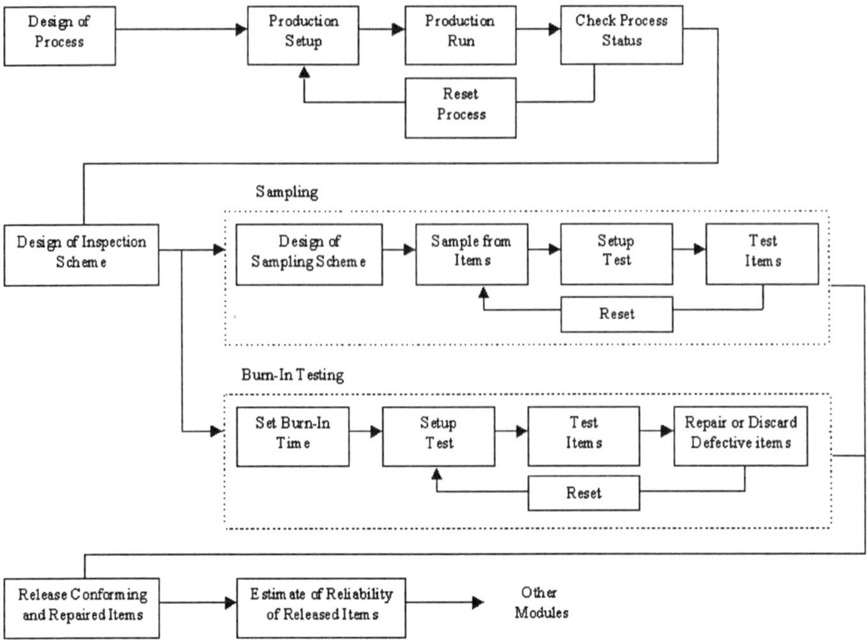

Figure 14.7 Flow of Activities in the Production Module

control charts can be used to monitor and maintain the process. The link to warranty, however, is the performance of released items. Estimation of this characteristic requires testing. Two types of tests that can be performed, Life Testing, and Burn-In or Infant Mortality Testing, identify different types of product failure. Life testing involves running a sample of items until a failure occurs, in order to estimate reliability and to identify the types of failure mode that can be expected. Burn-In Testing involves running all items to identify and weed-out any early "process-related" faults.

Life tests are necessary to estimate reliability; this information is used by the post-sale servicing module to estimate warranty costs. The sampling scheme defines how items are selected for testing since life testing is typically destructive, and the size of the sample influences the accuracy of the estimates made and the cost of the testing process. For continuous production, the sampling scheme may require selecting items at a pre-specified frequency (e.g. every 50th item) and adjusting the frequency and type of test based on the results. For lot production, a specified number of items are selected from each lot and tested. Depending on the number of successes and failures, successive samples may be required. An optimal sample size achieves a trade-off between accuracy and costs.

Burn-In Testing reduces the number of infant-mortality failures by testing all products for a specified duration before distribution. Burn-In often results in testing and scrap costs to the manufacturer.

Items that fail the tests can either be repaired or scrapped. Items subjected to life testing and subsequently released (regardless of whether they passed, or failed and were repaired) have had part of their life used up compared to items that have not been tested. Information gathered through testing and analysis can then be used to further improve the production process.

4.3 Marketing Module. The purpose of the marketing module is to manage market information, risk, and costs. Prior to product launch, the decision maker sets targets for price and warranty terms. To help select values that are achievable and satisfy certain criteria, models are an important and useful tool. In selecting the price and warranty terms, for example, the decision criterion might be to minimise the risk to the consumer. In the case of a product with negotiated price and warranty terms, the decision criteria could be to minimise the consumer's and producer's risks. Another criteria could be to achieve a certain level of sales or to maximise (or meet) a certain market share. Other criteria could involve costs, revenue, and profits.

Figure 14.8 illustrates some of the inputs and outputs to the Marketing Module and some of the models that make up the module. These models include risk models for measuring risks associated with the product, demand or sales models for estimating the numbers of items that will be sold, marketing cost models for estimating promotion, advertising and market research costs, market share models for evaluating market penetration, and revenue and profit models for commercial appraisal. Inputs include market information, the new product strategy, and targets for objectives. Outputs include price, reliability and cost targets, warranty terms, and sales forecasts.

The two main influences a warranty has on marketing are the reduction of risk to the consumer and the increase in demand due to the signalling of reliability. Purchasing a product with a warranty reduces, for the consumer, the risk of additional costs that result from item failure. As a result, better warranties lead to higher demand and sales. Estimates of demand lead to estimates of market share, revenue, and profits, based on the design of the warranty policy.

The marketing module is comprised of a number of stages as illustrated in Figure 14.9. Identification of product features requires knowledge of the types of products and features that can be designed by the organisation's resources, manufacturing information regarding the pro-

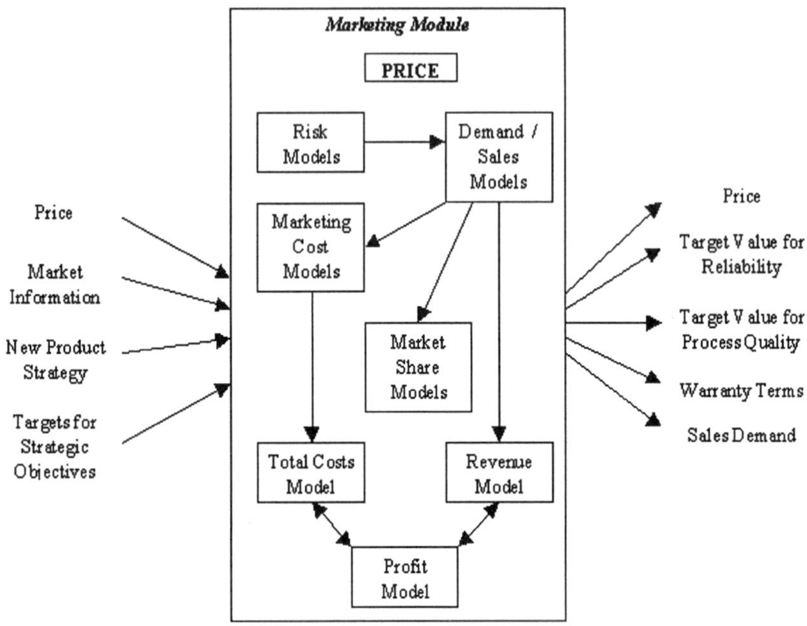

Figure 14.8 Marketing Module

duction facilities and plant equipment, and post-sale servicing information about types of warranties that have been offered by the organisation and the location and capacity of servicing facilities.

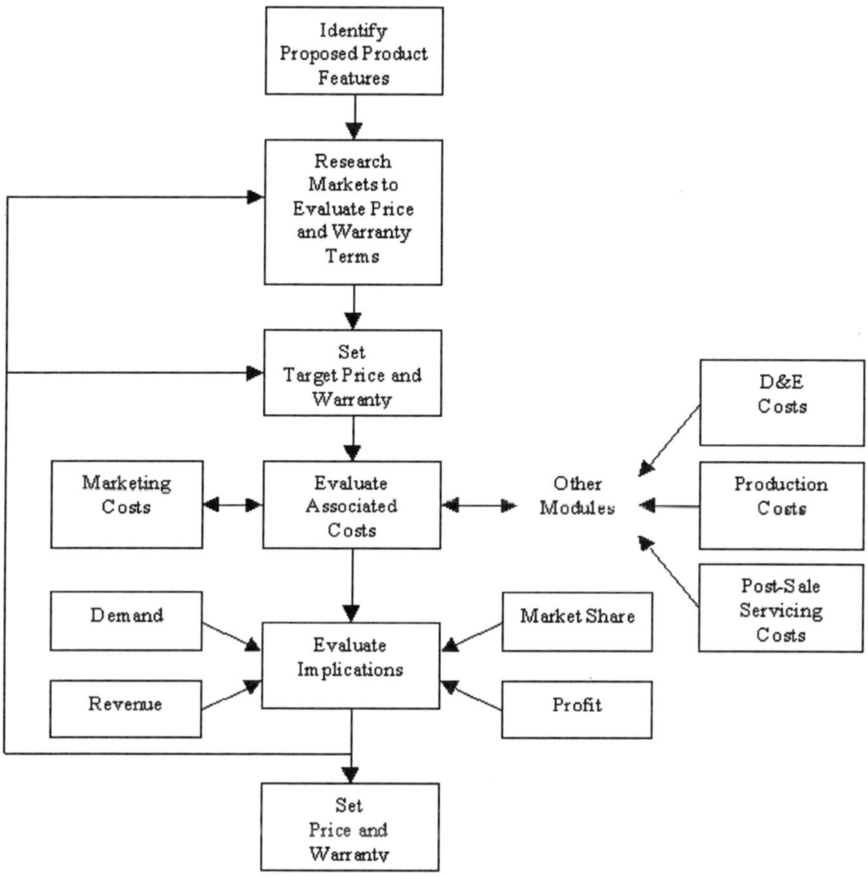

Figure 14.9 Flow of Activities in the Marketing Module

A preliminary outline of a new product's features provides the basis for the market research effort. These features are used to research the marketplace to estimate realistic limits for price and warranty terms. For a given product line, initial target values for price and warranty are set within these limits which are dictated by the market, but the actual price/warranty values that are eventually chosen are determined by the design, production, and servicing capabilities of the organisation.

An evaluation of the costs associated with these target values involves estimating the costs required to achieve the respective D&E, Production, and Post-Sale Servicing targets, as well as the Marketing costs associated with research, promotion, and advertising. Market research costs should be considered since they affect the accuracy of the price and warranty limits.

An evaluation of the marketing implications of these limits is also necessary to ensure that demand, market share, revenue, and profit objectives are met. If these targets are not met then further market research can be undertaken to ascertain what features, price, and warranty are desired, or the initial price and warranty targets can be adjusted to achieve the strategic objectives.

These evaluations progress throughout the pre-launch stage of the life cycle. Firm target values should be reached early on in the development cycle to allow other modules to take appropriate actions to meet the targets. When the launch stage begins, it becomes necessary to set the final price and warranty values. However, before these are set, the latest information from each of the other modules should be used in the cost and implication evaluations.

4.4 Post-Sale Servicing Module. The purpose of the Post-Sale Servicing module is to measure, improve, and control product performance, warranty service, and warranty costs. The cost implications of different warranty terms can be assessed and fed back throughout the development cycle. Following product release, Post-Sale Servicing acquires repair data that can be used to improve future designs or determine if other changes (e.g. to price/warranty terms) are needed.

Figure 14.10 illustrates some of the inputs and outputs to the Post-Sale Servicing Module and some of the types of models that make up the module. The module contains models for warranty costs, repair quality, warranty reserves for replacement, repair and refund warranty policies, repair limits, warranty execution, and product distribution. Inputs include the warranty terms, the reliability of products, and the usage and claim execution patterns. Outputs include expected warranty costs, spare parts requirements, and warranty funds requirements.

The Post-Sale Servicing Module is comprised of a number of stages illustrated in Figure 14.11. These stages include receiving claims, verifying their validity, rectifying problems and determining the cost of rectification, outputting reports, and updating systems.

When a claim is made, its validity must be ascertained. Typically, this requires determining the age of the item and identifying the failure modes. A tracking system allows details such as serial number or lot number to be used to identify the sale date, from which the age can be determined. If the claim is valid, then repairs are made at the manufacturer's (or agent's) expense. If the claim is invalid (e.g. due

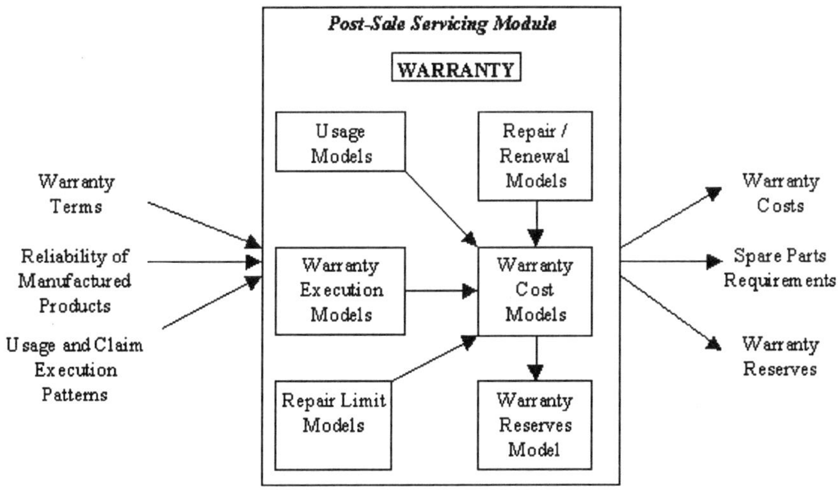

Figure 14.10 Post Sale Servicing Module

to violation of warranty terms), then repair at the customer's expense must be approved. If this is acceptable, the repair is made.

The cost of rectifying the problem should be estimated regardless of whether it is at the customer's or producer's expense since this data can then be used for product and process improvement, comparison of estimates with actuals, and subsequent revision of models.

From this information, a number of reports can be produced. Customer reports are provided with the return of the item, management reports summarise cost and quantity information, repair reports detail failure modes and frequencies, and periodic warranty assessment reports document the success of the warranty program including model estimates and actuals, forecasts, etc.

5. MODELS AND DATA MANAGEMENT

In the early stages of product development, much of the information will be subjective or based on the judgement of experts or come from historical records of similar systems or components, 'theoretical data' from design analyses, test data, and market information. As the development progresses, better data becomes available due to design definition, prototype testing, and trial manufacturing runs. This data is used to refine models, reducing the confidence limits of estimated parameters, and to assist the definition of target values for objectives and the development of models to measure these objectives.

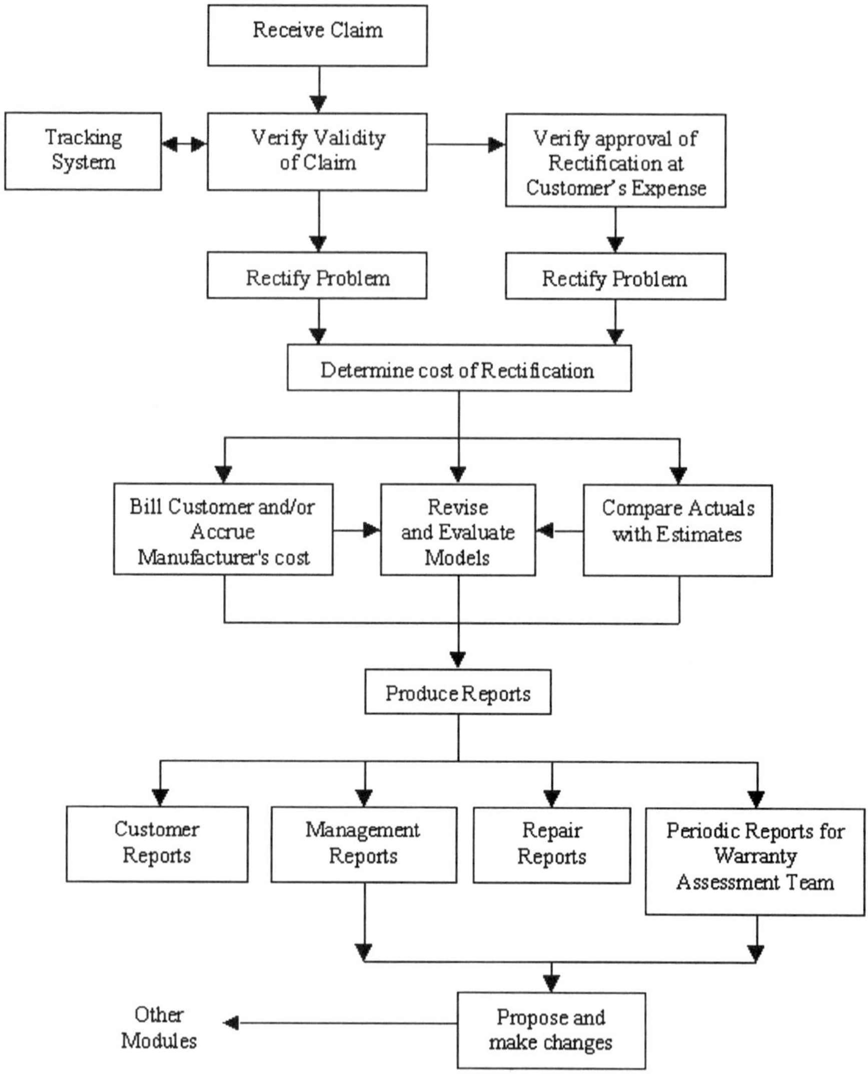

Figure 14.11 Flow of Activities in the Post-Sale Servicing Module

As product launch approaches, the warranty management system is used to finalise warranty terms. The warranty terms determine contractually the cost for each claim that is made. The design and production determine the number of failures that are likely to occur during the warranty period. Customers determine whether to make claims.

Once the product enters the market, models are used to evaluate actual data and compare parameters and characteristics with earlier forecasts. Analysts continue to refine models using actual production and repair data. Comparison will highlight invalid assumptions and allow for corrections to be made to the appropriate areas of the process. If the models are found to inadequately reflect field failure patterns or warranty costs, then a change in strategy, warranty terms, design, or manufacturing process may be required to correct the situation. Competitors actions also need to be observed and implications considered. If a competitor increases the warranty length on a product, the impact of increasing warranty length to match that of the competitor should be evaluated. If the warranty is increased without adequate consideration, the associated cost increase may outweigh the competitive advantage of increasing the warranty length.

5.1 D & E Models.

5.1.1 Failure / Reliability Models. Failure modeling for warranty cost analysis involves modeling the time to first failure and the time to failure after a repair has been made. First failure and subsequent failure are different because actions taken to rectify the first failure may alter the failure characteristics of the item. Failure/reliability models describe how failures are distributed over time. Reliability is the complement of failure and is the probability that an item will not fail.

Failure models can be categorised as black box or white box models (see Figure 14.12). A product is considered to be a system, a network of components. The time to failure of a system can be modeled by a failure distribution function or by the failure characteristics of the components that make up the system.

Black box models of both systems and components involve probability failure distributions. Various distributions have been used to model failure. The form of the probability distribution function F(t) is selected based on the analysis of failure data or mathematical convenience and can be supplemented with engineering judgement. The time to first failure can represent calender time, usage (e.g. mileage, cycles), or some other time scale.

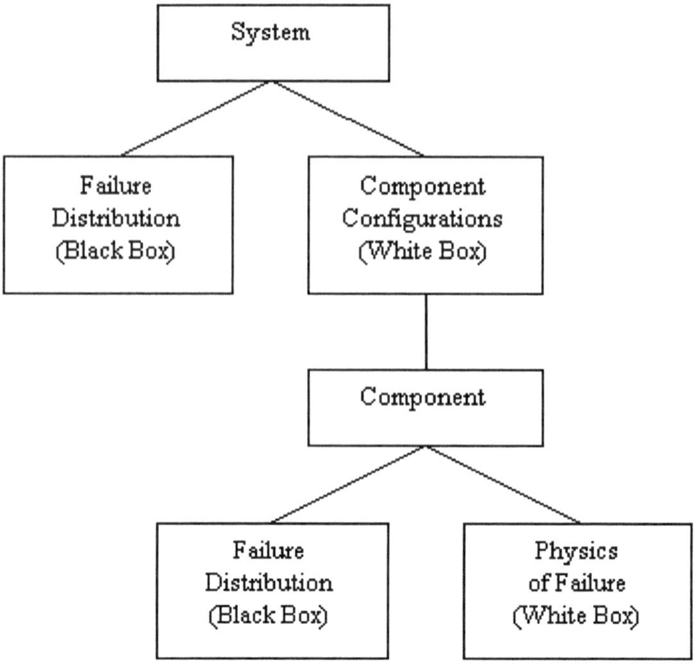

Figure 14.12 Failure Model Hierarchy

White box models of systems characterise the configuration and linking of components within the system through the mathematical structure function (Hoyland and Rausand (1994)). This approach uses techniques such as block diagrams of series; parallel and network system configurations; theory of coherent structures; analysis of redundancies such as hot, cold and warm standby; consideration of dependent failures; and estimates of component reliabilities to model system reliability.

White box models of components involve modeling the physics of failure (e.g. the accumulation of damage, the process of corrosion, etc). The time to failure is obtained by modelling the failure process, such as crack growth as a function of stress or temperature cycles, and is affected by different variables, such as timescales (i.e. calendar time, operating time), wear variables, and shock variables (Kordonsky and Gertsbakh (1995 a and b)). Because of its dependence on specific failure mechanisms and products, this approach is more difficult to generalise. Examples of white box failure models include stress-strength, damage-endurance, challenge-response, and tolerance-requirement models (Dasgupta and Pecht (1991)). For further details, see Murthy and Hussain (1995) and Nguyen and Murthy (1988).

5.1.2 Development Time (Reliability Growth) Models. Reliability growth models are mathematical formulae representing system reliability as a function of development time. Reliability growth modeling requires the estimation of reliability model parameters (as functions of usage or calendar time) and reliability growth model parameters (as functions of development time).

For further details, see Balaban (1978), Dhillon (1980), Crow (1972), Fries and Sen (1996), Mazzuchi and Soyer (1992), Mazzuchi and Soyer (1990), Mazzuchi and Soyer (1993), Xie and Zhao (1996), Meth (1992), and Duane (1964).

5.1.3 Development Cost Models. Development cost models describe how development costs relate to development time or product reliability. Development costs are accrued in the form of material costs, labour costs, and other costs including capital for test equipment, office costs, electricity, and so on. As the product's reliability improves, the costs increase because better components and more reliable materials come at a higher price and because the time for testing and estimating the reliability of the product increases as it improves.

Decision makers can evaluate the cost required to develop the product to a specified reliability or for a specified development time, allowing cash flow forecasts to be developed which can be linked to the target date for product release.

For more details, see Donelson (1977), Byers and Galli (1977), and Murthy and Nguyen (1987).

5.2 Production Models.

5.2.1 Production Cost Models. Inputs for production cost models are the cost of materials and parts that make up the product, the cost of labour and equipment required for assembly, the cost of indirect elements such as electricity, leasing or purchase of production facilities, and administration and management. To calculate an estimate of the production cost per item produced, an estimate of the production demand is required.

The cost of production affects the sale price of the product. Since warranty cost targets are often evaluated in relation to sale price, the manufacturing cost per item also affects warranty costs and the development of warranty terms.

There are two types of production cost models: production cost per item produced and production cost per item sold. If a fraction of the items produced are retained for warranty reserves, then the number of items for sale will be less than the number produced. The cost estimate, therefore, becomes the production cost per item sold. This will be more than the production cost per item produced since the same total production cost is distributed over fewer items. Since the cost per item affects the sale price per item, it is important that the cost per item sold is considered.

For more details, see Djamaludin *et al* (1994).

5.2.2 Production Quality Models. Production quality models describe how product reliability is affected by the process state. Process quality can change from a state of in-control to one of out of control as items are produced. A trade-off exists between the costs required to prevent non-conforming items being produced and reaching the customer and the costs that result from failures within the warranty period.

Lot sizing is based on the concept that smaller lot sizes result in a higher proportion of conforming items. As the lot size increases, there is a higher chance that the process will change from the initial in-control state to an out-of-control one. Producing items in small lots increases the cost of production since the process must be checked and reset more frequently. It is possible to trade off the size of each lot produced with the reliability of items manufactured to achieve an optimal lot size that minimises the associated costs.

For more details, see Leemis and Beneke (1990), Kuo and Kuo (1983), Jensen and Petersen (1982), Watson and Wells (1961), Lawrence (1966), and Murthy, Wilson and Djamaludin (1993).

5.2.3 Plant Maintenance Models. Plant maintenance models describe the reliability and operation of plant equipment used in the manufacture of the product. These models are used to plan maintenance activities within the production plant to minimise the occurrence of unplanned downtime. Typically, unplanned downtime is more costly than planned downtime.

Maintenance models depend on the maintenance strategy or policy that is adopted and the performance characteristics (reliability) of the equipment being considered. Examples of maintenance policies include the Age Policy whereby an item is replaced when it reaches a specified age or upon failure or the Block Policy whereby an item is repaired or replaced at specified time intervals. Condition monitoring is useful if an item's failure probability can be mapped to a measurable variable or condition.

The maintenance model is generally expressed as the expected cycle cost divided by the expected cycle time, where the cycle is defined by the chosen policy. The maintenance cost takes into account the costs associated with equipment failure and subsequent downtime, as well as the costs associated with maintenance activities. Using this model, an optimal interval can often be estimated which minimises the total downtime and maintenance costs. The maintenance interval becomes useful when considering a network of plant equipment where the maintenance activities must be scheduled.

5.3 Marketing Models.

5.3.1 Risk Models. Purchasing behaviour is influenced by the consumer's perception of risk and provides a starting point for warranty modelling. The utility function represents a measure of risk in terms of monetary value. It is used to gauge the attractiveness of competing choices and the attitudes to uncertainty, for example, the choice between purchasing the product with or without warranty coverage. Consumers (and producers) can be classified as risk-neutral, risk-taking, or risk-averse.

The consumer's risk model describes the utility or benefit to the consumer. The decision to purchase a product is influenced by this risk and is dependent on factors such as price, reliability, the value of having a working item, the level of warranty coverage, the consumer's income level, and the effort required by the consumer to maintain the product in working order. In situations where the consumer cannot evaluate the product reliability prior to sale, the moral hazard problem arises. This occurs when there is the potential for the manufacturer to "cheat" by claiming that the product is better than it really is. The consumer must rely on the warranty as an indicator of reliability.

The producer's risk model describes the producer's utility or producer's profit. A product sold with a warranty has the potential to generate costs for the producer, so there is a risk in offering the warranty. This risk depends on the price, the warranty coverage, the product reliability, the value of a working item to the customer, the producer's general level of quality, and the cost associated with maintaining that level of quality. In situations where the manufacturer cannot evaluate the usage of the product, there is the potential for customers to "cheat" by claiming to have maintained the product better than they really have. This situation also leads to the moral hazard problem since less maintenance leads to greater warranty costs.

For more details, see Varian (1984), Singpurwalla and Wilson (1993), Glickman and Berger (1976), Ritchken and Tapiero (1986), and Cooper and Ross (1985).

5.3.2 Demand / Sales Model. Both the sale price and the warranty length influence the consumer's decision to purchase and, thus, affect the level of sales. Demand models describe the effect that price and warranty length have on the magnitude of sales. Price elasticity and warranty length elasticity (or simply warranty elasticity) are parameters that control the size of the effect.

Sales models describe the number of items sold as a function of time. Warranty cost forecasts can be developed by combining sales models with warranty cost models.

5.3.3 Marketing Cost Models. Marketing cost models describe the cost incurred when advertising and promoting a product, including the costs involved in researching the market. It may be possible to identify relationships between marketing costs and factors such as market share or demand. Such relationships are likely to be difficult to model since data may be hard to obtain and may depend on the type of product and market.

5.3.4 Market Share Models. Market share models describe a product's penetration into the market. Market share is the fraction of demand to potential demand and lies between 0 and 1. The price and warranty length are used to estimate the sales demand, but the minimum price and maximum warranty length are used to estimate the potential sales demand. There will be a minimum price due to costs. Similarly, there can be a maximum warranty length, beyond which the costs of offering the warranty are too great.

For additional details, see Mitra and Patankar (1988, 1990, 1993, 1997), Patankar and Mitra (1989), and Glickman and Berger (1976).

5.3.5 Revenue Models. Revenue and pricing models describe the magnitude of revenue. At the most basic level, revenue is the product of demand and price; however, elements including taxes must also be considered.

Pricing models are used to develop the price of a product based on costs, value, or competition. Cost-plus pricing models, for example, involve estimating the total cost per unit and applying an additional margin. For more details, see Kotler and Armstrong (1996).

5.3.6 Profit Models. The profit or Gross Margin (GM) is derived from the total revenue and the total cost. Profit models can be developed on a total or per item basis. Profit models and revenue/pricing models are interrelated. For more details, see Chun and Tang (1995).

5.4 Post Sale Servicing Models.

5.4.1 Warranty Cost Models. Warranty cost models describe the cost of offering a warranty. They exist or can be built for each type of warranty policy that an organisation offers. Different policies can be used as part of different management strategies.

The most common warranty policies are the Free Replacement Warranty (FRW), the Pro-Rata Warranty (PRW) and Combination Warranties. Under the FRW policy, if failure occurs within the warranty period, the customer is provided with a replacement free of charge. Under the PRW policy, the customer gets a refund (or discount on another item), where the value depends on the duration of coverage remaining; if the item fails near the end of the coverage period, the refund amount is less than if it had failed earlier on. Combination warranties typically consist of a FRW period followed by a PRW period.

The basic format of a warranty cost model is the number of failures under warranty multiplied by the consequences of failure (cost and/or spares). These and other warranty cost models are found in Blischke and Murthy (1994).

See also Hill, Beall and Blischke (1991), Kao and Smith (1993), and Kalbfleisch, Lawless and Robinson (1991).

5.4.2 Repair / Renewal Models. Repair/renewal models describe the quality of repair processes. A failed repairable item is subjected to repair. The type of repair affects the distribution of future failures. Poor repair can shorten subsequent life; very good repair can extend product life. Repaired items are considered to be as-good-as-new (AGAN), minimal (or as-good-as-old AGAO), or imperfectly repaired. Failures occur randomly as points on a time and/or usage axis and can be modelled by a point process.

AGAN repair corresponds to the replacement of non-repairable items (e.g. components) or the repair of repairable items (e.g. systems), where the item after rectification is assumed to be as good as the failed one was when it was new. An ordinary renewal process can be used to model failures; if the rectification is AGAN, the time to complete the rectification action (commonly termed 'time to repair') is assumed to be negligible compared to the time to failure, and failures are considered to be independent and identically distributed (IID).

Minimal repair is often assumed when the failed system is repaired to a state that was apparent just before failure. Minimal repair assumes that complex systems are made of many parts and that the repair or replacement of the failed part does not affect the failure rate of the rest of the system. In effect, the system's age is the same as it was just before failure. Therefore, the product is assumed to have the same failure distribution. If the time to repair is negligible compared to the life of the product, then a non-homogeneous Poisson process with a time dependent intensity function is used.

Imperfect repair returns the system to some state that is better than just before failure but not as good as new. Imperfect repair can be modelled using a modified (delayed) renewal process (Ross (1970)).

See also Tuckwell (1995), Grimmett and Stirzaker (1992), and Murthy (1991).

5.4.3 Warranty Execution Models. Warranty execution models describe the likelihood of customers making claims in the event of product failure. A primary assumption when modeling warranty costs is that each failure within the warranty period results in a claim. This may not be a correct assumption in all cases. If a customer is dissatisfied with the product or the organisation's service, or the product is at the end of it's warranty period, then the customer may decide not to have the item repaired even if it is under warranty. Alternatively, false claims may be made.

For further details, see Patankar and Mitra (1995).

5.4.4 Usage Models. Product usage models describe the degree of usage that the product has undergone. Product usage at its most basic level can be characterised as normal or abnormal. Usage models can be developed for each of the types of consumer: governmental, industrial/commercial, and domestic.

Typically, industrial/commercial users will have a higher usage rate than domestic users. Usage models can be developed to measure reliability-related metrics such as loads, number of cycles, number of starts, effect of environmental factors, etc.

Depending on the type of product and customer, this may have an effect on the probability of failure within the warranty period. Consequently, different users may require different warranty terms.

5.4.5 Warranty Reserves Models. Warranty reserves models describe the amount of funds, spare parts, or items that need to be set aside to cover future warranty requirements. Models for these amounts apply to rebate, repair, and replacement warranties, respectively (Thomas (1989), Menke (1969), and Amato and Anderson (1976)).

Rebate warranties require that the manufacturer refunds a monetary amount specified by the warranty terms. To ensure that these payouts are met, a fraction of the sale price of each item must be set aside by the manufacturer as warranty reserves. Only costs are refunded. This fraction is set aside at the time of sale, but the warranty payouts actually occur throughout the warranty length. Investment of the warranty reserves, for example, in short term interest bearing deposits, generates income which can be used to lower the amount required for the reserves.

For non-repairable products covered by a non-renewing warranty, the manufacturer may be interested in the number of spare units to set aside as replacements. Under a non-renewing warranty, when a replacement is provided, the coverage only lasts until the end of the original warranty period.

When products are repairable, failure can be rectified by a repair or a replacement. The type of repair influences the performance of the product after repair and, hence, determines the demand rate for repairs.

If repair can be considered AGAN, the expressions for the non-repairable case are applicable. If repair is considered minimal (i.e. the failure rate after repair equals the failure rate immediately prior to failure), then a non-stationary Poisson process describes the failures.

5.4.6 Warranty Servicing Models. Warranty servicing models are used to develop servicing strategies. When a claim is made on a FRW for a repairable product, the manufacturer can either repair the item or replace it with a new one. Depending on the claimed fault, estimates of repair cost and repair time can be made. If these estimates are above certain limits (repair cost limits and repair time limits respectively), it may be more cost effective to replace the item rather than repairing it. For more details, see Blischke and Murthy (1994 and 1996), Chung (1994).

5.5 Data for Model Building.

A database is necessary for the storage, maintenance, and organisation of data for a warranty management system. Problems to be overcome include the timeliness and consistency of the data used. The proper use of a database ensures that data can be shared between users, that duplication is minimised, that there is consistency or integrity of data, and that the data is secure (Beynon-Davies and Hutchings (1993)).

Technical data encompasses data that affects the reliability of the product or process including material strengths, component specifications, system configurations, test data, environmental measurements, and production equipment specifications and records, repair times, failure times, failure modes and mechanisms, and operating conditions.

Commercial data includes data that has a marketing impact such as price and warranty terms, sales volume (demand), risk, and profits. Marketing data focuses on consumers (potential customers) while post-sale support deals with customers.

Commercial data includes repair costs, warranty registration card details, and customer complaints with respect to many different aspects of the product. These include technical features, packaging, design, image, functional performance, quality, availability, reputation, before-sale service, after-sale service, size, colour, running costs, ease of maintenance, ease of use, the provision of training or simple instructions, compatibility with existing equipment, safety, pollution hazard reduction, and so on.

Data can be collected at the system level or the component level. The hierarchy can be separated into as many levels as required - System, Subsystem, Component, Part, Material, etc.

Cost data can be broadly categorised as direct and indirect. Direct costs can be described as the cost of materials and labour that result in the final product, and indirect costs are those that are required to support the production process. For example, the cost of the metal, plastic, oil, components, technical labour, etc. that make up the product are direct costs; the cost to design the product, set up and maintain the production plant, perform administrative tasks, service, and market the product, and manage the entire process are indirect costs.

Historical data includes failure and sales records from previous products and components. Failure records from tests or the field provide time-to-failure data that is useful for reliability analysis and subsequent warranty cost analysis. Sales data can be used to forecast market penetration and can also be a useful source for data mining, in which the database is 'mined' for patterns that can indicate consumer buying patterns (Berry and Linoff (1997)).

Vendor data includes components, materials, and/or sub-systems that are purchased from outside the manufacturing organisation. Test data from vendors can be obtained and verified by in-house testing, if necessary, and used the same way as historical data.

Test data allows performance to be quantified and reliability to be estimated. Experiments should ideally be designed and carried out under controlled conditions so that the information obtained is meaningful. If the product is complex and expensive, testing may be required at several levels: material, component, and system. For simple products, it may be adequate to test only the completed product. As the complexity of the product increases, so do the data and analysis required for aggregation of meaningful information.

Handbook data includes specifications and calculations obtained from technical publications. Data of this type may typically include labour costs in certain regions, formulae for various technical relationships, market indices for commodities, and so on.

Various computer applications are used when designing a product. These produce various data such as results from Finite Element Analyses, Fatigue calculations, reliability estimates, and so on. This data should be incorporated into the warranty management system to augment other sources.

Information from experts can be incorporated into various analyses to provide judgemental data. Past experience will allow personnel to have a subjective "feel" for certain situations. This can be a valuable source of data when no other is available.

Random samples are often taken when inspecting items from the production line. The data are used to estimate production quality and other metrics of interest.

Environmental data includes temperatures, stresses, etc. that are encountered during tests or during operational life (if recorded). These variables can be used to evaluate the effect of different environments on product operation.

For more details, see Blischke and Murthy (2000).

5.6 Database Management.

A database must be maintained and managed for it to be helpful for users. A database management system (DBMS) provides the procedures and mechanisms for maintaining and managing a database. Three main functions of a database management system are file maintenance, information retrieval, and database administration. The relationship between these functions is illustrated in Figure 14.13.

5.6.1 File Maintenance.

File maintenance involves adding records to tables, updating data in tables, and deleting records from tables. The users of the database are responsible for maintaining the quality of the data that is in the database.

5.6.2 Information Retrieval (Inputs and Outputs).

The system must allow the user to input their requirements (targets, objectives, models, etc) and view the results. Typically, users will enter or update data in the database tables through a Graphical User Interface (GUI) or "Front End". Applications that enter or update data may do this directly, or also be controlled through a GUI.

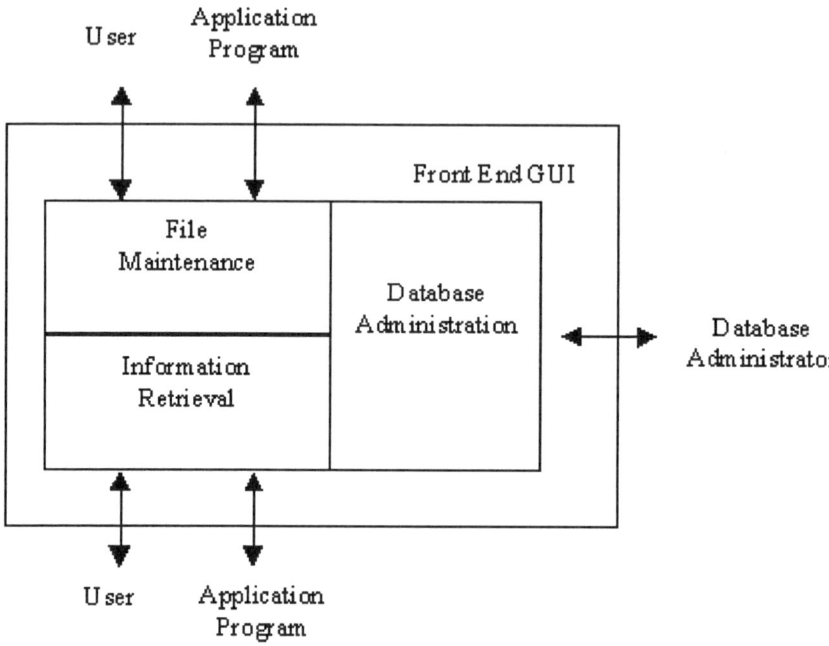

Figure 14.13 Database Management System (adapted from Beynon-Davies and Hutchings (1993))

The main inputs to a warranty management system are operational data from each module and values for decision variables (e.g. target values, model types, etc.). These should be displayed to the user via 'forms' so that the analysis is consistent and understandable. For example, the decision maker selects a failure distribution, warranty type, and warranty length(s) as inputs, then the system estimates the parameters of failure distribution from underlying data and estimates the expected warranty cost and bounds.

The main outputs are objective values for criteria (e.g. estimated warranty cost, number of failures per item sold, sales demand (total, or as a function of time), manufacturing cost, or market share. If the system is being used to evaluate optimal values of variables, then these are also outputs. These outputs can be displayed graphically through the GUI or by using reports.

Each estimated value must be accompanied by a measure of its variability, for example, as confidence limits, otherwise the estimated value is of limited use. The quality of the estimates and models depends on the quality of the data entered into the system.

5.6.3 Database Administration. A database requires maintenance so that it performs as intended. Database administration involves the creation, deletion, and restriction of tables. During development of the database, tables are created for storing data. As the design of the database changes, inevitably some tables become obsolete and require deletion. It is important that only certain personnel have the ability to make such changes.

5.6.4 Database Implementation. Brennan (1994) summarised the procedure for implementing a warranty database outlining the following elements.

- Determine key data elements
- Analyse existing databases and systems for adequacy
- Evaluate future requirements for other uses or projects
- Make cost trade-offs of performance requirements versus cost (of computer systems)
- Select hardware and software which meets requirements
- Develop the I/O tools: sources of inputs and types of outputs
- Continue maintenance, updating, and programming

6. CONCLUSION

Product warranty is affected by decisions made throughout a product's life, particularly its early stages. In turn, warranty decisions have an impact on activities and decision making at the different stages of manufacturing. A warranty management system provides a framework for effective decision making where warranty decisions are linked to other manufacturing decisions from a product life perspective. In this chapter, we proposed a structure for such a warranty management system and discuss the models and data related issues for effective decision making.

Future research could focus on reviewing the models suggested in Section 5 and developing a comprehensive library for use by programmers or software developers. Furthermore, a GUI could be developed that is generic enough to be incorporated into the product development process of manufacturing organisations.

References

[1] Amato, H.N. and Anderson, E.E. Determination of Warranty Reserves: An Extension. *Management Science*, **22** (1976) 1391-1394.

[2] Armistead, C. and Clark, G. A Framework for Formulating After-Sales Support Strategy:. *International Journal of Operations and Production Management*, **11**(3) (1991) 111-124.

[3] Balaban Reliability growth models, *Journal of Environmental Science*, **21** (1978) 11-18.

[4] Berry, M.J.A. and Linoff, G. *data mining techniques: for marketing, sales, and customer support.* John Wiley & Sons, Inc., New York, 1997.

[5] Beynon-Davies, P. and Hutchings, A.R. Modelling and databases, *European Journal of Operational Research*, **64** (1993) 327-337.

[6] Blischke, W.R. and Murthy, D.N.P. *Warranty Cost Analysis.* Marcel Dekker, New York, 1994.

[7] Blischke, W.R. and Murthy, D.N.P. *Product Warranty Handbook.* Marcel Dekker, New York, 1996.

[8] Blischke, W.R. and Murthy, D.N.P. *Reliability.* John Wiley & Sons, Inc., New York, 2000.

[9] Blischke, W. and Murthy, D.N.P. Product warranty management - I: A taxonomy for warranty policies. *European Journal of Operational Research*, **62** (1992) 127-148.

[10] Brennan, J.R. *Warranties: planning, analysis, and implementation.* McGraw-Hill Logistics Series, (1994) 167-184.

[11] Byers, J.K. and Galli, D.H. Reliability growth apportionment. *IEEE Transactions on Reliability*, **R26** (1977) 242-244.

[12] Cohen, L. *Quality Function Deployment: How to Make QFD Work for You.* Addison-Wesley Publishing Company, Reading, Massachusetts 1995.

[13] Chun, Y.H. and Tang, K. Determining the optimal warranty price based on the producer's and customer's risk preferences. *European Journal of Operational Research*, **85** (1995) 97-110.

[14] Chung, K. Optimal repair-cost limit for a consumer following expiry of a warranty, *Microelectronics and Reliability*, **34**(10) (1994) 1689-1692.

[15] Cooper, R. and Ross, T.W. Product warranties and double moral hazard. *Rand Journal of Economics*, **16**(1)Spring (1985) 103-113.

[16] Crow, L.H. *Reliability Growth Modeling, Technical Report No. 55.* U.S. Army Materiel Systems Analysis Agency, Aberdeen Proving Ground, Maryland, 1972.

[17] Dasgupta, A. and Pecht, M. Material Failure Mechanisms and Damage Models; *IEEE Transactions on Reliability*, **40**(5) December (1991) 531-536.

[18] Dhillon, B.S. Reliability Growth: A Survey; *Microelectronics and Reliability*, **20** (1980) 743-751.

[19] Djamaludin, I., Murthy, D.N.P., and Wilson, R. Quality control through lot sizing for items sold with warranty. *International Journal of Operational Research*, **33** (1994) 97-107.

[20] Donelson, J. Cost model for testing program based on nonhomogeneous Poisson failure model. *IEEE Transactions on Reliability*, **R26** (1977) 189-194.

[21] Duane, J.T. Learning curve aroach to reliability monitoring. *IEEE Transactions on Aerospace*, **2** (1964) 563-566.

[22] Fussel, J.B. *Fault Tree Analysis - Concepts and Techniques in Generic Techniques in Reliability Assessment*, Henley, E.J. and Lynn, J. (Eds.), Nordoff, Leyden, Holland, 1976.

[23] Fries, A. and Sen, A. A Survey of Discrete Reliability-Growth Models. *IEEE Transactions on Reliability*, **45**(4) December (1996) 582-604.

[24] Glickman, T.S. and Berger, P.D. Optimal price and protection period decisions for a product under warranty. *Management Science*, **22** (1976) 1381-1390.

[25] Grimmett, G.R. and Stirzaker, D.R. Probability and Random Processes. *Oxford University Press*, New York, 1992.

[26] Hill, V.L., Beall, C.W. and Blischke, W.R. A simulation model for warranty analysis. *International Journal of Production Economics*, **22** (1991) 131-140.

[27] Hoyland, A. and Rausand, M. System Reliability Theory: Models and Statistical Methods; *John Wiley and Sons, Inc.*, New York, 1994.

[28] IEEE Std. 352 IEEE Guide for General Principles of Reliability Analysis of Nuclear Power Generating Station Protection Systems. *IEEE Inc.*, New York, 1982.

[29] Ives, B., and Vitale, M.R. After the sale: Leveraging maintenance with information technology. *MIS Quarterly*, **12** (1988) 7-20.

[30] Jensen, F. and Petersen, .E. Burn-In: An Engineering Approach to the Design and Analysis of Burn-In Procedures. *John Wiley & Sons, Inc.*, New York, 1982.

[31] Kalbfleisch, J.D., Lawless, J.F. and Robinson, J.A. Methods for the analysis and prediction of warranty claims. *Technometrics*, August, **33**(3) (1991) 273-285.

[32] Kao, E.P.C. and Smith, M.S. Discounted and per unit net revenues and costs of product warranty: the cast of phase-type lifetimes. *Management Science*, **39**(10) October (1993) 1246-1254.

[33] Kordonsky, Kh.B. and Gertsbakh, I.B. System state monitoring and lifetime scales - I. *Reliability Engineering and System Safety*, **47** (1995 a) 1-14.

[34] Kordonsky, Kh.B. and Gertsbakh, I.B. System state monitoring and lifetime scales - II. *Reliability Engineering and System Safety*, **47** (1995 b) 147-154.

[35] Kotler, Philip and Armstrong, Gary *Principles of Marketing; 7th Ed.;* International Ed.; Prentice Hall Int., Englewood Cliffs, N.J., (1996) 351-358.

[36] Kuo, W. and Kuo, Y. Facing the Headaches of Early Failures: A State-of-the-Art Review of Burn-In Decisions, *Proceedings of the IEEE*, **71**(11) (1983) 1257-1266.

[37] Lawrence, M.J. An investigation of the burn-in problem, *Technometrics*, **8**(1) (1966) 61-71.

[38] Leemis, L.M. and Beneke, M. Burn-In Models and Methods: A Review, *IIE Transactions*, **22**(2) June (1990) 172-180.

[39] Lele, M.M. and Karmarkar, U.S. *Good product support is smart marketing.* Harvard Business Review, November-December (1983) 124-132.

[40] Mazzuchi, T.A. and Soyer, R. Reliability assessment and prediction during product development. *Proceedings annual reliability and maintainability symposium*, (1992) 468-474.

[41] Mazzuchi, T.A. and Soyer, R. A Bayes method for assessing product-reliability during development testing, *IEEE Transactions on Reliability*, **42**(3) September (1993) 503-510.

[42] McGuire, E.P. *Industrial Product Warranties: Policies and Practices.* The Conference Board Inc., New York, 1980.

[43] Menke, W.W. Determination of Warranty Reserves. *Management Science*, **15** (1969) B-542-B-549.

[44] Meth, M.A. Reliability-Growth Myths and Methodologies: A Critical View. *Proceedings Annual Reliability and Maintainability Symposium*, (1992) 337-342.

[45] MIL-STD 882. *System Safety Program Requirement.* U.S. Department of Defense, Washington, DC, 1984.

[46] Mitra, A. and Patankar, J.G. Warranty cost estimation: a goal programming approach. *Decision Sciences*, **19** (1988) 409-423.

[47] Mitra, A. and Patankar, J.G. A multi-objective model for warranty estimation. *European Journal of Operational Research*, **45** (1990) 347-355.

[48] Mitra, A. and Patankar, J.G. An integrated multicriteria model for warranty cost estimation and production. *IEEE Transactions on Engineering Management*, **40**(3) August (1993) 300-311.

[49] Mitra, A. and Patankar, J.G. Market share and warranty costs for renewable warranty programs. *International Journal of Production Economics*, **50** (1997) 155-168.

[50] Murthy, D.N.P. A note on minimal repair. *IEEE Transactions on Reliability*, **40**(2) June (1991) 245-246.

[51] Murthy, D.N.P. and Blischke, W.R. Strategic Warranty Management: A life cycle approach. *IEEE Transactions on Engineering Management*, **47** (2000) 40-44.

[52] Murthy, D.N.P. and Chattophadhyay, G. Warranties for second hand products, *Proceedings of Ninth International FAIM Conference*, Tilburg, Netherlands, (1999) 1145-1159.

[53] Murthy, D.N.P and Hussain, A.Z.M.O. Warranty and optimal redundancy design. *Engineering Optimization*, **23** (1995) 301-314.

[54] Murthy, D.N.P. and Nguyen, D.G. Optimal development testing policies for products sold with warranty. *Reliability Engineering*, **19** (1987) 113-123.

[55] Murthy, D.N.P., Wilson, R.J., and Djamaludin, I. Product warranty and quality control. *Quality and Reliability Engineering International*, **9** (1993) 431-443.

[56] Nguyen, D.G. and Murthy, D.N.P. Optimal reliability allocation for products sold under warranty. *Engineering Optimization*, **13** (1988) 35-45.

[57] O'Leary, D.J. Managing warranty policies, ASQC 48th Annual Quality Congress Proceedings. Milwaukee, Wis., *American Society for Quality Control* , (1994) 517-521.

[58] Patankar, J.G. and Mitra, A. A multi-objective model for warranty cost estimation using multiple products. *Computers Operations Research*, **16**(4) (1989) 341-351.

[59] Patankar, J.G. and Mitra, A. Effects of warranty execution on warranty reserve costs. *Management Science*, **41**(3) March (1995) 395-400.

[60] Ritchken, P.H., Chandramohan, J. and Tapiero, C.S. Servicing, quality design and control. *IIE Transactions*, **21**(3) September (1989) 213-219.

[61] Ritchken, P.H. and Tapiero, C.S. Warranty design under buyer and seller risk aversion. *Naval Research Logistics Quarterly*, **33** (1986) 657-671.

[62] Ross S.M. *Applied Probability Models with Optimization Applications.* Holden Day, San Francisco, 1970.

[63] Singpurwalla, N.D. and Wilson, S. The warranty problem: its statistical and game theoretic aspects. *SIAM Review*, **35**(1) March (1993) 17-42.

[64] Thomas, M.U. A Prediction model for Manufacturer Warranty Reserves. *Management Science*, **35** (1989) 1515-1519.

[65] Tuckwell, H.C. *Elementary applications of probability theory.* London, Chapman & Hall, London, 1995.

[66] Varian, H.R. *Microeconomic analysis.* W.W. Norton., New York, 1984.

[67] Watson, G.S. and Wells, W.T. On the possibility of improving the mean useful life of items by eliminating those with short lives. *Technometrics*, **3**(2) (1961) 281-298.

[68] Xie, M. and Zhao, M. Reliability growth plot - an underutilized tool in reliability analysis. *Microelectronics and Reliability*, **36**(6) (1996) 797-805.

Chapter 15

WARRANTY AND QUALITY

D.N.P. Murthy and I. Djamaludin
The University of Queensland
St. Lucia, Brisbane,
Q 4072, Australia

Abstract Warranty is an important element in the marketing of new products, and warranty results in additional cost to the manufacturer. The warranty servicing cost is influenced by design quality, manufacturing quality, and service quality. This paper deals with this link and reviews the literature linking warranty and quality. It concludes with a brief discussion of topics for future research.

Keywords: Warranty, design quality, manufacturing quality, service quality.

1. INTRODUCTION

Modern manufacturing is characterised by (i) rapidly changing technologies, (ii) global markets, (iii) fierce competition, (iv) nearly identical products due to very similar components and technology used and, (v) educated and more demanding customers. This has serious implications for both buyers and manufacturers.

In the purchase decision of a product, buyers typically compare characteristics of comparable models of competing brands. When competing brands are nearly identical, comparison is difficult. In many instances, customers choose a particular product solely on the basis of some product related characteristics such as product price and perceived product quality. In such situations, warranty plays an important role in the product choice, and manufacturers offer many different types of warranties.

A warranty of any type, since it involves an additional service associated with a product, will lead to potential warranty servicing costs beyond those associated with the design, manufacture, and sale of the product. These future costs, are unpredictable and typically range from

2% to as much as 15% of net sales (McGuire (1980)). As a result, warranty has a significant impact on the total profits for a manufacturing business.

In the context of a new product the three different notions of quality are as follows:

- Design Quality (defined through product reliability)
- Manufacturing Quality (defined through conformance to design)
- Service Quality (defined in terms of customer satisfaction).

Each of these has an impact on the warranty servicing cost. By improving quality, one can reduce the warranty servicing cost. However, this is worthwhile only if the reduction in the expected warranty servicing cost exceeds the additional cost associated with quality improvement.

There are several papers that deal with warranty and quality. The bulk of them deal with a single notion of quality, and optimal decisions with regard to quality improvement achieve a trade-off between the improvement cost and the savings in the expected warranty servicing cost.

In this paper, we review this literature and focus on the quality improvement aspects. For some simple models we present the results which characterise the optimal strategies. For more complex models, we discuss the salient features and omit giving the results for the sake of brevity. Following this, we discuss some topics for future research.

The outline of the paper is as follows. Sections 2 and 3 give a brief overview of warranty and quality, respectively. Section 4 deals with a review of models linking warranty and quality. We conclude with a brief discussion of topics for future research in Section 5.

Notation:

FRW	: Free replacement warranty
PRW	: Pro-rata warranty
W	: Warranty period
$F(t)$	: Failure distribution of new item
$f(t)$	: Failure density function associated with $F(t)$
$r(t)$	: Failure rate associated with $F(t)$
$\bar{F}(t)$	: Survival function associated with $F(t)$
$F_c(t)$	: Failure distribution of conforming item
$F_n(t)$	: Failure distribution of nonconforming item
$F_m(t)$	: Failure distribution of a module (with redundancy)
$F_s(t)$	: Failure distribution of partially energised item (warm standby)
$N(t)$	: Number of failures over the interval [0,t)

$M(t)$	: Renewal function associated with $F(t)$
$s(x)$	: Rebate function
τ	: Development period [for reliability improvement]
	: Burn-in period [for quality control]
θ	: Parameter of the failure distribution [$\theta > 0$] [For exponential failure distribution, θ is the failure rate and for Weibull and Gamma, θ is the scale parameter]
θ_0	: Initial value of θ
θ_m	: Limiting minimum value of θ after development for an infinite time
$\tilde{\theta}(\tau)$	: Value of θ after development for period τ
$C_d(\tau)$	: Development cost with development for period τ
C_h	: Cost of handling each claim
$C_i(T)$	: Cost of inspection per unit for a period of time T
C_m	: Manufacturing cost per unit
C_r	: Expected cost of each minimal repair
C_s	: Setup cost
C_w	: Warranty cost per item for the manufacturer
P	: Unit sale price
p_1	: Fraction of conforming items in a lot.
p_2	: Fraction of non-conforming items in a lot.
p	: Probability of produced item conforming
$p_{in}[p_{out}]$	: Probability of a produced item conforming when the process is in control[out of control]
q	: Probability of process staying in control at the end of producing an item [lot production]
q_s	: Probability that the switch works satisfactorily [cold standby redundancy]
Q	: Lot size

2. PRODUCT WARRANTY

2.1 Concept and Role. A warranty is a manufacturer's assurance to a buyer that a product or service is or shall be as represented. It is considered to be a contractual agreement between the buyer and manufacturer entered into upon the sale of the product or service. A warranty may be implicit or it may be explicitly stated. In broad terms, the purpose of a warranty is to establish liability among the two parties (manufacturer and buyer) in the event that an item fails. An item is said to fail when it is unable to perform satisfactorily its intended function when properly used. The contract specifies both the performance that is to be expected and the redress available to the buyer if a failure occurs.

In recent years, manufacturers and third parties (dealers, financial institutions in the credit business) have been offering extended warranties. These can be viewed as maintenance service contracts that buyers can choose by paying an extra amount. The normal warranty is an integral part of the product sale and is factored into the sale price.

2.2 Product Categories. New products can be divided into following three categories:

1. Consumer durables such as household appliances, cars, and personal computers bought by individual households as a single item.

2. Industrial and commercial products bought by businesses for the provision of services (for example, equipment used in a hospital to provide medical care, aircraft used by airlines, or products such as computers or copy machines. These are bought either individually (for example, a single x-ray machine bought by a hospital) or as a batch, or lot, of $Q(Q > 1)$ items (for example, batteries bought by a car manufacturer, fleet of trucks bought by a car rental agency). We need to differentiate "standard" off-the-shelf products from "custom-built" products built to buyer's specifications. We shall call the latter category products as "specialised" products.

3. Government acquisitions such as a new fleet of tanks or jet fighters. These are usually "custom built" and are products involving new and evolving technologies. As such, they are characterised by a high degree of uncertainty in the product development process. (Note: A government is also a large buyer of "standard" industrial and commercial products but these do not involve product development as part of the warranty.)

2.3 Warranty Policies. Warranties can be divided into different sub-groups based on dimensionality (one-dimensional involve only age or usage, two-dimensional involve both age and usage) and whether the warranty is renewing or not. In a renewing warranty, the repaired or replacement item comes with a new warranty identical to the initial warranty. Blischke and Murthy (1992) develop taxonomy for new product warranty.

Warranties Policies for Standard Products Sold Individually The most important characteristic of a warranty is the form of compensation to the customer on failure of an item. The two most common are the following two basic warranties:

One-Dimensional Non-renewing Free Replacement Warranty (FRW) Policy

The manufacturer agrees to repair or provide replacements for failed items free of charge up to a time W from the time of the initial purchase. The warranty expires at time W after purchase.

One-Dimensional Non-renewing Pro-Rata Rebate Warranty (PRW) Policy

The manufacturer agrees to refund a fraction of the purchase price should the item fail before time W from the time of the initial purchase. The buyer is not constrained to buy a replacement item.

The combination warranty is obtained by combining these two basic policies. An example is a combination policy that provides free replacement up to a specified time and a replacement at pro-rated cost during the balance of the warranty period.

Warranty Policies for Standard Products Sold in Lots

Under this warranty, an entire batch of items is guaranteed to provide a specified total amount of service, without specifying a guarantee on any individual item. For example, rather than guaranteeing that each item in a batch of 100 will operate without failure for 2000 hours, the batch as a whole is guaranteed to provide at least 200,000 hours of service. If after the last item in the batch has failed, the total service time is less than 200,000 hours, items are provided as specified in the warranty (e.g., free of charge or at pro rata cost) until such time as the total of 200,000 hours is achieved.

Warranty Policies for Specialised Products

In the procurement of complex military and industrial equipment, warranties of a certain type play a very different and important role, that of being an incentive to the seller to increase the reliability of the items after they are put into service. This is accomplished by requiring that the contractor service the items in the field and make design changes as failures are observed and analysed. The incentive is an increased fee paid to the contractor if it can be demonstrated that the reliability of the item has, in fact, been increased. Warranties of this type are called Reliability Improvement Warranties (RIW).

2.4 Review of Warranty Literature. A three-part review paper, Blischke and Murthy (1992), Murthy and Blischke (1992a) and Murthy and Blischke (1992b) deals with the concept and taxonomy for warranties, a framework for the study of warranties and warranty cost analysis, respectively. The paper by Blischke (1990) deals with warranty cost analysis. A more recent review paper is by Thomas and Rao (1999).

Over the last six years, four books have appeared on the subject. Blischke and Murthy (1993) deal with cost analysis of over 40 different warranty policies for new products. Blischke and Murthy (1995) is a collection of research papers dealing with warranty from many different perspectives. Sahin and Polatoglu (1998) deal with the cost analysis of some basic one-dimensional warranty policies. Brennan (1994) deals with warranty administration in the context of defence products.

Finally, Djamaludin *et al.*(1995) list over 1500 papers on warranties divided into different categories.

2.5 Warranty Cost Analysis. The simplest models for warranty cost analysis are based on the following assumptions:

1. Whenever a failure occurs, it results in an immediate claim.
2. All claims are valid.
3. The time to rectify a failed item (either through repair or replacement) is sufficiently small in relation to the mean time between failures that it can be approximated as being zero.
4. The manufacturer has the logistic support (spares and facilities) needed to carry out the rectification actions without any delays.

Relaxing one or more of these assumptions results in more complex models, and these can be found in Blischke and Murthy (1993) and (1995).

Several different notions of warranty costs have been studied. In this paper, we focus our attention on the manufacturer's expected warranty cost per unit. Because of various uncertainties in manufacturing, X, the time to first failure, is a random variable. Let $F(x)$ denote the distribution function for X and be called the failure distribution. Let $f(x)$ and $r(x)$ denote the failure density and failure rate associated with $F(x)$. Let θ denote the parameters of the distribution; we have not indicated this explicitly for notational convenience. Finally, let μ denote the expected value of X.

Expected Warranty Cost [FRW Policy]

In the case of a non-repairable product, failures under warranty are rectified instantaneously through replacement by new items. As a result, the number of failures over the period $[0, t)$, $N(t)$, is characterized by a renewal process with time between renewals distributed according to $F(t)$. The cost C_w to the manufacturer over the warranty period W is a random variable given by

$$C_w = [C_m + C_h]N(W), \tag{15.1}$$

where C_m is the unit manufacturing cost, and C_h is the handling cost of each warranty claim.

The expected number of failures during warranty, $E[N(W)]$ is given by

$$E[N(W)] = M(W), \tag{15.2}$$

where $M(.)$ is the renewal function given by

$$M(t) = F(t) + \int_0^t M(t-x)f(x)dx. \tag{15.3}$$

As a result, the expected warranty cost per unit to the manufacturer is given by

$$E[C_w] = [C_m + C_h]M(W). \tag{15.4}$$

In the case of a repairable product, the manufacturer has the option to repair a failed item. If the failed items are repaired minimally (see, Barlow and Hunter (1960)), then failures over the warranty period occur according to a non-homogeneous Poisson process with intensity function equal to the failure rate $r(t)$. Let C_r denote the average cost of each repair. Thus, the manufacturer's expected warranty cost, per unit sold, is given by

$$E[C_w] = [C_r + C_h]\int_0^W r(x)dx. \tag{15.5}$$

Cost expressions for other types of repairs can be found in Blischke and Murthy (1993).

Expected Warranty Cost[PRW Policy]

We confine our attention to the linear rebate function given by

$$s(x) = \begin{cases} (1 - x/W)P & 0 \leq x < W \\ 0 & otherwise. \end{cases} \tag{15.6}$$

The warranty cost to the manufacturer is given by

$$C_w = C_h + s(X), \tag{15.7}$$

when X (item life time) $< W$. As a result, the expected warranty cost per unit to the manufacturer is given by

$$E[C_w] = \int_0^W [C_h + s(x)]f(x)dx. \tag{15.8}$$

Using (15.6) in (15.8) and carrying out the integration, we obtain

$$E[C_w] = P\left[F(W) - \frac{\mu_W}{W}\right] + C_h F(W), \tag{15.9}$$

where

$$\mu_W = \int_0^W x f(x)dx \tag{15.10}$$

is the partial expectation of X.

3. QUALITY

The different stages for a new product are as follow.

Design $\rightarrow$ Manufacture $\rightarrow$ Market $\rightarrow$ Post-sale Service

There are several different notions of quality for each stage. We focus our attention on design, manufacturing, and service quality.

3.1 Design Quality.

3.1.1 Quality Notion. Garvin (1987) defines many different notions of product quality, and these include transcendent, product based, user based, manufacturing based and value based. Several of these are influenced by the decisions made during the design stage. As such, we will call these "design quality". In the context of product warranty, the design quality of relevance is product performance over time. This can be described in terms of reliability $R(x)[= 1 - F(x)]$, failure rate $[r(x)]$, or mean time to first failure $[\mu]$. Higher value for $R(W)$, lower value for $r(x)$ for $0 \leq x \leq W$, or higher value for μ are different indicators of design quality.

3.1.2 Quality Improvement. There are basically two approaches to improving product reliability. These are (i) using redundancy and (ii) reliability growth through a development program.

Redundancy

Here one or more components are replicated to improve the reliability of the product. As such, for a component that is replicated, we use a module of $k(\geq 2)$ components instead of a single component. The three different types of redundancy that have been used are (i) hot standby, (ii) cold standby, and (iii) warm standby (see Hussain (1997)). Let $H(x)$ and $F(x)$ denote the failure distribution for the module and component, respectively.

In hot standby all k replicates are connected in parallel and in use so that the module failure time is the largest of the k component failure times. As a result, $H(x)$ is given by

$$H(x) = [F(x)]^k. \tag{15.11}$$

In cold standby, only one component is in use at any given time. When it fails, it is replaced by a working component (if available) through a switching mechanism. If the switch is perfect and components do not degrade when not in-use, then the module failure time is the sum of the k component failure times. As a result, $H(x)$ is given by

$$H(x) = F(x) * F(x) * \cdots * F(x), \tag{15.12}$$

where * is the convolution operator.

In warm standby, the component in use has a failure distribution $F(x)$, and the remaining non-failed components (in partially energised state as opposed to fully energised state when put into operation) have a failure distribution $F_s(x)(< F(x)$ for $x > 0)$. As a result, a component can fail before it is put into use. This implies that the failure distribution of a component put into use is a complicated function of $F(x)$, $Fs(x)$, and the duration for which it was partially loaded before being put into use. In the case $k = 2$, the distribution function for the module is given by

$$H(x) = F(x) - \int_0^x \bar{F}(x - y + \bar{F}^{-1}(\bar{F}_s(y)))f(y)dy. \tag{15.13}$$

For general k, the expressions for $H(x)$ are more complex and can be found in Hussain (1997).

For all three types of redundancy, $H(x) < F(x)$, implying that a module is more reliable than a single component. The reliability increases with k, the number of replicates in a module. If the switch is not perfect, then $H(x)$ (given by (15.12) for cold standby and by (15.13) for warm standby) gets modified. For further details, see Hussain (1997).

Note that cost of a module depends on k (the number of replication) and the type of redundancy (as both cold and warm standby redundancies involve a switching mechanism). The cost of a module is kC_m in the case of hot standby and $C_m(k+\delta)$, $0 < \delta \leq 1$ in the case of cold and warm standby, where δC_m is the additional cost of the switching mechanism.

Reliability Growth

This involves a research and development (R&D) effort where the product is subjected to an iterative process of test, analyse, and fix cycles. During this process, an item is tested for a certain period of time or until a failure occurs. Based on the analysis of the test run and failure mode, design and/or engineering modifications are made to improve the reliability. The process is repeated, resulting in reliability improvement (or growth) of the product. The reliability growth models can be broadly categorised into two groups: continuous and discrete models. Fries and Sen (1996) gives a comprehensive review of the discrete models, and a discussion of continuous time models can be found in Hussain (1997).

In continuous time models, the improvement in reliability can be modelled through a parameter (θ) of the distribution which decreases with development. Usually, θ is the inverse of the scale parameter so that in the case where $F(x)$ is exponential, it is the failure rate associated with $F(x)$.

In one of the well-known continuous time models, developed by Duane (1964), the parameter corresponds to the failure rate for an exponential distribution. After development for a period τ, the failure rate $\theta(\tau)$ is given by

$$\theta(\tau) = \lambda\beta\tau^{\beta-1}, \tag{15.14}$$

where λ is the failure rate before the development program is initiated, and $0 < \beta < 1$. Note that this is a deterministic function that decreases with τ increasing.

Crow (1974) models the number of modifications, $N(\tau)$, carried out during the development period as a non-homogeneous Poisson process with the intensity function given by (15.14), and τ replaced by t. The failure rate $\theta(\tau)$ after development for a period τ is given by (15.14), and the expected number of design modification is given by

$$E[N(\tau)] = \int_0^\tau \theta(t)dt. \tag{15.15}$$

In real life, the reliability improvement at the end of the development period is uncertain. Murthy and Nguyen (1987) build on the Crow's model and treat changes in the failure rate after each modification as random variables. As a result, the failure rate at the end of the development period is uncertain.

Hussain and Murthy (1999) propose another model where the outcome of the development process is uncertain. Their model is as follows. As the development program continues, θ decreases, thereby improving the reliability of the product; however, the outcome $\tilde{\theta}(\tau)$ after a development (testing) period of τ is uncertain. Let θ_o denote the initial value of θ, that is, the value before the development program is initiated, and let θ_m be the limiting minimum value after development for an infinite time. The term $\tilde{\theta}(\tau)$ is obtained by

$$\tilde{\theta}(\tau) = \theta_o - (\theta_o - \theta_m)Z_\tau, \tag{15.16}$$

where Z_t is a random variable which is a function of the development time τ and is modelled as

$$\mathrm{Z}_\tau = [1 - \exp(-\rho\tau)]\mathrm{Y}, \tag{15.17}$$

where $Y \in [0, 1]$ is a random variable distributed according to a beta distribution with parameters α and β. The effect of τ, the development period, is modelled in a deterministic manner where $\rho > 0$. This implies that conditional on Y, $E[Z_t]$ increases, and $E[\tilde{\theta}(\tau)]$ decreases as τ increases.

The cost of development, $C_d(\tau)$, is an increasing function of the development period τ, and this increases the unit manufacturing cost. The increase is given by $C_d(\tau)/TQ$, where TQ is the total number of items manufactured.

3.2 Manufacturing Quality.

3.2.1 Quality Notion. The process used for manufacturing a product depends on the demand for the product and is determined by economic considerations. If the demand is high, then it is economical to use a continuous production process. If the demand is low to medium, then it is more economical to use a batch production process, where items are produced in lots (or batches).

Due to variability in the manufacturing process, some of the items do not conform to design specifications, and these are termed "non-conforming" in contrast to the remaining which are termed "conforming". The characteristics (e.g., reliability) of a non-conforming item are inferior to a conforming item. We call this "manufacturing quality", and higher quality implies fewer non-conforming items being produced in a probabilistic sense.

Let $F_c(x)$ and $F_n(x)$ denote the failure distributions of conforming and non-conforming items, where $F_c(x) < F_n(x)$ implying that the reliability of a non-conforming item is smaller than that for a conforming item. One needs to differentiate two types, Types I and II, of non-conforming items.

A Type I non-conforming item occurs where $F_n(x) = 1$ for $x > 0$. This implies that the item is nonfunctional and is detected immediately after it is put in use. This type of non-conformance is usually due to defects in assembly (e.g., a dry solder joint). For a Type II non-conforming item, the mean to first failure (μ) is greater than zero and hence cannot be detected easily as a Type I non-conforming item.

The probability that an item produced is non-conforming depends on the state of the manufacturing process. In the simplest characterisation, the state can be modelled as being either in control or out of control. When the state is in control, all the assignable causes are under control, and, although non-conformance cannot be avoided entirely, the probability that an item produced is non-conforming is very small. When the state changes to out of control, this probability increases significantly. The process starts in control and changes to out of control after a certain (random) number of items are produced, and it stays there until some action is initiated to change it back to in control.

3.2.2 Modelling Occurrence of Non-conforming Items. Let p_{in} and p_{out} denote the probability that an item produced is conforming when the process is in control and out of control, respectively. In general, $p_{in} >> p_{out}$. In the extreme cases, $p_{in} = 1$, implying that all items produced are conforming when the state is in control, and $p_{out} = 0$, implying that all items produced are non-conforming when the process is out of control.

Continuous Production

The manufacturing process starts in control, and, after a random length of time, it changes to out of control. When the process is in control, the probability that an item produced is conforming is p_{in} and that it is non-conforming is $(1 - p_{in})$. Since the failure distributions of the two types of items are $F_c(x)$ and $F_n(x)$, respectively, the failure distribution of an item, $F(x)$, given that the process is in control, is a mixture of these two distributions, namely

$$F_{in}(x) = p_{in}F_c(x) + (1 - p_{in})F_n(x). \tag{15.18}$$

Similarly, given that the process is out of control, the item failure distribution is given by a mixture distribution of the form

$$F_{out}(x) = p_{out}F_c(x) + (1 - p_{out})F_n(x). \tag{15.19}$$

Once the process state changes from in control to out of control, it remains in that state until it is brought back to an in control state through some corrective action.

Batch Production

Here the items are produced in lots of size Q. At the start of each lot production, the state of the process is checked to ensure that it is in control. If the process state is in control at the start of the production of an item, it can change to an out of control state with probability $(1 - q)$ or continuc to be in control with probability q. Once the state changes to out of control, it remains there until completion of the lot. As mentioned previously, an item produced with the state in control [out of control], is conforming with probability $p_{in}[p_{out}]$ where $p_{in} >> p_{out}$.

Let N_c denote the number of conforming items in a lot. The expected value of this (for details, see Djamaludin (1993)) is given by

$$E[N_c] = \frac{q(p_{in} - p_{out})(1 - q^Q)}{(1 - q)} + p_{out}\, Q, \tag{15.20}$$

and the expected fraction of conforming items in a lot of size Q, $\phi(Q)$, is given by

$$\phi(Q) = \frac{q(p_{in} - p_{out})(1 - q^Q)}{(1 - q)Q} + p_{out}. \tag{15.21}$$

It is easily seen that $\phi(Q)$ is a decreasing sequence in Q. This implies that the expected fraction of conforming items in a batch decreases as the batch size (Q) increases.

3.2.3 Quality Improvement. The occurrence of non-conforming items during manufacturing can be controlled using one or more of the following approaches: weeding, prevention, and process improvement. The first two are "on-line" approaches and the third one is an "off-line" approach to controlling reliability degradation. We briefly discuss each of these approaches.

3.2.4 Weeding. The aim of weeding essentially, is to detect non-conforming items before they are released for sale. This requires testing of items produced. For Type I non-conformance, testing takes very little time since a non-conforming item is detected immediately after it is put into operation. In contrast, the detection of non-conforming items for Type II non-conformance involves testing for a significant length of time. In either case, testing involves additional costs.

In life testing, the testing is often terminated before all the items fail. This leads to two types of censoring called Types I and II, respectively. In Type I censoring, all items in the sample are tested for a fixed period of time T, and in Type II censoring, life testing is terminated when a fixed number of items on the test fail. Note that in Type I censoring, the number of failed items at the end of the testing period is a random number, and, in Type II censoring, the total duration of testing is a random variable.

Another issue that needs to be addressed is the level of testing. One can either test 100% or less than 100%. Again, cost becomes an important factor in deciding on the level of inspection. The quality of testing is still another variable that must be considered. If testing is perfect, then every non-conforming item tested is detected. With imperfect testing, not only are non-conforming items not detected, but also a conforming item may be classified as non-conforming. As a result, the outgoing quality (the fraction of conforming items) depends on the level of testing and the quality of testing. The cost of inspection per unit, $C_i(T)$ is an increasing function of the testing period T and, as a result, increases the total manufacturing cost per unit. For 100% testing, the increase is given by $C_i(T)$, and, for less than 100% testing, the increase is $nC_i(T)/Q$, where n is the number of items tested in a lot and Q is the lot size.

Burn-in is used for weeding out Type II non-conforming items. As indicated earlier, the failure distribution of an item is given by a mixture distribution of the form

$$F(x) = pF_c(x) + (1 - p)F_n(x), \tag{15.22}$$

where p is the probability that a produced item is conforming.

Burn-in involves testing items for a period τ. Those that fail during testing are scrapped. The rationale for this is that non-conforming items are more likely to fail than conforming items and, hence, are weeded out.

The probability that a conforming [non-conforming] item will fail during testing for a period τ is given by $F_c(\tau)$ $[F_n(\tau)]$. As a result, the probability that an item that survives the test is conforming is given by

$$p_1 = \frac{p\bar{F}_c(\tau)}{p\bar{F}_c(\tau) + (1-p)\bar{F}_n(\tau)} \quad . \tag{15.23}$$

Since $\bar{F}_n(\tau) < \bar{F}_c(\tau)$, we obtain $p_1 > p$. The failure distribution of an item that survives the test is given by

$$G_1(x) = p_1 F_{c1}(x) + p_2 F_{n1}(x), \tag{15.24}$$

where $F_{c1}(x)$ and $F_{n1}(x)$ are given by

$$F_{c1}(x) = \frac{F_c(x+\tau) - F_c(\tau)}{1 - F_c(\tau)} \quad , \tag{15.25}$$

and

$$F_{n1}(x) = \frac{F_n(x+\tau) - F_n(\tau)}{1 - F_n(\tau)} \tag{15.26}$$

for $x \geq 0$. Note that as τ increases, p_1 (the probability that an item released is conforming) increases, and, hence, the outgoing quality is improved. However, this is achieved at the expense of the useful life of conforming items released being reduced by an amount τ.

Prevention

When the process is in control, the occurrence of non-conforming items is due to uncontrollable factors. When the process is out of control, the occurrence of non-conforming items increases, and the reason for the change in the process state is due to one or more of the controllable factors deviating significantly from their set values.

In the case of continuous production, the process starts in control and changes to out of control with the passage of time. If the change is detected immediately, then it can be brought back into control at once, and the high occurrence of non-conforming items when the state is out of control would thereby be avoided. This is appropriate for Type I non-conformance. For Type II non-conformance, this is not appropriate because testing would involve putting items on a test bed and testing to failure. For Type I failure, attribute control charts can be used to detect the change in the state of the process and for appropriate actions to be initiated.

In the case of batch production, the process starts in control and can change to out of control during the production of a lot. This affects the number of non-conforming items in the lot. As indicated earlier, the expected fraction of non-conforming items in a lot increases with the lot size Q. This implies that the outgoing quality increases as Q decreases. However, this has implications with regard to unit manufacturing cost since setting up each batch production results in additional set-up cost.

The manufacturing cost per unit C_m is given by the equation $C_m = (C_s + QC_p + \psi C_b)/Q$, which increases as Q decreases. Depending on the process being out of control or in control at the end of production of a lot, $\psi = 1$ or 0. C_p is the cost of material and labor per unit, and C_b is the cost to bring the process state back to in control if it was initially out of control.

Process Improvement

The design of the manufacturing process has a significant impact on p_{in}, the probability that an item is conforming when the process is in control. Ideally, one would like to have this probability equal to one so that no item produced is non-conforming. This involves proper design of the experiment to determine the optimal settings for the various controllable factors.

3.3 Service Quality.

3.3.1 Notion of Quality. Service quality has received a lot of attention, and there are several dimensions to assessing customer satisfaction. Some of these are relevant in the context of product warranty. However, in this sub-section, we focus our attention on the quality of repair service to rectify a failure.

An important quality notion is the quality of repair. This can be described in terms of the number of attempts before a failure is rectified. There are many reasons for this to occur. These include inherent difficulty in locating the root cause for the failure, lack of experience on the part of the person carrying out the repair, and so on.

Another variable of interest is the time needed to carry out the repair. This is the sum of the waiting and repair times. The former is influenced by factors such as availability of spares and repair personnel. These are closely linked to the warranty servicing logistics. Better quality of service implies shorter mean times for carrying out repairs.

For a repairable product, when a failure occurs within the warranty period, the manufacturer has the choice of either repair or replacement. The choice influences future failures over the remaining warranty period. Obviously, replacing the failed item by a new one lowers the probability of a subsequent claim and increases customer satisfaction; however this increases the warranty servicing cost.

3.3.2 Quality Improvement. Service quality can be improved through better training of repair personnel and by improving the logistic management. The literature on logistics is vast, and the concepts and techniques discussed can be used for improving service quality in the context of servicing warranty.

4. WARRANTY AND QUALITY: REVIEW OF MODELS

The warranty servicing cost can be reduced by improving quality (design, manufacturing, and servicing). However, improving quality is achieved by additional costs. Hence, quality improvement is worthwhile only if the savings in the expected warranty cost exceed the additional cost incurred.

Many models have been proposed to determine optimal decisions with regard to quality, taking into account the impact of quality on warranty servicing cost. The bulk of these models deal with either design quality or manufacturing quality. Very few papers deal with models that take into account both these qualities and/or service quality. In this section, we review these models.

4.1 Warranty and Design Quality.

4.1.1 Redundancy. As indicated earlier, redundancy involves replication of one or more components. This is possible only for certain products where incorporation of such replication is permissible by the functional design of the item.

Murthy and Hussain (1994) deal with an optimal decision based on the total expected cost (which is the sum of the warranty servicing cost and the manufacturing cost). The results for $k = 2$ are as follow.

Hot Standby

The following characterises whether building in redundancy is worthwhile or not for a non-repairable product sold with FRW policy.

Proposition 1: [Murthy and Hussain (1994)]

Building in redundancy is worthwhile if

$$\frac{C_h}{C_m} > \frac{1 + 2M_H(W) - M_F(W)}{M_F(W) - M_H(W)}, \tag{15.27}$$

where $M_F(t)$ and $M_H(t)$ are the renewal functions associated with distribution functions $F(x)$ and $H(x)$ (given by (15.11) with $k = 2$), respectively. An implication of this is that redundancy is worthwhile if C_h/C_m is high.

Cold Standby

The switch is imperfect so that qs is the probability that the switch will function properly when needed ($q_s = 1$ corresponds to a perfect switch). As a result, the failure density function for the module is given by

$$h(x; q_s) = (1 - q_s)\, f(x) + q\,_s f(x) * f(x), \tag{15.28}$$

where * is the convolution operator.

The following proposition characterises whether building in redundancy is worthwhile or not for a nonrepairable product sold with a *FRW* policy.

Proposition 2: [Murthy and Hussain (1994)]

Building in redundancy is worthwhile if

$$\frac{C_h}{C_s} > \frac{(2+\delta)[1 + M_H(W; q) - [1 + M_F(W)]}{M_F(W) - M_H(W; q)}, \tag{15.29}$$

where $M_H(x)$ is the renewal function associated with the failure density given by (15.28). The implication of this is that redundancy is worthwhile if C_h/C_m and q are large, and δ is small.

Warm Standby

Proposition 2 is applicable where $M_H(x)$ is the renewal function associated with $H(x)$ given by (15.13). For further details, see Hussain (1997).

4.1.2 Reliability Growth . Murthy and Nguyen (1987) develop three different models, Models I-III, for determining optimal reliability improvement, taking into account the impact of reliability on the expected warranty cost. Optimal reliability improvement is obtained by minimising the sum of the expected warranty servicing cost and the development cost. Model I is based on the Crow's model, and the decision variable is the development period τ. The number of developments carried out is a random variable with the expected value given by (15.15). In Model II, the development is stopped after the nth modification is carried out. As a result, the development time and the value of $\theta(\tau)$ are random variables. In Model III, the failure rate of the product changes after each modification, and the amount of reduction in the failure rate is uncertain. The development is stopped after time τ or after the nth modification, whichever occurs first.

Hussain and Murthy (1999) deal with a model where the outcome of the development process is uncertain and is given by (15.16) and (15.17). They assume that Y is distributed according to a beta distribution with parameters α and β so that

$$\mathrm{E[Y]} = \frac{\alpha}{\alpha + \beta}. \tag{15.30}$$

They consider a repairable product sold with a *FRW* policy and with failures under warranty repaired minimally. In the case where the item failure distribution is given by a Weibull distribution with scale parameter $\tilde{\theta}(\tau)$ and shape parameter γ, the following proposition indicates if development is worthwhile or not.

Proposition 3: [Hussain and Murthy (1999)]
If $\mathrm{C_r}(\theta_o^\gamma - \theta_\mathrm{m}^\gamma)\mathrm{T}^\gamma < \mathrm{C_s}/\mathrm{Q}$, then no development is the optimal strategy.

4.1.3 Reliability Growth and Redundancy. Hussain and Murthy (2000) combine both development and redundancy to improve reliability. The development is carried out for a period τ (a decision variable to be selected optimally). The reliability of the product depends on $\tilde{\theta}(\tau)$, which is a random variable as in Hussain and Murthy (1999). If the reliability improvement through development is small, then redundancy can be used as a means of reducing the expected warranty cost. This translates into the following decision problem. If $\tilde{\theta}(\tau) > \theta_C$ (implying reliability improvement through development is small), then use redundancy. On the other hand, if $\tilde{\theta}(\tau) < \theta_C$ (implying reliability improvement through development is large) then do not use redundancy. The decision variable θ_C that needs to be determined optimally, taking into account the option of building in redundancy.

4.2 Warranty and Manufacturing Quality.

A non-conforming item is less reliable than a conforming item. This has serious implications for both buyers and manufacturers. From a buyer's perspective, it leads to increased maintenance costs, and it affects the operational cost because of increased down time. This in turn affects the buyer's satisfaction. From the manufacture's point of view, it leads to increase warranty servicing costs when items are sold with warranty. In addition, items that fail early in their life can seriously affect the reputation of the product and its subsequent sales. This implies that the manufacturer must control reliability degradation during manufacturing.

4.2.1 Weeding. Type I Non-conformance Maruchek (1987) deals with an optimal single sampling plan to prevent the resulting liability caused by non-conforming items (of Type I non-conformance) being released to the customer. Items are produced in lots of size Q and the fraction defective (p_2) in each lot is a random variable distributed according to beta distribution. A sample of n items is inspected: if the number of defectives in each sample is less than c, the lot is released for sale; if not, the lot is rejected. Two cases are considered: (i) rejected lots are 100% inspected and (ii) rejected lots are scrapped. According to the model it is assumed that the expected percentage of consumers receiving a non-conforming unit who demand settlement and the average settlement per claim is known. She derives conditions for inspection or no inspections to be the optimal strategy.

Mamer (1987) deals with the trade-off between product reliability and warranty length. Reliability varies from item to item and is modelled by a parameter θ which is a random variable. For a conforming item $\theta = \theta_c$; and for a non-conforming item, $\theta = \theta_n$ where $\theta_c < \theta_n$. In his model, two types of failures are considered, one due to wear (modelled by either $F_c(x)$ or $F_n(x)$) and the other due to accidental damage (which is assumed to occur according to a Poisson process). As a result, item replacements occur according to a renewal process with renewal times given by $Z = \min\ (X, Y)$, where X is distributed according to $F_c(x)$ or $F_n(x)$, and Y is exponentially distributed with parameter λ. A consumer may not replace a failed item after the expiration of warranty with probability α.

Mamer considers three types of warranty policies: (i) pro-rata, (ii) renewing free replacement and (iii) non-renewing free replacement. He derives the optimal warranty length for each policy by equating the expected discounted cost of ownership with some target level of ownership cost.

Ritchken *et al.*[1989] deal with the optimal design of quality control plans based on the long run average profit maximization. Their model incorporates the following costs: (i) production cost which is an increasing function of p (the probability that an item is conforming), (ii) cost of inspection, (iii) cost of rework, and (iv) cost of service capacity for repairing failed items returned by consumers.

The long run average profit per unit time depends on the quality control plan ϕ, the servicing capacity, k, and the manufactured quality (or the targeted design) θ, and is obtained by,

$$\hat{\pi}(\phi, k, \theta) = \frac{E\{\pi_j\}}{E\{\tau_j\}}, \tag{15.31}$$

where τ_j , $j = 1, 2, ..$ is the number of time periods between successive instances in which servicing capacity k is exceeded, and π_j is the net profit over period τ_j. The expected profit over a renewal period of length t is the expected income minus the expected cost. The expected cost is the sum of inspection and rework costs.

Tapiero and Lee (1989) deal with a problem similar to that of Ritchken *et al.*(1989). They consider two cases: non-degenerate and degenerate probability distributions for the probability that an item is non-conforming, and a simple single sampling plan where the decision to inspect a lot is based on an outcome of an inspection of a sample.

Chen (1991) models the quality of products as a function of the process design, where θ_U and θ_L are the upper and lower tolerance limits of a possible range of consumer satisfaction and of the average (μ) and the variance (σ) production process. The items are produced in batches, and p_1 is the proportion of conforming items in a lot, that is, items with characteristics within the two limits $[\theta_L, \theta_U]$. The failure time of both non-conforming and conforming items is distributed according to the Weibull distribution. He studies various inspection schemes such as single, double, and sequential sampling plans. He uses a model proposed by Balcer and Sahin (1986) to calculate the expected warranty cost and studies the effect of an inspection scheme on the total cost.

Type II Non-conformance

Mann and Saunders (1969) and Mann (1970) assume that the parameter of the failure distribution (assumed Weibull) varies from lot to lot. They consider the optimal warranty period W to be offered for each lot. Their decision is based on taking a sample of n items from each lot and testing them at a fixed level of stress until a number of items in the sample fails (Type II censoring).

Lie and Chun (1987) deal with quality improvement through a single sample attribute plan with non-destructive testing. They consider three types of warranty policies: (i) failure free for non-repairable items, (ii) failure free for repairable items, and (iii) a rebate refund policy for non-repairable items. The optimal sample size n and the acceptance number c are found by minimising the total cost comprised of the quality control cost and warranty servicing cost. An optimum sample size n^* for a given c and fraction of non-conforming items in a lot (p_2) are given by the following proposition.

Proposition 4: [Lie and Chun (1987)]

The optimal n^* satisfies $\chi^2_{1-\beta}/(2p_1) \leq n* \leq \chi^2_{\alpha}/(2p_0)$ where the degrees of freedom are $2c+2$. n^* is the lower bound of n if $(C_i + C_r p_2)/(k_1 p_2) \geq (\bar{C}_W)_c - (\bar{C}_W)_n$ and the upper bound of n if $(C_i + C_r p_2)/(k_1 p_2) \leq (\bar{C}_W)_c - (\bar{C}_W)_n$, where

$$k_1 \equiv \begin{cases} 1 & \text{for repairable items} \\ 1 - p_2 & \text{for non - repairable items} \end{cases}$$

and $\bar{C}_W$ is the average warranty cost per item with subscripts c for the conforming item and n for the non-conforming item. The acceptable quality level is p_0 , and the limiting quality level is p_1. Their model assumes zero testing time, which is an unrealistic assumption for Type II non-conforming item.

Murthy *et al.*(1993) study the optimal testing period (with Type I censoring) for items sold with warranty. The production process is assumed to be in steady state. The occurrence of non-conforming items is purely due to variations in the uncontrollable factors. Therefore, each item produced is conforming (non-conforming) with probability p_1 (p_2). The inspection cost per item for testing period T is $C_i(T)$, which is an increasing function of T. The optimal T^* is obtained by minimising the asymptotic total cost per item sold, which is the sum of the asymptotic manufacturing cost and the asymptotic warranty cost per item sold. They consider three different warranty policies and consider conditions to determine if testing is the optimal strategy or not. In the case where conforming and non-conforming items have exponential failure distributions with failure rate λ_1 and λ_2, respectively, the results are given by the following propositions.

Proposition 5: [Murthy et al.(1993)]

For items sold with a *FRW* policy and failures under warranty repaired minimally, no life testing is the optimal strategy if

$$(p\lambda_1 + (1-p)\lambda_2)[C_m + C_d + C_i(0^+)] \\ -p(1-p)(\lambda_2 - \lambda_1)^2 W C_R \geq 0, \tag{15.32}$$

or if

$$(p\lambda_1 + (1-p)\lambda_2)[C_m + C_d + C_i(0^+)] \\ +C_i'(0^+) - p(1-p)(\lambda_2 - \lambda_1)^2 W C_R \geq 0 \tag{15.33}$$

and

$$C_i'(T) \geq C_i'(0^+)(\lambda_2 e^{-\lambda_2 T} - \lambda_1 e^{-\lambda_1 T})/(\lambda_2 - \lambda_1) \tag{15.34}$$

for all $T \geq 0$, where $C_i(O^+) > 0$ corresponds to a non-zero set-up cost for life testing.

Proposition 6: [Murthy et al.(1993)]

For items sold with a *FRW* policy, failures under warranty repaired minimally, and where $C_i(O^+) = 0$, then life testing is the optimal strategy if

$$(p\lambda_1 + (1-p)\lambda_2)(C_m + C_d) + C_i'(0^+) \\ -p(1-p)(\lambda_2 - \lambda_1)^2 W C_R < 0. \tag{15.35}$$

Proposition 7: [Murthy et al.(1993)]

For items sold with a *PRW* policy and unconditional refund, no life testing is the optimal strategy if

$$(p\lambda_1 + (1-p)\lambda_2)[C_m + C_d + C_i(0^+)]$$
$$-p(1-p)(\lambda_2 - \lambda_1)[\psi_2(W) - \psi_1(W)] \geq 0 \qquad (15.36)$$

or if

$$(p\lambda_1 + (1-p)\lambda_2)[C_m + C_d + C_i(0^+)]$$
$$+C_i'(0^+) - p(1-p)(\lambda_2 - \lambda_1)[\psi_2(W) - \psi_1(W)] \geq 0 \qquad (15.37)$$

and (15.34) holds for all $T \geq 0$, where

$$\psi_2(W) - \psi_1(W) = [kS(1-\beta) + C_h](e^{-\lambda_1 W} - e^{-\lambda_2 W})$$
$$+[kS\beta/(\lambda_1\lambda_2 W)][\lambda_2(1 - e^{-\lambda_1 W}) - \lambda_1(1 - e^{-\lambda_2 W})] \geq 0 \qquad (15.38)$$

for all $W > 0$.

Proposition 8: [Murthy et al.(1993)]

For the *PRW* policy with unconditional refund and where $C_i(O^+) = 0$, then life testing is the optimal strategy if

$$(p\lambda_1 + (1-p)\lambda_2)(C_m + C_d) + C_i'(0^+)$$
$$-p(1-p)(\lambda_2 - \lambda_1)[\psi_2(W) - \psi_1(W)] < 0. \qquad (15.39)$$

Proposition 9: [Murthy et al.(1993)]

For non-repairable items sold with *FRW* policy (so that replacements under warranty involve new items), no life testing is the optimal strategy if, and only if ,

$$\alpha(T)\,[J(T) - J(0)] = \{[1 - \alpha(T)]\,(C_m + C_d) + C_i(T)\}\,[1 + M_T(W)]$$
$$-(C_m + C_h)\alpha(T)\,[M_0(W) - M_T(W)] \qquad (15.40)$$

is non-negative for all $T > 0$. Expressions of $M_O(W)$ and $M_T(W)$ can be obtained analytically using Laplace transforms (see Barlow and Proschan (1965) for details).

Kwon (1996) deals with a method to find the optimal sampling plans that minimise the expected average cost per lot. The accepted lots are sold under pro-rata warranty. The item failure distribution is Weibull with a known shape parameter and an unknown scale parameter. A sample of n items is drawn at random and life tested simultaneously. The test is terminated when $r (r \leq n)$ items have failed (Type II censoring). The expected average cost per lot of size Q is a function of (n, r, c). The optimal sampling plan is obtained by minimising the expected average cost per lot.

Environmental Stress Screening (ESS)

Environmental Stress Screening (ESS) is sometimes confused with burn-in since they both reduce early field failures. The difference is that burn-in is conducted under ambient conditions, while ESS is conducted under accelerated environmental conditions. The environmental stresses are more severe than the normal operating conditions and, in some cases, different than the operational conditions.

Coleman (1990) suggests the use of ESS to improve product reliability. The screening process can be applied at various stages during production (e.g. component, assembly or on a complete system). The type of stress used depends on the type of faults to be detected, and should be selected for maximum effectiveness on weak units without causing damage to good units. The ESS facilities usually represent a significant capital investment and increased operating costs which can be justified only if the field servicing and repair costs are high.

Kar and Nachlas (1997) study a net-profit model to optimally select the warranty period, burn-in , and stress variables to minimise the expected net-profit, Z_T, given by

$$Z_T = Z(W) - C_1(S^{(a)}, \tau) - C_\tau(\tau) - E(C_w(W)), \tag{15.41}$$

where $Z(W)$ is the expected revenue function, $C_1(S^{(a)}, \tau)$ is the burn-in cost, τ is the burn-in time, $(S^{(a)}$ is a vector of burn-in variables with $^{(a)}$ implying acceleration, and $C_\tau(\tau)$ is the failure cost during burn-in. The optimal τ , W, and $(S_i^{(a)}$ are obtained by minimising Z_T, subject to constraints that reflect the extent to which the system components can endure various stress variables.

Pohl and Dietrich (1995) deal with the optimal screening duration and develop a three level ESS model for a complex electronic system (an electronic printed circuit board, a PCB). They investigate the possibility that ESS may actually reduce the screened population useful life. Screening is performed at three levels: (i) component, (ii) board, and (iii) system levels. The system consists of i different types of components which are assembled into l different types of PCBs using j different types of PCB connections. Failure distribution at the component is assumed to be a mixture of exponential distribution and a mixture of Weibull distributions at the board and system levels.

The system life cycle cost is the sum of fixed screening cost at the component level, cost of screening components for a specified duration, cost of replacing the component during PCB screening, system screening, and the warranty servicing cost. The PCB cost depends on how much screening is done at each assembly level. The PCB life cycle cost is the sum of fixed screening cost per PCB, cost of screening PCB for a specified duration, cost of repairing PCB during system screening, and repair cost during warranty period. The optimal screening duration is obtained by minimizing the system life cycle cost.

Pohl and Dietrich (1999) deal with a model that is an extension of their earlier model. Here, detection of failures is imperfect so that a failed item may escape to the next level of assembly before being detected. The system studied is similar to that in Pohl and Dietrich (1995); however, they examine multi-level multi-component structures, and the lifetimes of components and connections follow a phase-type probability distribution.

4.2.2 Burn-In. Nguyen and Murthy (1982) study optimal burn-in to minimise the manufacturer's expected cost, defined as the sum of the expected manufacturing and warranty servicing costs. They consider both repairable and non-repairable products.

For repairable items sold with FRW policy with failures during burn-in repaired minimally, the expected cost of manufacturing per item, $C_M(\tau)$ is given by

$$C_M(\tau) = C_m + c_1 + c_2\tau + C_r \int_0^\tau r(t)dt, \tag{15.42}$$

where τ is the burn-in duration, C_m is the manufacturer cost per item, c_1 is the fixed setup cost of burn-in per unit, c_2 is the cost per unit time of burn-in per unit, C_r is the expected cost of each repair, and C_h is the handling cost for each warranty claim. For further details, see Nguyen and Murthy (1982). The manufacturer's expected cost per unit, $E[C_s(W;\tau)]$, is given by,

$$E[C_s(W;\tau)] = C_M(\tau) + (C_r + C_h) \int_0^W r(\tau + t)dt \tag{15.43}$$

For non-repairable items sold with FRW policy, the expected cost of manufacturing per item, $C_M(\tau)$ is given by,

$$C_M(\tau) = \frac{C_m + c_1 + c_2 \int_0^\tau \bar{F}(t)dt}{\bar{F}(\tau)}. \tag{15.44}$$

Therefore, the manufacturer's expected cost per unit, $E[C_s(W;\tau)]$, is given by

$$E[C_s(W;\tau)] = C_M(\tau) + [C_M(\tau) + C_h]M_\tau(W), \tag{15.45}$$

where $M_\tau(W)$ is the renewal function associated with the distribution function $F_\tau(t)$, and $C_M(\tau)$ is given by (15.44).

For non-repairable items sold with a PRW policy with a rebate function given by

$$s(t) = \begin{cases} \vartheta P(1 - \varsigma t/W), & if \quad 0 \le t \le W \\ 0, \quad if \quad t \ge W \end{cases} \tag{15.46}$$

where $0 < \vartheta \le 1$, $0 \le \varsigma \le 1$, and P is the sale price per item, the manufacturer's expected cost per unit is given by

$$E[C_s(W;\tau)] = \frac{C_m + c_1 + c_2 \int_0^\tau \bar{F}(t)dt}{\bar{F}(\tau)} +$$

$$\vartheta P \frac{W\bar{F}(W) - (1-\varsigma)W\bar{F}(W+\tau) - \varsigma \int_0^W \bar{F}(\tau + t)dt}{W\bar{F}(W)}. \tag{15.47}$$

The optimal burn-in , τ^*, is obtained by minimising (15.43), (15.45), or (15.47) depending on the warranty policy. If the failure rate $r(t)$ is bathtub shaped with $r(t)$ strictly decreasing over $0 \le \tau \le \tau_m (\tau_m > 0\ )$, then $\tau^* < \tau_m$.

4.2.3 Preventing Occurrence of Non-conforming Items. Lot size production

Djamaludin *et al.*(1994) deal with quality control through lot sizing. They consider a model where the process starts in control. The state can change anytime during the production of an item to an out of control state with probability $(1 - q)$ or remain in control with probability q. Once the process changes, it stays in that state until the last item in the lot is produced. $p_{in}[p_{out}]$ is the probability of that item produced when the process state is in control [out of control] is conforming. The asymptotic manufacturing cost per item, $C_M(Q)$, is given by

$$C_M(Q) = C_m + \left[C_s + (1 - q^Q)C_b\right]/Q. \tag{15.48}$$

When items are sold with a FRW policy and failures under warranty are repaired minimally, the asymptotic warranty cost per item, $C_M(Q)$, is given by

$$C_w(Q) = C_r\left\{p_1 \int_0^W r_c(t)dt + p_2 \int_0^W r_n(t)dt\right\}, \tag{15.49}$$

where p_1 [p_2] is the fraction of conforming [non-conforming] items in a lot. The optimal lot size is obtained by minimising the total cost given by the sum of the above two costs. In general, one needs to use a numerical procedure to obtain the optimal solution. If $q \approx 1$, then one can use the following approximation:

$$q^Q \approx 1 + Qln(q) + \{Qln(q)\}^2/2. \tag{15.50}$$

Using this approximation and treating Q as a continuous variable, an approximate optimal lot size Q^{**} can be obtained analytically and is given by:

$$Q^{**} = q\left[\frac{2C_s}{(1-q)\{C_r(p_{out}-p_{in})qA - C_b(1-q)\}}\right]^{1/2}. \tag{15.51}$$

where A is given by

$$A = \left[\int_0^W r_n(t)dt - \int_0^W r_c(t)dt\right]$$

4.2.4 Preventing Occurrence of Non-conforming Item and Weeding. Djamaludin *et al.*(1995) deal with a model similar to that in Djamaludin *et al.*(1994), except that it also involves testing a fraction of items in some lots. At the end of each lot production of Q items, the state of the process is assessed. If it is found to be in control, then the lot is released without testing. However, if the state is found to be out of control, the last K items in the lot are tested for a period of T. Items which fail during testing are scrapped, and the others which are not tested and those which survive the test, are released for sale. An item released for sale can be one of the four types: Type A (conforming and not tested), Type B (non-conforming and not tested), Type C (conforming and tested) and type D (non-conforming and tested).

The asymptotic manufacturing cost per item is given by

$$C_M(Q,K,T) = \frac{QC_m + C_s + (1-q^Q)B + E[\varepsilon_i]C_{sc}}{L - E[\varepsilon_i]}, \tag{15.52}$$

where B is given by

$$B = \{C_b + KC_i(T)\}$$

where C_{sc} is scrapping cost, C_b is the cost of bringing the process in control if it is out of control, and $E[\varepsilon_i]$ is the expected number of items scrapped in lot i.

The asymptotic warranty cost per item (FRW for repairable product) is given by

$$C_w(Q,K,T) = C_r[\gamma_1(W)\{\frac{q}{1-q}(1-q^Q)\}$$

$$+\gamma_2(W)\{Q - \{\frac{q}{1-q}(1-q^Q)\} + \{\beta_1(\mathrm{T},\mathrm{W}) - \gamma_1(\mathrm{W})\}\mathrm{E}_{\mathrm{KQ}}$$

$$+\{\beta_2(\mathrm{T},\mathrm{W}) - \gamma_2(\mathrm{W})\}\{\mathrm{K}(1-\mathrm{q}^{\mathrm{Q}}) - \mathrm{E}_{\mathrm{KQ}}\}] \qquad (15.53)$$

where

$$E_{KQ} = \{q/(1-q)\}\{q^{Q-K} - q^Q\} - Kq^Q,$$

$$\gamma_{\mathrm{j}}(\mathrm{W}) = \mathrm{p_j}\int_0^{\mathrm{W}} \mathrm{r_c(t)dt} + (1-\mathrm{p_j})\int_0^{\mathrm{W}} \mathrm{r_n(t)dt},$$

and

$$\beta_{\mathrm{j}}(\mathrm{T},\mathrm{W}) = \mathrm{p_j}\bar{F}_c(T)\int_0^{\mathrm{W}} \hat{\mathrm{r}}_{\mathrm{c}}(\mathrm{t})\mathrm{dt} + (1-\mathrm{p_j})\bar{\mathrm{F}}_{\mathrm{n}}(\mathrm{T})\int_0^{\mathrm{W}} \hat{\mathrm{r}}_{\mathrm{n}}(\mathrm{t})\mathrm{dt}.$$

Subscript $j = in$ or out, depending on the process state, is in control or out of control at the end of the production of a lot; $\hat{r}_c$ [$\hat{r}_n$] is the failure rate of conforming [non-conforming] item which survives the test and is released for sale (for further details, see Djamaludin et al.(1995)).

The optimal values for Q^*, K^* and T^* are obtained by minimizing the asymptotic total cost per item which is the sum of the above two costs.

Djamaludin *et al.*(1997) deal with a model similar to that in Djamaludin *et al.*(1995), except that, at the end of production of a lot, the process state is not known. The testing scheme is as follows. Items are numbered sequentially (1 through Q) in the order in which they are produced in a lot. The item Q (last item produced in a lot) is life tested for a period of T. If it survives, then it and the remaining items in the lot are released with no further testing. If it fails, then it is scrapped, and item K is tested for a period of T. If item K survives, then it and the remaining items in the batch are released with no further testing. However, if it fails, then item $(K+1)$ to $(Q-1)$ are life tested for a period of T. Those survive the test are released with the first $(K-1)$ items which are released with no testing. An item released for sale can be one of the four types, Type A, B, C and D, as in Djamaludin *et al.*(1995). They consider three warranty policies: (i) FRW for a repairable product, (ii) PRW for a non-repairable product, and (iii) FRW for a non-repairable product. The optimal Q^*, K^* and T^* is obtained by minimising the asymptotic total cost per item released, which is the sum of the asymptotic manufacturing cost and the asymptotic warranty servicing cost.

4.2.5 Releasing with No Testing. Murthy *et al.*(1995) deal with the case where the manufacturer is unable to carry out life testing before releasing the items for sale. As a result, both conforming and non-conforming items are released for sale. Since non-conforming items have a higher failure rate, they tend to fail early, and this affects consumer satisfaction. One way of overcoming this is through a consumer incentive warranty policy. Under this policy, should the first failure occur within a period $T(T \leq W)$, the customer is offered the following two options. Option 1 is a total refund (money back guarantee), and Option 2 is a new replacement item with a new warranty identical to the original warranty and a lump sum C_{ls} as compensation for the inconvenience caused due to the early failure. All failures beyond T and within warranty are repaired minimally at no cost to the consumer. The choice of Option 1 implies the loss of a customer and this is captured through a penalty cost C_g to the manufacturer (for the loss of future sales and goodwill). Let T_0 denote the consumer's expectation of minimum reliability for the product. The manufacturer's expected total warranty cost per item for items sold with a FRW policy is given by

$$C(T;W) = F(T)\left\{[F(T_0)/F(T)](S + C_h + C_g) + D\right\}$$
$$+F(T)\{(C_r + C_h)E\} \tag{15.54}$$

where D and E are given by

$$D = [1 - (F(T_0)/F(T))]\,[C_m + C_h + C_{ls} + C(T;W)]$$

$$E = [p_1\bar{F}_c(T)\int_T^W r_c(t)dt + (1 - p_1)\bar{F}_n(T)\int_T^W r_n(t)dt]$$

and $F(T) = pF_c(T) + (1-p)F_n(T)$. They prove the following:

Proposition 10: [Murthy et al.(1995)]
$T^* < W$; if

$$F(T)\left[\int_{T_0}^W r_1(t)dt - (S + C_h + C_g)/(C_r + C_h)\right] >$$
$$[(C_m + C_h + C_{ls})/(C_r + C_h) - 1], \tag{15.55}$$

then $T^* > T_0$.

4.3 Warranty and Service Quality. The manufacturer must service all claims for items sold with a FRW policy. As a result, the servicing strategy has an impact on subsequent claims and costs over the remainder of the warranty period. If a failed item is always replaced by a new one, then fewer subsequent failures result, which is better from the buyer's point of view. However, this strategy increases the warranty servicing cost since repair, in general, is cheaper than replacement by a new item. Nguyen and Murthy (1989) deal with the choice between replacement and repair and take into account the different costs associated with each action. They look at several alternate decision rules for choosing between these two actions.

When the time to repair is not negligible, the buyer is deprived of the usage of a failed item which is undergoing repair. One way of improving buyer satisfaction is through better quality of service, where quality is defined in terms of measures such as mean repair time or the probability of repair being completed within a specified time period. Another approach is the use of loaners. Here a failed item is replaced by a working item on loan till the failed item is repaired and returned. Karmarkar and Kubat (1983) deal with this topic.

Maintainability deals with repair subsequent to failure at the product design stage. Through proper design (such as modular structure and built-in diagnostics), the repair time and cost can be significantly reduced, thus, improving the quality of service. Hegde and Kubat (1989) deal with this topic.

For certain products (for example, a large pump in a chemical plant), the warranty servicing requires the manufacturer (or his agent) to carry out the servicing on site. This implies a travelling repairman as opposed to a fixed repair facility. In this case, the service quality is influenced by a number of factors such as the number of repairmen, scheduling rules, and so on. Agnihothri and Karmarkar (1992) deal with this topic.

4.4 Integrated Models. In this section, we briefly review the papers which deal with two or more notions of quality in an integrated manner.

Hussain and Murthy (1998) consider a model that integrates design quality and manufacturing quality. An improvement in reliability is achieved through the use of redundancy (where a component is replaced by a module involving two or more components) and through inspection and testing. As a result, one needs to choose between the following two options: (i) testing at component level to weed out non-conforming items and (ii) testing at module level without any testing at the component level. Hussain and Murthy (1998) derive some results for different types of redundancy for Type I non-conformance. Hussain and Murthy (1996) deal with the case where the non-conformance is Type II, and testing involves burn-in.

Murthy and Kumar (2000) consider a model that integrates the three different types of quality: design, manufacturing, and service. Their model involves three players: manufacturer, retailer, and buyers. The design quality is a decision variable under the control of the manufacturer. The quality of manufacturing is dependent on the design quality and the type of manufacturing process involved. The retailer carries out the inspection and testing of each item prior to its sale and rectifies any non-conforming item. The retailer also services the warranty claims. The quality of service is uncertain in the sense that not all repairs are successful and, as a result, a failed item might need to be repaired more than once to rectify the failure. They use a game-theoretic approach to determine the optimal strategies for the manufacturer and the dealer and present analytical results for a simple case where the failure distribution is exponential and failures are repaired minimally.

5. CONCLUSION AND TOPICS FOR FUTURE RESEARCH

As can be seen from the previous section, the bulk of the papers dealing with warranty and quality deal with either design quality or manufacturing quality. Papers dealing with warranty and servicing quality or integrated models involving two or more types of quality are very limited in number. In this section, we briefly discuss some topics for future research.

The models in sections 4.1 and 4.2 assume that all buyers are similar in their usage pattern. In real life, this is not the case, and the usage intensity varies across the population. Kim *et al.* (2000) deal with this topic, review the literature and proposing some new models. Quality improvement strategies (during design and/or manufacturing) take into account the effects of variations in usage intensity have not been studied and offer considerable scope for research in the future.

The models for reliability growth are simplistic and do not model the real world very effectively. There is a need for building new stochastic models that capture the uncertain outcome of reliability in a more realistic manner. Another factor is the assessment of the reliability achieved at the end of the development period. Since testing is often limited and done in an accelerated mode, the assessed reliability can differ significantly from the true reliability. This uncertainty needs to be captured in the model in order to derive optimal design quality for items sold with a warranty.

The models for quality variations in manufacturing discussed in section 3.2.2 are simple. The process starts in control and changes to out of control with time and/or production. In real life, the degradation is more complex, involving both gradual and sudden changes. This implies that one needs more complex dynamical models to model the changes to the state and the output quality of the process. See Murthy and Djamaludin (1990) for more on this topic.

For Type II non-conformance, weeding out non-conforming items requires burn-in. This takes time; however, the time can be reduced by accelerated testing. Most models for accelerated testing assume a deterministic relationship between the failure times at the normal level and the higher stress level. In real life, this is not the case. Often, the relationship is uncertain, and testing at higher stress levels can introduce new modes of failure (see Murthy and Tang (2000)). The implication is that these issues need to be taken into account in models dealing with warranty and quality.

Service quality in the context of warranty is still a largely unexplored area. There is considerable opportunity to build new models that incorporate various logistic related issues (for example, spares) and guaranteed service performance measures.

Finally, the design of more integrated models, in which all three different quality notions are addressed in an effective manner, is an open and challenging area for future research. However, one would need to take into account the interaction between manufacturer and retailer (or third party service agent) and various options for cooperation and competition. This raises several other issues, such as moral hazard and incentive contracts etc., that also need to be incorporated into the models.

References

[1] Agnihothri, S.R. and Karmarkar, U.S., Performance evaluation of service territories. *Operations Research*, **40** (1992) 355-366.

[2] Balcer, Y. and Sahin, I., Replacement costs under warranty: cost moments and time variability. *Operations Research*, **34** (1986) 554-559.

[3] Barlow, R.E. and Hunter, L.C., Optimum preventive maintenance policies. *Operations Research*, **8** (1960) 90-100.

[4] Barlow, R.E. and Proschan, F., *Mathematical Theory of Reliability.* Wiley, New York, 1965.

[5] Blischke, W.R., Mathematical models for analysis of warranty policies. *Mathematical Computational Modelling*, **13** (1990) 1-16.

[6] Blischke, W.R. and Murthy, D.N.P., Product warranty management - I: A taxonomy for warranty policies. *European Journal of Opera-*

tional Research, **62** (1992) 127-148.

[7] Blischke, W.R., and Murthy, D.N.P., *Warranty Cost Analysis.* Marcel Dekker, Inc., New York, 1993.

[8] Blischke, W.R. and Murthy, D.N.P. [Eds.], *Product Warranty Handbook.* Marcel Dekker, Inc., New York, 1995.

[9] Brennan, J.R., *Warranties, Planning, Analysis and Implementation.* McGraw-Hill, Inc., New York, 1994.

[10] Chen, X., *Product Quality and Warranty Cost.* Ph.D. Dissertation, The University of Wisconsin-Milwaukee, 1991.

[11] Coleman, G., *Stress screening for product reliability.* Electronic Production, March, (1990) 65-68.

[12] Crow, L.H., *Reliability analysis for complex, repairable system, Reliability and Biometry.* edited by F. Proschan and R.J. Serfling, SIAM, Philadelphia, (1974) 379-410.

[13] Djamaludin, I., *Quality Control Schemes for Items Sold with Warranty.* Ph.D. Dissertation, The University of Queensland, Brisbane, Australia, 1993.

[14] Djamaludin, I., Murthy, D.N.P. and Blischke, W.R., *Bibliography on warranties.* Product Warranty Handbook, edited by W.R. Blischke and D.N.P. Murthy, Chapter 33, Marcel Dekker, Inc., New York, 1995.

[15] Djamaludin, I., Murthy, D.N.P. and Kim, C.S., *Warranty and preventive maintenance. Proceedings of the First Western Pacific and Third Australia-Japan Workshop on Stochastic Models in Engineering.* Technology and Management, September, Christchurch, (1999) 81-90.

[16] Djamaludin, I., Murthy, D.N.P. and Wilson, R.J., Quality control through lot sizing for products sold with warranty. *International Journal of Production Economics*, **33** (1994) 97-107.

[17] Djamaludin, I., Wilson, R.J. and Murthy, D.N.P., Lot sizing and testing for items with uncertain quality. *Mathematical and Computer Modelling*, **22**(10-12), (1995) 35-44.

[18] Djamaludin, I., Wilson, R.J. and Murthy, D.N.P., *Lot sizing and life testing for quality improvement of items sold with warranty.* Optimization in Quality Control, edited by K.A. Al-Sultan and M.A. Rahim, Kluwer Academic Publishers, Boston, 1997.

[19] Duane, J.T., Learning curve approach to reliability monitoring. *IEEE Transactions on Aerospace*, **2** (1964) 563-566.

[20] Fries, A. and Sen, A., A survey of discrete reliability-growth models. *IEEE Transactions on Reliability*, **45** (1996) 582-604.

[21] Garvin, D.A., *Competing on the eight dimensions of quality*. Harvard Business Rev., July, (1987) 101-109.

[22] Hussain, A.Z.M.O., *Warranty and Product Reliability*. Ph.D Dissertation, The University of Queensland, Brisbane, Australia, 1997.

[23] Hussain, A.Z.M.O. and Murthy, D.N.P., Warranty and optimal redundancy with uncertain quality. *Proceedings of the Second Australia-Japan Workshop on Stochastic Models*, Gold Coast, September, 226-235, (1996) (Also to appear in the International Journal of Mathematical and Computer Modelling).

[24] Hussain, A.Z.M.O. and Murthy, D.N.P., Warranty and redundancy design with uncertain quality. *IIE Transactions*, **30** (1998) 1191-1199.

[25] Hussain, A.Z.M.O. and Murthy, D.N.P., Warranty and reliability improvement through product development. *Proceeding of the First Western Pacific and Third Australia-Japan Workshop, Stochastic Models in Engineering*, Technology and Management, September, Christchurch, (1999) 166-175.

[26] Hussain, A.Z.M.O. and Murthy, D.N.P., Warranty and redundancy design with uncertain quality. *IIE Transactions*, **30** (2000) 1191-1200.

[27] Kar, T.R. and Nachlas, J.A., Coordinated warranty and burn-in strategies. *IEEE Transactions on Reliability*, **46**(4), (1997) 512-518.

[28] Karmarkar, U.S. and Kubat, P., The value of loaners in product support. *IIE Transactions*, **15** (1983) 5-11.

[29] Hegde, G.G. and Kubat, P., Diagnostic design: A product support strategy. *European Journal of Operational Research*, **38** (1987) 35-43.

[30] Kim, C.S., Djamaludin, I. and Murthy, D.N.P., Warranty cost analysis with heterogeneous usage intensity. *Accepted for publication in International Transactions of Operations Research*, 2000.

[31] Kwon, Y.I., A Bayesian life test sampling plan for products with Weibull lifetime distribution sold under warranty. *Reliability Engineering and System Safety*, **53** (1996) 61-66.

[32] Lie, C.H. and Chun, Y.H., Optimum single-sample inspection plans for products sold under free and rebate warranty. *IEEE Transactions on Reliability*, **R-36**(5) (1987) 634-637.

[33] Mamer, J.W., Discounted and per unit costs of product warranty. *Management Science*, **33**(7) (1987) 916-930.

[34] Mann, N.R., Warranty periods based on three ordered sample observations from a Weibull populations. *IEEE Transactions on Reliability*, **R-19**(4) (1970) 167-171.

[35] Mann, N.R. and Saunders, S.C., On evaluation of warranty assurance when life has a Weibull distribution. *Biometrika* **56**(3) (1969) 615-625.

[36] Marucheck, A.S., On product liability and quality control. *IIE Transactions*, **19**(3) (1987) 355-360.

[37] McGuire, E.P., *Industrial Product Warranties: Policies and Practices.* The Conference Board Inc., New York, 1980.

[38] Murthy, D.N.P. and Blischke, W.R., Product warranty management - II: An integrated framework for study. *European Journal of Operational Research*, **62** (1992a) 261-281.

[39] Murthy, D.N.P. and Blischke, W.R., Product warranty management - III: A review of mathematical models. *European Journal of Operational Research*, **63** (1992b) 1-34.

[40] Murthy, D.N.P., Djamaludin, I. and Wilson, R.J., A consumer incentive warranty policy and servicing strategy for products with uncertain quality. *Quality and Reliability Engineering International*, **11** (1995) 155-163.

[41] Murthy, D.N.P. and Hussain, A.Z.M.O., (1994), Warranty and optimal redundancy design. *Engineering Optimization*, **23** (1994) 301-314.

[42] Murthy, D.N.P. and Kumar, K.R., Total product quality. International Journal of Production *Economics*, **67** (2000) 253-267.

[43] Murthy, D.N.P. and Nguyen, D.G., Optimal development testing policies for products sold with warranty. *Reliability Engineering*, **19** (1987) 113-123.

[44] Murthy, D.N.P. and Tang, L.C., Stochastic stress life models. *Proceedings of ESREL 2000, SARS and SAR Europe Annual Conference*, (2000) 307-312.

[45] Murthy, D.N.P., Wilson, R.J. and Djamaludin, I., Product warranty and quality control. *Quality and Reliability Engineering International*, **9** (1993) 431-443.

[46] Nguyen, D.G. and Murthy, D.N.P., Optimal burn-in time to minimize cost for products sold under warranty. *IIE Transactions*, **14** (1982) 167-174.

[47] Pohl, E.A. and Dietrich, D.L., Environmental stress screening (ESS) strategies for complex systems: A 3-level mixed distribution model. *Microelectronics and Reliability*, **35**(4) (1995) 637-656.

[48] Pohl, E.A. and Dietrich, D.L. Optimal stress screening strategies for multi-component systems sold under warranty: The case of phase-type lifetimes. *Annals of Operations Research*, **91** (1999) 137-161.

[49] Ritchken, P.H., Chandramohan, J. and Tapiero, C.S. Servicing, quality design and control. *IIE Transactions*, **21**(3) (1989) 213-220.

[50] Sahin, I. and Polatoglu, H. *Quality, Warranty and Preventive Maintenance*. Kluwer Academic Publishers, Boston, 1998.

[51] Tapiero, C.S. and Lee, H.L. Quality control and product servicing: A decision framework. *European Journal of Operational Research*, **39** (1989) 61-73.

[52] Thomas, M.U. and Rao, S.S. Warranty economic decision models: A summary and some suggested directions for future research. *Operations Research*, **47** (1999) 807-820.

VI

QUALITY

Chapter 16

SPC-BASED DIAGNOSIS OF PROCESSES WITH MULTIPLE CONCURRENT DYSFUNCTIONAL STEPS

John R. English,
Department of Industrial Engineering
University of Arkansas
Fayetteville AR, USA

Mohammed Alsein, and
Transplace.com
Lowell AR, USA

Michael H. Cole
Department of Industrial Engineering
University of Arkansas
Fayetteville AR, USA
mcole@uark.edu

Abstract This paper develops a statistical process control (SPC) approach to process diagnosis for scenarios in which multiple sub-processes can be concurrently dysfunctional. MDS (Multiple concurrent Dysfunctional Steps) is an extension of SDS (Single Dysfunctional Step), an SPC approach for scenarios in which only one step can be dysfunctional at any given time. The SDS framework was developed in Krishnamurthi *et al.* (1993) and analyzed in Melton *et al.* (1997). Analysis shows the MDS approach does an excellent job diagnosing complex process disturbances as long as the monitored variables are limited to a "vital few." The MDS approach integrates existing SPC knowledge into the field of process diagnosis and enables QC personnel to diagnose systems using SPC data that is already being collected for other purposes.

Keywords: Process diagnosis, special cause diagnosis, multiple concurrent dysfunctional steps, process diagnosis, statistical process control.

1. INTRODUCTION AND MOTIVATION

Most modern manufacturing processes are complex compositions of multiple interdependent process steps. Each process step can be measured by multiple process variables. Typically, each process variable is monitored using statistical process control (SPC) techniques such as control charts. These variables interact with each other based on the condition of the underlying process steps. When a process step becomes dysfunctional, it is desirable to diagnose the process as quickly as possible. That is, when a process variable (or set of variables) goes out of control, we want to correctly identify the dysfunctional process step (or set of steps) in a timely efficient manner. The multidimensional aspects of process diagnosis require expert knowledge on the shop floor. However, an appropriate statistical perspective on process diagnosis exceeds reasonable expectations for even the most qualified operators without applicable SPC tools.

This paper develops an SPC approach (MDS, Multiple concurrent Dysfunctional Steps) for cases in which multiple process steps can be concurrently dysfunctional. This extends the SPC approach developed by Krishnamurthi *et al.* (1993) and Melton *et al.* (1997), which is known as SDS (Single Dysfunctional Step). The MDS approach integrates existing SPC knowledge into the field of process diagnosis and enables QC personnel to diagnose systems using SPC data that is already being collected for other purposes.

The remainder of this paper includes a review of the process diagnosis literature, a mathematical exposition of MDS, a set of experimental results comparing MDS to SDS, and conclusions.

2. RELATED LITERATURE

Diagnosis techniques are of two kinds: rule-based expert systems (Table 16.1) and model-based systems (Table 16.2). Expert systems are computer programs that solve problems using a knowledge base and inference procedures (Kalpakjian, 1992).

Model-based diagnosis systems seek to develop a general theory of diagnosis based on the behavior of the process components.

The SPC approach of Krishnamurthi *et al.* (1993) and Melton *et al.* (1997) is much simpler to employ than more complex approaches such as parsimonious covering theory (Peng and Reggia (1987), Reggia and

Table 16.1 Rule-based Expert Systems for Process Diagnosis.

Reference	Approach	Domain
Cook *et al.* (1992)	SPC, heuristics	Manufacturing
Kim and May (1995)	Neural networks	Integrated circuit fabrication equipment
Pham and Oztemel (1995)	Neural network for SPC pattern recognition	Manufacturing

Peng (1989)) or expert systems (Pham and Oztemel (1995)). One of the advantages of the SPC approach is that it relies partly on control chart data that is likely already being collected.

3. SPC APPROACH TO PROCESS DIAGNOSIS

This section details the mathematics behind the MDS approach to process diagnosis. Consider an M-step process that is characterized by N process variables (see Figure 16.1). Based on the relationship between the N monitored process variables and the M existing process steps, the SPC approach incorporates general laws of probabilities with existing methods of process control (i.e., control charts) to determine the most probable dysfunctional process step(s).

N : number of process variables
y_i : value of process variable i $(i = 1, 2, ..N)$
LCL$_i$: lower control limit of process variable i
UCL$_i$: upper control limit of process variable i
x_i : $= 0$ if process variable i is in control
$(P[x_i = 0] = P[LCL_i < y_i < UCL_i])$
$= 1$ if process variable i is not in control
$(P[x_i = 1] = 1 - P[x_i = 0])$
$\mathbf{X}_k$: binary vector of variable states $\{x_1, x_2, ...x_N\}$
$(k = 0, 1, ...(2^N - 1))$
M : number of process steps
z_j : $= 0$ if process step j is functional,
$= 1$ if dysfunctional $(j = 1, 2, ...M)$
$\mathbf{Z}_b$: system state diagnosis vector $\hat{z}_1, \hat{z}_2, ...\hat{z}_M$ $(b = 0, 1, ...(2^M - 1))$

Table 16.2 Model-based Systems for Process Diagnosis.

Reference	**Approach**	**Domain**
Hudlicka and Lesser (1987)	-	-
Gwee *et al.* (1993)	Fuzzy logic for SPC pattern recognition neural networks	Frequency trimming of crystal resonators
Kumar *et al.* (1995)	Parsimonious covering theory	Networks
Manivannan and Banks (1990)	Real-time simulation	CNC machines
Olive *et al.* (1995)	Statistical filtering, state estimation	-
Roth *et al.* (1995)	Causal tree	Galvanizing processing line
DeKleer (1990)	Kernel diagnosis	-
DeKleer *et al.* (1992)	Minimum entropy technique	-
Peng and Reggia (1987), Reggia and Peng (1989)	Connectionist	-
Krishnamurthi *et al.* (1993) Melton *et al.* (1997)	SPC	-

$\hat{Z}_b$: system state vector ($b = 0, 1, ...(2^M - 1)$)
$P[\mathbf{X}_k|\hat{Z}_\text{b}]$: known probability that process variable state $\mathbf{X}_k$ results given $\hat{Z}_b$
$\max_b\{P[\mathbf{X}_k|\hat{Z}_\text{b}]\}$ = maximum $P[\mathbf{X}_k|\hat{Z}_\text{b}]$ over all process variable states
= most probable process variable state given system state $\hat{Z}_b$
$I(b,k)$: $= 1$ if $P[\mathbf{X}_k|\hat{Z}_\text{b}] = max_b\{P[\mathbf{X}_k|\hat{Z}_\text{b}]\}$,
$= 0$ otherwise

The problem is to diagnose the system (i.e., determine which steps are out of control, if any) given the values of the various process variables. Assume an upper and lower control limit (UCL_i and LCL_i) have been set for each process variable y_i. The state of the process variables can then be characterized by a vector $\mathbf{X} = \{x_1, x_2, ...x_N\}$, where x_i is 0 if

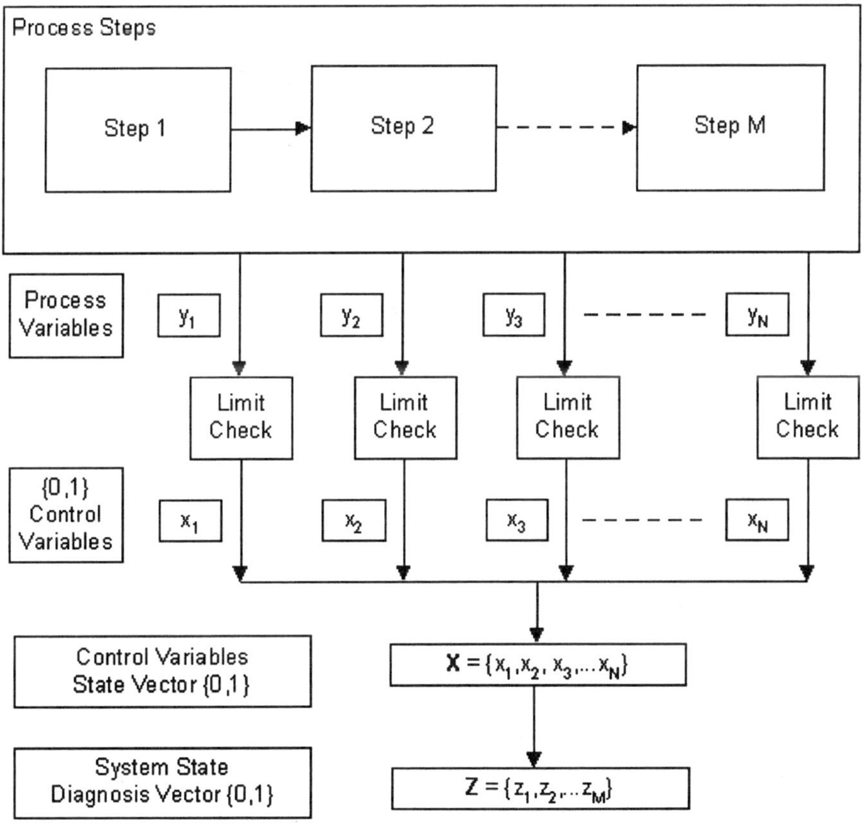

Figure 16.1 M Step Process characterized by N Process Variables.

y_i is in control ($\text{LCL}_i \leq y_i \leq \text{UCL}_i$), and 1 if y_i is out of control ($y_i <$ LCL_I or $y_i >\text{UCL}_I$). Note that X can assume any of 2^N values since each process variable is either in control or out of control. Diagnosing the process is equivalent to constructing a system state diagnosis vector $\mathbf{Z} = \{z_1, z_2, ...z_M\}$, where $z_j = 0$ if Step j is diagnosed as functional, and $z_j = 1$ if Step j is diagnosed as dysfunctional. Note that $\mathbf{Z}$ can assume any of 2^M values since each process step can be diagnosed as either functional (0) or dysfunctional (1). The diagnosis is correct if the system state diagnosis vector $\mathbf{Z}$ is equal to the (unknown) actual system state vector $\hat{Z}$.

For example, a possible case for a system with 3 process steps and 6 process variables is $\mathbf{X} = \{1,1,0,0,0,1\}$ (process variables 1, 2, and 6 are out of control), and a possible diagnosis is $\mathbf{Z} = \{0,1,0\}$ (only process step 2 is dysfunctional).

Now, assume $P[\mathbf{X}_k|\hat{Z}_\text{b}]$, the probability that process variable state $\mathbf{X}_k (k = 0, 1, ...(2^N - 1))$ results given system state $\hat{Z}_b$ $(b = 0, 1, ...(2^M - 1))$, is known based on knowledge of the real-world system.
Let $max_b P[\mathbf{X}_k|\hat{Z}_\text{b}]$ = maximum $P[\mathbf{X}_k|\hat{Z}_\text{b}]$ over all process variable states (= probability of the most probable system state $\hat{Z}_\text{b}$ when $\mathbf{X}_k$ is observed).

Let $I(b, k) = 1$ if $P[\mathbf{X}_k|\hat{Z}_\text{b}] = max_b\{P[\mathbf{X}_k|\hat{Z}_\text{b}]\}$, 0 otherwise. Thus,

$$P[\text{correct diagnosis}|\mathbf{X}_k] = \sum_{k=0}^{2^N-1} (P[\mathbf{X}_k|\hat{Z}_\text{b}]^* I(b, k))$$

4. RESULTS OF COMPARISONS BETWEEN MDS AND SDS

This section compares MDS to SDS. The research results are based on process scenarios in which the cause and effect relationships between the process steps and the process variables are generated by Bernoulli trials as in Melton *et al.* (1997). Process variables are assigned to process steps in accordance with the Bernoulli distribution to construct the process scenarios considered, and the generated process scenarios are used to evaluate the ability of the SPC approach to correctly diagnose multiple dysfunctional process steps. Specifically, the Bernoulli probability (p) associated between a process variable and process step is an experimental factor. That is, the probability that a process variable i is related to a process step j is set at a controlled level, in this case either .1 or .5. The low level (.1) indicates a low dependency between a process step and a monitored variable, while the high level (.5) indicates a higher dependency between the process variable and the underlying process

Table 16.3 Process Scenarios.

Scenario	p	M (# steps)	N (# process variables)	**Shift** shift size (σ)
1	0.1	3	3	1
2	0.1	3	3	3
3	0.1	3	6	1
4	0.1	3	6	3
5	0.1	5	3	1
6	0.1	5	3	3
7	0.1	5	6	1
8	0.1	5	6	3
9	0.5	3	3	1
10	0.5	3	3	3
11	0.5	3	6	1
12	0.5	3	6	3
13	0.5	5	3	1
14	0.5	5	3	3
15	0.5	5	6	1
16	0.5	5	6	3

step. Larger values of p lead to a larger number of process variables dependent on a given process step.

This research considers three additional experimental factors: the number of process steps (3 or 5), the number of process variables (3 or 6), and the size of possible process shifts in the mean (1σ or 3σ) of the monitored process variables. As shown in Table 16.3, 16 possible process scenarios are considered.

Each process condition is replicated 3 times giving a total of 48 experimental process scenarios for data analysis. The Type I error rates and whether MDS is better able than SDS to make correct diagnoses

Table 16.4 Data Sets.

Data Set	# Steps	# of Possible Dysfunctional Steps
1	3	0
2	5	0
3	3	0, 1, 2, 3
4	5	0, 1, 2, 3, 4, 5

are evaluated. The comparison between MDS and SDS quantifies the difference in the ability to make correct process diagnosis when multiple dysfunctional process steps occur. Analysis of variance (ANOVA) is performed to determine the effect of the number of process steps, the number of process variables, the values of p, and the sigma shift values on the ability to make correct process diagnoses.

The data are divided into two sets of eight process scenarios, each depending upon the number of process steps (3 and 5) considered. If the data are not divided based on the number of process steps, the ANOVA results would be biased towards the scenarios with the larger numbers of process disturbance combinations since there is an increasing number of ways that multiple steps can be dysfunctional as the number of steps increases. In addition to being divided by process steps, the data are also categorized based upon the type of disturbance to be detected. Specifically, we consider the scenarios in which no process steps are dysfunctional and in which any/none of the process steps can be dysfunctional. This division results in 4 data sets as described in Table 16.4.

4.1 Data Sets #1 and #2. Data sets #1 and #2 provide data for the analysis of Type I errors since, as shown in Table 16.2, the conclusion that any process step is dysfunctional when the process is undisturbed is a Type I error. The design examines the ability of MDS to make correct process diagnosis when no steps are dysfunctional. Specifically, the P[correct diagnosis $|Z_a]$ values are recorded for the 3 replications for data set #1 (3 process steps) and the 3 replications for data set #2 (5 process steps). Table 16.5 shows the resulting ANOVA for data set #1.

Table 16.5 Data Set #1 ANOVA.

Source	**DF**	**SS**	**MS**	**F Value**	**Pr > F**
Model	7	.0004	.00006	7.21	.0005
Error	16	.00012	.00000		
Corrected Total	23	.00052			
Source	**DF**	**SS**	**MS**	**F Value**	**Pr > F**
Shifts	1	.00000	.00000	.0200	.8852
Var	1	.0002	.0002	27.8700	.0001
Pvalue	1	.0001	.0001	15.6800	.0011
Shifts*Var	1	.0000	.0000	.0200	.8852
Shifts*Pvalue	1	.0000	.0000	.3400	.5657
Var*Pvalue	1	.00005	.00005	6.2200	.0240
Shifts*Var*Pvalue	1	.00000	.00000	.3400	.5657

The interaction between the number of process variables and the value of p (Var*Pvalue) is statistically significant. Since the interaction is significant, the significance of the main factors in isolation cannot be judged.

Figure 16.2 is an interaction plot of the number of process variables and the p value.

The y-axis represents the probability of a correct diagnosis that no process steps are dysfunctional. The probability of a Type I error grows as the number of process variables related to a given process step increases (value of p).

To maximize the probability of correct diagnosis when no steps are dysfunctional and the number of process steps are small, the number of process variables and the p values should be held at a minimum. Increasing the number of process variables decreases the ability to make correct process diagnosis for larger values of p. The SPC system driving MDS should be designed to measure only the vital few process variables. Table 16.6 shows the ANOVA for data set #2.

As with data set #1, the only 2-way interaction found to be significant is the interaction between the number of process variables and the

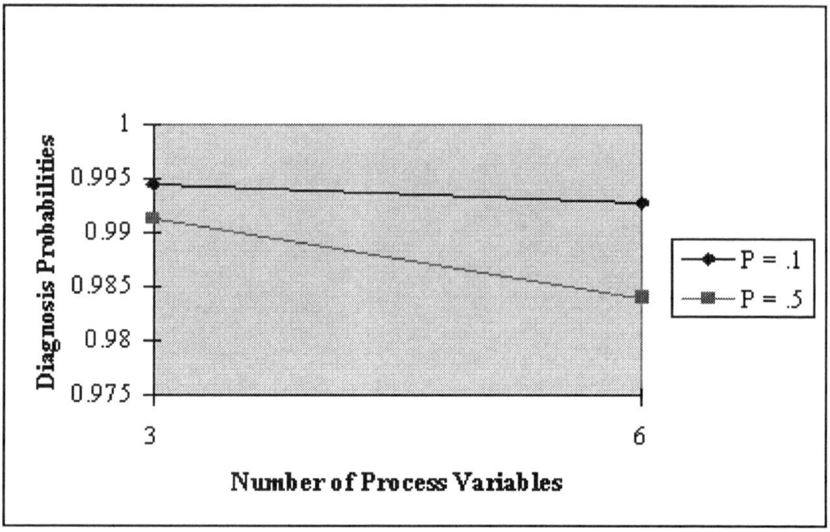

Figure 16.2 Interaction plot of number of process variables vs. p.

p values (Var*Pvalue). Due to this interaction, the significance of the associated main factors cannot be judged. Figure 16.3 shows the interaction between the number of process variables and the p values for the 5 process step scenarios.

Data set #2 indicates results similar to data set #1 with respect to interaction effects between the value of p and the number of process variables. Type I error increases as the number of process variables increase, and this increase is greater when the number of dependent variables is greater. Again, there is evidence that supports the use of the vital few in monitoring or diagnosing processes. The shift factor has no effect on the results for data sets #1 and #2 since no process steps are dysfunctional.

4.2 Data Set #3. Data set #3 is used to examine the increase in correct diagnoses as a result of considering multiple dysfunctional process steps for cases with 3 process steps. In particular, the average probability of correct diagnosis using MDS minus the probability of correct diagnosis using SDS for the same experimental scenario is determined for each of the 8 process scenarios when only 3 process steps are considered. Averaging is necessary in this case since 0 process steps,

Table 16.6 Data Set #2 ANOVA.

Source	DF	SS	MS	F Value	Pr > F
Model	7	.0002	.00003	17.6900	.0005
Error	16	.00003	.00000		
Corrected Total	23	.0002			
Source	**DF**	**SS**	**MS**	**F Value**	**Pr > F**
Shifts	1	.0000	.0000	.0000	1.0000
Var	1	.0002	.0002	92.4800	.0001
Pvalue	1	.00003	.00003	15.6800	.0011
Shifts*Var	1	.0000	.0000	.0000	1.0000
Shifts*Pvalue	1	.0000	.0000	.0000	1.0000
Var*Pvalue	1	.00003	.00003	15.68	.0011
Shifts*Var*Pvalue	1	.0000	.0000	.0000	1.0000

1 process step, 2 process steps (steps 1&2, 1&3 or 2&3), or 3 process steps can be concurrently dysfunctional. This situation gives rise to the complication addressed in this research: the real possibility of multiple dysfunctional process steps. As a point of analysis, the number of dysfunctional process steps is considered as a factor with 4 levels (0, 1, 2, or 3 simultaneous dysfunctional process steps). Therefore, 4 average probabilities of correct diagnosis using MDS minus the probability of correct diagnosis using SDS are generated for the 8 process scenarios for each of the 3 replications yielding 96 observations. Table 16.7 shows the ANOVA for data set #3.

One main factor and two 2-way interactions are found to be significant. The 2-way interaction between the number of dysfunctional process steps and the sigma shift values (Steps*Shifts) is significant. Since the interaction is significant, the significance of these main factors in isolation cannot be judged.

Figure 16.4 shows the interaction between the number of dysfunctional process steps and the shift values.

Regardless of the shift value, no observable difference in the diagnosis probabilities between MDS and SDS exists when the number of dysfunctional process steps is equal to 0 or 1 as desired. When more than one step is dysfunctional, MDS provides better diagnoses than SDS for larger disturbances in the processes.

The interaction between the number of dysfunctional process steps and number of process variables (Steps*Var) is significant. Figure 16.5 shows the interaction plot.

MDS provides improved diagnosis abilities when the number of dysfunctional process steps is larger than 1. The number of process variables affects the diagnosis probabilities. The advantage of MDS is higher when the number of process variables is relatively low. For the case with a large number of process variables (6) and dysfunctional steps (3), the relative performance of MDS degrades, but is still better than that of SDS. The practical significance of the degraded performance would depend on the real-world application.

Only one main factor, the p value, is significant. The modified approach gives better diagnosis probabilities when $p = 0.1$ than when $p = 0.5$. This result supports the need in diagnosis applications to identify the vital few monitored process variables.

4.3 Data Set #4. Data set #4 is constructed like data set #3 except that the number of dysfunctional process steps can be 0, 1, 2, 3, 4, or 5, yielding a factor with 6 levels. This data set is used to

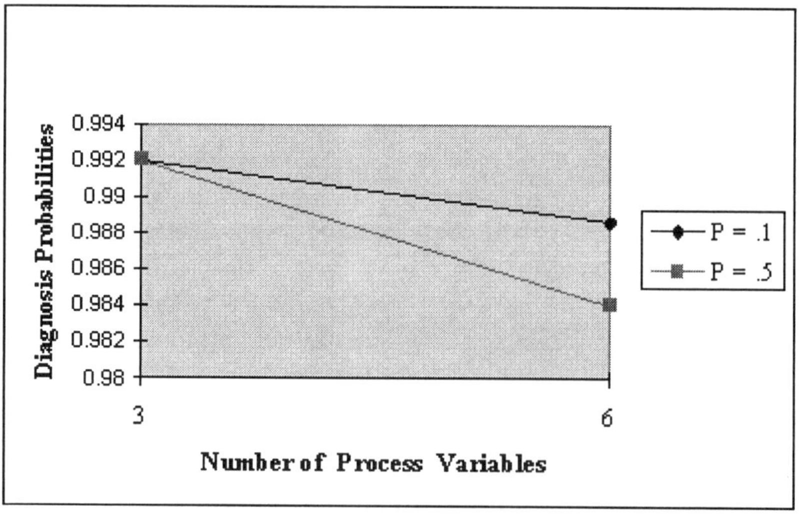

Figure 16.3 Interaction plot of number of process variables vs. p value.

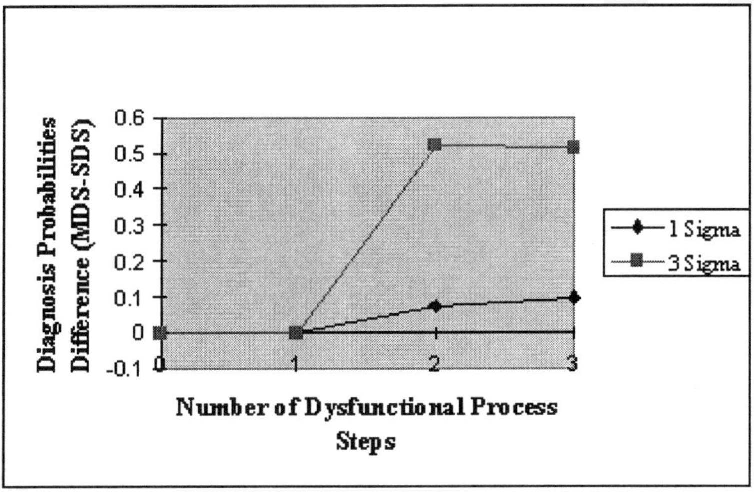

Figure 16.4 Interaction plot of Steps vs. Shifts.

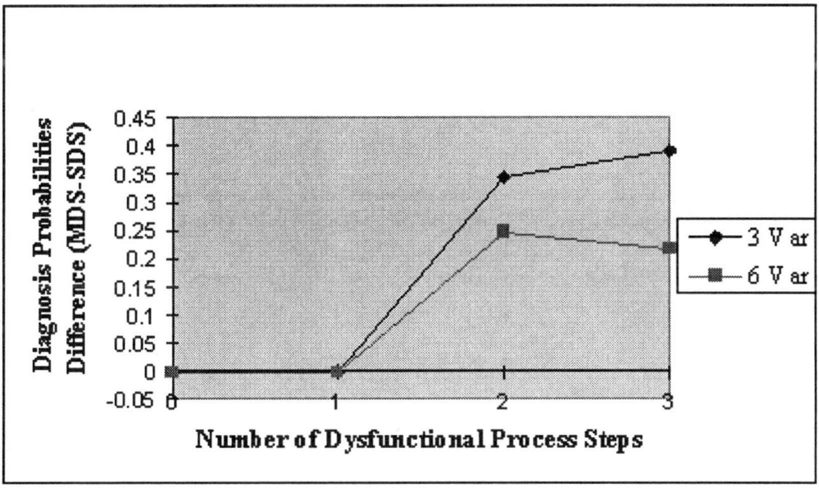

Figure 16.5 Interaction plot of Steps vs. Number of Process variables (Var).

Table 16.7 Data Set #3 ANOVA.

Source	**DF**	**SS**	**MS**	**F Value**	**Pr > F**
Model	31	5.0176	.1619	10.56	.0001
Error	64	.9810	.0153		
Corrected Total	95	5.9986			
Source	**DF**	**SS**	**MS**	**F Value**	**Pr > F**
Steps	3	2.1932	.7311	47.70	.0001
Shifts	1	1.1236	1.1236	73.31	.0001
Var	1	.1122	.1122	7.32	.0087
Pvalue	1	.0872	.0872	5.69	.0200
Steps*Shifts	3	1.1259	.3753	24.49	.0001
Steps*Var	3	.1266	.0422	2.75	.0496
Steps*Pvalue	3	.1022	.0340	2.22	.0940
Shifts*Var	1	.0000	.0000	.0000	.9555
Shifts*Pvalue	1	.0035	.0035	.23	.6319
Var*Pvalue	1	.0454	.0454	2.96	.0900
Steps*Shifts*Pvalue	3	.0046	.0015	.10	.9588
Steps*Var*Pvalue	3	.0479	.0159	1.04	.3806
Steps*Shifts*Var	3	.0019	.0006	.04	.9883
Shifts*Var*Pvalue	1	.0122	.0122	.80	.3740
Steps*Shifts*Var*Pvalue	3	.0307	.0102	.67	.5749

examine the improvements in diagnosis for cases with 5 process steps. Table 16.8 shows the ANOVA.

No 4-way or 3-way interactions are found to be significant. A number of 2-way interactions were found to be significant: Var*Pvalue, Shifts*Pvalue, Steps* Pvalue, and Steps*Shifts. Since their interactions are significant, the significance of the associated main factors cannot be judged in isolation. Note that the F-value for Var is 0.0, as compared to

Table 16.8 Data Set #4 ANOVA.

Source	**DF**	**SS**	**MS**	**F Value**	**Pr > F**
Model	47	5.2968	.1127	12.04	.0001
Error	96	.8982	.0093		
Corrected Total	143	6.1951			
Source	**DF**	**SS**	**MS**	**F Value**	**Pr > F**
Steps	5	1.6953	.3390	36.24	.0001
Shifts	1	1.7096	1.7096	182.70	.0001
Var	1	.0000	.0000	.00	.9698
Pvalue	1	.3780	.3780	40.40	.0001
Steps*Shifts	5	.8685	.1737	18.56	.0001
Steps*Var	5	.0226	.0045	.4800	.7878
Steps*Pvalue	5	.3068	.0613	6.56	.0001
Shifts*Var	1	.0052	.0052	.56	.4564
Shifts*Pvalue	1	.0921	.0921	9.84	.0023
Var*Pvalue	1	.0532	.0532	5.69	.0191
Steps*Shifts*Pvalue	5	.0786	.0157	1.68	.1465
Steps*Var*Pvalue	5	.0354	.0070	.76	.5825
Steps*Shifts*Var	5	.0037	.0007	.08	.9951
Shifts*Var*Pvalue	1	.0175	.0175	1.87	.1745
Steps*Shifts*Var*Pvalue	5	.0299	.0059	.64	.6702

7.32 in the table for data set #3. However, recall that Var is involved in a 2-way interaction in data set #3 and that its significance as a main factor could not be judged in that case.

The interaction between the number of process variables and the p values (Var*Pvalue) is significant. Figure 16.6 shows the interaction.

The MDS approach provides better diagnosis capabilities than the SDS approach. MDS outperforms SDS, especially when the number of

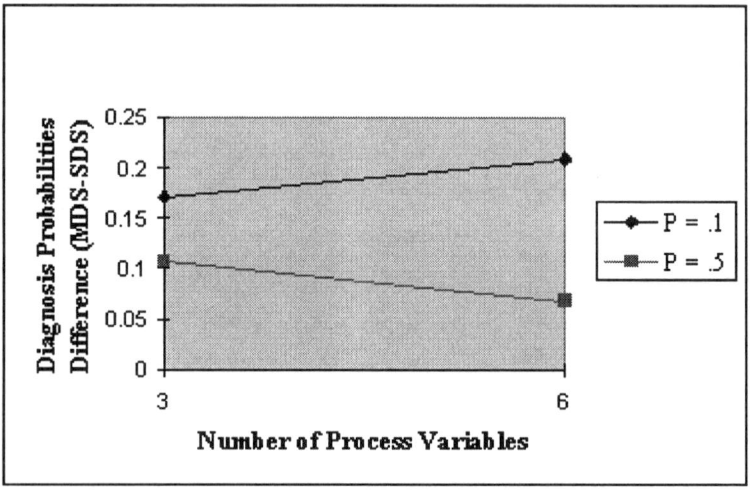

Figure 16.6 Interaction plot of Var vs. value of p.

process variables is large and the p value is small. A larger number of process variables coupled with large values of p decreases the ability to make correct process diagnoses. A larger number of process variables coupled with small values of p provide the best diagnosis capability.

The 2-way interaction between the sigma shift and the p values (Shifts*Pvalue) is found to be significant. Figure 16.7 shows the interaction plot.

The greatest difference between the MDS and SDS is realized when the shift value is 3 and the value of p is .1. The improved performance is greatest for larger shifts in the effected process variables.

The 2-way interaction between the number of dysfunctional process steps and the p values is found to be significant. Figure 16.8 shows the interaction.

The probability of making a correct diagnosis is identical for MDS and SDS when zero or one step is dysfunctional. When more than one step is dysfunctional, MDS provides improved diagnosis probabilities. The advantage of MDS over SDS is greater for smaller p values. Thus, MDS works best when few process variables are related to each process step. When more than one process step is dysfunctional, the advantage of MDS over SDS tends to decrease as the number of dysfunctional steps

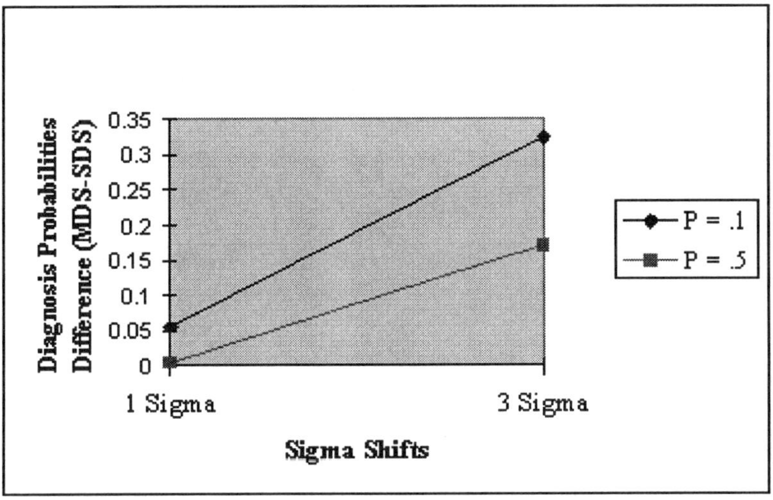

Figure 16.7 Interaction plot of Shifts vs. value of p.

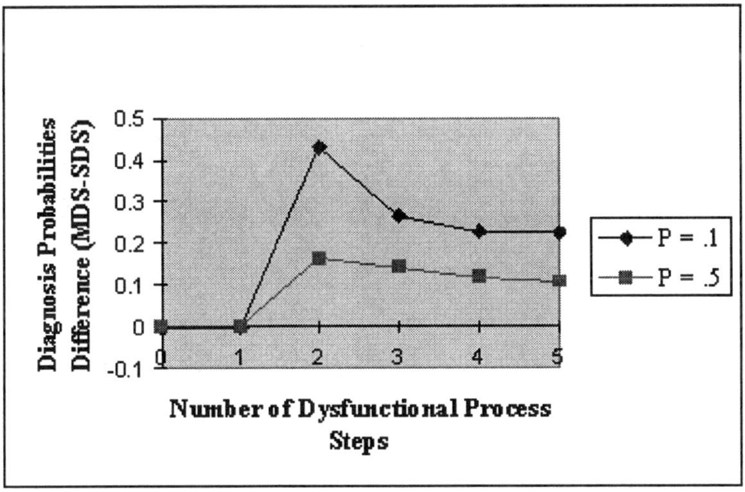

Figure 16.8 Interaction plot of Steps vs. value of p.

increases. The advantage of MDS over SDS does not change much as the number of dysfunctional process steps increases from 4 to 5.

The interaction between the number of dysfunctional process steps and the sigma shift (Steps*Shifts) is found to be significant. Figure 16.9 shows the interaction.

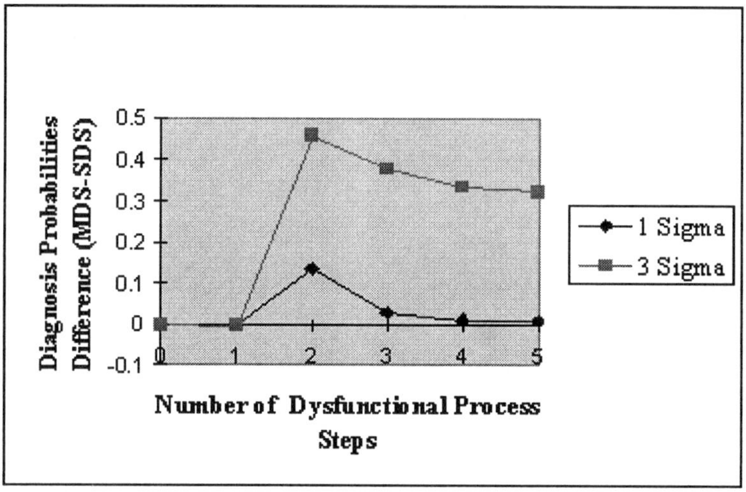

Figure 16.9 Interaction plot of Steps vs. Shift.

The probability of making a correct diagnosis is identical for MDS and SDS when zero or one step is dysfunctional. When more than one step is dysfunctional MDS provides improved diagnosis probabilities. The advantage of MDS over SDS is greater for larger shifts in variable values. When the shift is small and the number of dysfunctional variables is high, MDS offers only marginal improvements over SDS.

5. CONCLUSIONS

The MDS approach provides better diagnosis results than the SDS approach for cases where multiple dysfunctional process steps can exist concurrently. The MDS approach performs the same as the SDS approach for cases in which zero or one step is dysfunctional. The MDS approach performs best when the number of multiple dysfunctional process steps is small (i.e., two or three). Increasing the number of process steps leads to a decrease in the ability to make correct process diagnoses, particularly when the number of dependent variables is large and the process variable shifts are small.

The MDS approach appears to perform best with a minimum number of process variables (e.g., monitoring three process variables gives better results than monitoring six process variables). Increasing the number of process variables decreases the ability to make correct diagnoses. The practitioner should be careful to establish the vital few process variables to be monitored. If the process consists of a large number of process variables, the process should be divided into sub-operations. Each sub-operation should be measured with the fewest reasonable number of process variables.

The MDS approach appears to perform best when relatively few process steps are related to each variable. In the experiments above, the p values affect the number of process variables related to the process steps. Large p values lead to a large number of process variables that are dependent on a given process step and decrease the diagnosis probabilities. This observation gives support to the establishment of the vital few process variables.

References

[1] Cook, D.M., Massey, J.G., and Mckinney, C. Knowledge-based Approach to Statistical Process Control. *Computers and Electronics in Agriculture*, **7**(1) (1992) 13-22.

[2] DeKleer, J. Using Crude Probability Estimates to Guide Diagnosis. *Artificial Intelligence*, **45**(3) (1990) 381-391.

[3] DeKleer, J., Mackworth, A.K., and Reiter, R. Characterizing Diagnosis and Systems. *Artificial Intelligence*, **56**(2-3) (1992) 197-222.

[4] Gwee, B.H., Lim, M.H., and Soong, B.H. Self-Adjusting Diagnostic System for the Manufacture of Crystal Resonator. *Proceedings of the 28th Annual Meeting of the IEEE Industry Application Conference*, **3** 2014-2020, 1993.

[5] Hudlicka, E., and Lesser, V. Modeling and Diagnosing Problem Solving System Behavior. *IEEE Transactions on Systems, Man and Cybernetics*, **17**(3) (1987) 407-419.

[6] Kalpakjian, S. *Manufacturing Engineering and Technology*. 2nd ed., Addison-Wesley Publishing Company, New York, 1992.

[7] Kim, B., and May, G.S. Real-Time Diagnosis of Semiconductor Manufacturing Equipment using Neural Networks. *Proceedings*

of the 17th IEEE/CPMT International Electronics Manufacturing Technology Symposium, 224-231, 1995.

[8] Krishnamurthi, M., English, J., and Taylor, G.D. Diagnosis of Discrete Part Manufacturing Processes. *Institute of Industrial Engineers: 2nd Industrial Engineering Research Conference Proceedings*, 564-568, 1993.

[9] Kumar, G. P., and Venkataram, P. Network Fault Diagnosis Using a Realistic Abductive Reasoning Model. *Engineering Application of Artificial Intelligence*, **8**(6) (1995) 703-708.

[10] Manivannan, S., and Banks, J. Towards a Real-Time Knowledge-Based Simulation System for Diagnosing Machine Failures. *1990 Winter Simulation Conference Proceedings*, 603-608, 1990.

[11] Melton, K. D., English, J.R., and Taylor, G.D. A Statistical Process Control Approach to Process Diagnosis. *International Journal of Quality Sciences*, **2**(2) (1997) 87-105.

[12] Olive, J.M., Hmida, F., and Egea, M. Detection and Diagnosis Approach by State Estimation Applied to Production Machines. *Proceedings of the 1995 34th IEEE Conference on Decision And Control*, **3** (1995) 2403-2405.

[13] Peng, Y., and Reggia, J.A. Probabilistic Causal Model for Diagnostic Problem Solving - II Diagnostic Strategy. *IEEE Transactions on Systems, Man and Cybernetics*, **17**(3) (1987) 395-406.

[14] Pham, D.T., and Oztemel, E. Integrated Neural Network and Expert System Tool for Statistical Process Control. *Proceedings of the Institution of Mechanical Engineers*, **209**(B2) (1995) 91-97.

[15] Reggia, J.A., and Peng, Y. A Connectionist Model for Diagnostic Problem Solving. *IEEE Transactions on Systems, Man and Cybernetics*, **19**(2) (1989) 285-297.

[16] Roth, S., Sirou, F., Bloch G., and Cecchin, T. Towards an Object Approach for Quality Monitoring and Diagnosis on a Galvanizing Line. *Proceedings of the 1994 IEEE Conference on Control Applications* **2** (1994) 1139-1143.

Chapter 17

STEPWISE-PROGRAMMED REGULATION OF MANUFACTURING QUALITY

John J. Liu
University of Wisconsin-Milwaukee
Milwaukee, Wisconsin, USA

Abstract In the process of attaining manufacturing targets (e.g., 6-σ quality targets), regulatory actions are often restricted to the so-called stepwise-programmed (or dispersal feedback) regulation, which contains a countable set of feedback-based decision-making points. For example, quality auditing of Toyota production systems calls for a series of feedback reviews (or audits) guided by a feedback strategy, which usually leads to disruptive stepwise changes in regulatory actions. The disruptive nature of dispersal feedback causes the quality regulation to be non-smooth (i.e., non-smooth Hamiltonian), with the result that classical solutions are no longer suitable. As shown in modern control theory, the so-called generalized (or viscosity) solutions need to be considered in this case. However, non-smooth quality regulation problems are predominantly treated as if they were smooth.

In this paper, we show the non-smoothness of quality regulation by means of a series of quality audits in a stochastic *quality diffusion* system developed by Liu and Nam (1998). Then, we develop a dispersal quality regulation model, from which generalized solutions applicable to the non-smooth Hamiltonian can be obtained, using the *discrete approximation* of Subbotin (1995).

Keywords: Quality regulation, quality audits, stochastic quality diffusion. system

1. INTRODUCTION

It is known, to both researchers and practitioners, that quality diffusion is continuous, while its regulation is discrete. This can be seen from 6-Sigma manufacturing, where continuous improvement is pursued by a series of regulatory actions facilitated via quality audits and reviews.

As typically applied in 6-Sigma manufacturing (Liu and Nam, 1999), the regulation is stepwise programmed. That is, the control is adjusted (programmed) at a discrete series of feedback points and must remain as programmed between feedback points (referred as dispersal feedback control). Let us look at some real world examples.

Example 17.1 *[Quality Liaison in 6-σ manufacturing (Liu, 2000)] In 6-sigma manufacturing, quality liaison is a commonly adopted mechanism for coordination, consisting of a series of* quality audits *to be exercised at an accountable set of review points. At each review point, collective feedbacks are obtained first and then used to determine (or to program) necessary changes in regulatory actions (e.g., target re-positioning and control adjustment). The regulation will remain as programmed between consecutive audits.*

Example 17.2 *[Sausage Moisture Control (Liu and Nam, 1999)] The moisture level is one of the most critical factors affecting the quality of sausage production. The regulation of moisture (e.g., control of temperature and ventilation) must follow a specific cooling procedure, which requires that the temperature be adjusted in multiple steps based on the feedback about actual moisture contents. The cooling procedure prohibits the use of simple bang-bang control since a sudden switch in temperature would harden the sausage casing (the surface) and block the moisture evaporation.*

The regulation of manufacturing quality is modeled in this paper as a regulated *quality diffusion* system (Liu and Nam 1998), which is a non-linear stochastic system. As we proceed, it will become clear that the dispersal of the stepwise-programmed regulation causes the lack of smoothness in the associated Hamiltonian. The regulation under such non-smooth *quality diffusion* is singular, and its Hamiltonian is non-smooth in general. It is known from control theory that the uniqueness and existence of a classic solution to a non-smooth HJB equation is no longer ensured. Thus, the so-called generalized (or viscosity) solutions must be considered when non-smoothness is present in the Hamiltonian. As far as we can ascertain, quality regulation problems, smooth or non-smooth, are all treated as if they were smooth.

In this paper, we first show that the Hamiltonian of a quality regulation problem is non-smooth when stepwise regulation is permissible, as in 6-σ manufacturing. Then, we develop a generalized solution method using the *discrete approximation* of Subbotin (1995). The key to the discrete approximation is the identification of a so-called positional (feedback) regulation strategy, which is characterized by a general measurable real function. Thus, the approximation is obtained by partitioning the regulation strategy into a large enough number of steps. To do this, we first convert the stochastic quality regulation problem to a deterministic dispersal regulation problem of first order partial differential equations (PDEs).

2. DISPERSAL QUALITY REGULATION

2.1 Regulated Quality Diffusion. Suppose that a certain manufacturing quality is to be attained over a finite horizon $[0, \tau]$. Let $y(t)$ be the manufacturing quality process, with an open-loop time-variant target $v(t)$. A target $v(t)$ is said to be *antagonistic* if it is independently controllable by an opponent, otherwise it is *non-antagonistic.* In general, manufacturing targets are non-antagonistic.

Given the target, control $u(t)$ is to be applied to drive the process $y(t)$ towards the target $v(t)$, causing changes (corrections) in $y(t)$ at a rate of $u(t)[v(t) - y(t)]$ (the *drift*). In addition, there is a random *disturbance* introduced onto the process. Therefore, $y(t)$ entails the *quality diffusion* process of Liu and Nam (1999):

$$dy(t) = u(t)[v(t) - y(t)]dt + \sigma y(t)d\xi(t), \qquad (17.1)$$
$$t \in [0, \tau], \ \ y^0 = y(0) \text{ (known)}$$

where $\xi(t)$ is a 1-dimensional Wiener process. Rearranging the terms, we write

$$dy(t) = [A(u(t))y(t) + M(u(t), v(t))]dt + \sigma y(t)d\xi(t), \quad (17.2)$$
$$t \in [0, \tau], \ \ y^0 = y(0) \text{ (known)}$$

where $A(u(t)) = -u(t)$, and $M(u(t), v(t)) = u(t)v(t)$.

Let $u(t) \in \mathcal{P} \subset R$ be stepwise programmed (explained further later), and $v(t) \in \mathcal{Q} \subset R$ be measurable. Both $\mathcal{P}$ and $\mathcal{Q}$ are *compact.* The mapping $A : \mathcal{P} \to R$ is assumed to be continuous and can be non-linear. Thus, equation (17.2) is a non-linear stochastic differential system. Note that the control $u(t)$ can be discontinuous in t, which is directly multiplied with the stochastic process $y(t)$ in (17.1).

2.2 Stepwise-Programmed Regulation. Suppose that the desired regulation of system (17.2) is restricted to the aforementioned programmed regulation under an a priori feedback schedule. A feedback schedule ∂S consists of a given number of m feedback reviews within the horizon $(0, \tau)$, defined as

$$\partial S = \{t_j \in (0, \tau) : 0 < t_j < t_{j+1} < \tau, j = 1, \cdots, m - 1\}, \qquad (17.3)$$

where t_j is the j-th review point. The time taken by each review is negligible. Following each review point t_j, the control remains as programmed over each regulation interval $S_j = (t_j, t_{j+1})$. Thus, the total control time domain excluding all the review points can be written as

$$I = (0, \tau) - \partial S = \{t \in (t_j, t_{j+1}) : j = 0, \cdots, m; t_0 = 0, t_{m+1} = \tau\}. \tag{17.4}$$

That is, total control time domain I has ∂S as its boundaries with $\bar{I} = [0, \tau]$.

Recall that the control $u(t)$ and the target $v(t)$ can only be changed (or *programmed*) at a feedback *review point* t_j and will remain as *programmed* during the regulation interval $S_j = (t_j, t_{j+1})$. Now, let us formally define the term *programmed regulation* introduced by Krasovskii and Subbotin (1988).

Definition 2..1 (programmed regulation) *Let $(x(t), t)$ denote the system* position *at t. A regulation $u(t)$ is said to be* programmed *over an interval $[\tau_1, \tau_2]$ if for any partition Δ with an arbitrary integer m, i.e.,*

$$\Delta : (\tau_1 =) t_0 = t_1 < \cdots < t_{m+1} (= \tau_2), \tag{17.5}$$

there exists a positional regulation strategy $\tilde{u}(x(t), t) : R \times [\tau_1, \tau_2] \to \mathcal{P}$, such that

$$u(t) = \tilde{u}(x(t), t), \forall t \in (t_j, t_{j+1}), j = 0, 1, \cdots, m. \tag{17.6}$$

A programmed regulation allows feedback at a countable set of review points $\{t_j : j = 1, 2, \cdots\}$. By definition, a *positional regulation strategy* $\tilde{u}(t)$ gives a *measurable* feedback policy, which can be itself *discontinuous* (e.g., stepwise). Note that in Definition 2..1 the regulation at each review point, $u(t_j)$, need not to be specified (i.e., can be free) because the actions at each review point are negotiable and, therefore, indefinite. It is such discreteness and indefiniteness that cause the so-called singular characteristics in quality regulation problems.

Since the target $v(t)$ is entirely dependent on the market, the target can be updated at each t_j and then fixed at $v(t_j)$ over the regulation period S_j. In other words, the target $v(t)$ will be stepwise, just like the regulation $u(t)$ (see Figure 17.1). The stepwise-programmed regulation is of particular importance to quality regulation problems, and is defined as follows.

Definition 2..2 (stepwise-programmed regulation) *A programmed regulation $u(t)$ is said to be stepwise-programmed over an interval $[\tau_1, \tau_2]$*

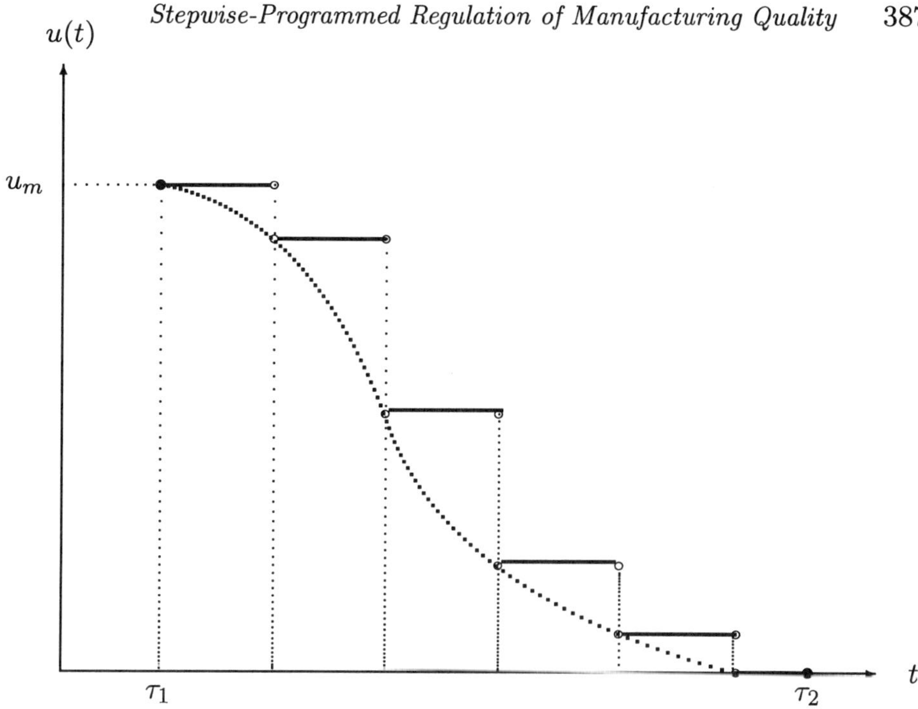

Figure 17.1 Stepwise-programmed regulation

if $u(t)$ is fixed at a constant during each regulation period S_j, as follows:

$$u(t) = \tilde{u}(x(t_j), t_j), \forall t \in (t_j, t_{j+1}), j = 0, 1, \cdots, m. \qquad (17.7)$$

2.3 Dispersal Quality Regulation Model. Denote the regulation-incurred cost by $h(u(t))$, and denote the quality deviation cost by

$$c(y(t) - v(t))^2, \qquad (17.8)$$

where c is a unit deviation cost associated with $y(t)$. Recall that the targets are variable within a bounded region (i.e., $v(t) \in \mathcal{Q}$); therefore, the so-called minimax deviation, as typically termed in pursuit theory, needs to be considered (Subbotin 1995). Simply, a minimax deviation is a minimized deviation under maximum disturbance (i.e., the worst target variation in this case). We then consider the following total expected cost as the minimax objective function with a known initial state y^0 :

$$\min_{u(t)\in\mathcal{P}} \max_{v(t)\in\mathcal{Q}} J(y^0, u(t), v(t)) = \int_0^T E\{(h(u(t)) + c(y(t) - v(t))^2\}dt, \qquad (17.9)$$

where $h : \mathcal{P} \to R$ is differentiable. At this point, the quality regulation problem is to drive the quality process $y(t)$ via a stepwise-programmed control so as to best attain its target $v(t)$ within a final period $[0, \tau]$. Then, a stochastic dispersal quality regulation (DQR) model can be constructed as follows:

DQR-stochastic Model

$$\min_{u(t)\in\mathcal{P}} \max_{v(t)\in\mathcal{Q}} J(y^0, u(t), v(t)) = \int_0^{\tau} E\{(h(u(t)) + c(y(t) - v(t))^2\}dt,$$

s.t.

$$dy(t) = [A(u(t))y(t) + M(u(t), v(t))]dt + \sigma y(t)d\xi(t),$$
$$t \in [0, \tau], \;\; y^0 = y(0) \text{ (known)}$$

Using stochastic control theory, the solution can be determined by solving a boundary problem of the HJBI equation, which is typically a second-order partial differential equation (PDE). As shown in control theory, the Hamiltonian, under dispersal feedback control, is in general non-smooth, and the uniqueness and existence of a classical solution to the HJBI equation is no longer ensured (Flemming and Soner, 1993). In this paper, we develop an alternative method by which the original *DQR-stochastic* problem is converted to an equivalent deterministic dispersal regulation system of first order PDEs. Next, we elaborate on the proposed alternative method.

It is easy to verify that the expected deviation cost in the *DQR-stochastic* model can be expressed in terms of $E\{y(t)\}$ and $E\{y^2(t)\}$ as follows:

$$\begin{aligned} &E\{h(u(t)) + c(y(t) - v(t))^2\} \\ &= h(u(t)) + c(v^2(t) - 2v(t)E\{y(t)\} + E\{y^2(t)\}). \quad (17.10) \end{aligned}$$

Introduce a column vector variable:

$$\begin{aligned} x(t) &= (E\{y(t)\}, E\{y_0^2(t)\})^t \quad (17.11) \\ &= (x_1, x_2)^t, \end{aligned}$$

where $x^0 = x(0)$. The first element in $x(t)$ is the mean of the quality process, and the last element is connected with the variance of the quality process. We can then write

$$L(x(t), u(t), v(t)) = E\{h(u(t)) + c(v(t) - y(t))^2\}. \quad (17.12)$$

Note from (17.10) that $L(\cdot)$ is linear in x, and thus

$$\frac{\partial L(x, u, v)}{\partial x} = D(u, v) = (L_{x_1}, L_{x_2})^t = (-2cv(t), c)^t, \quad (17.13)$$

where $D(u,v) \in R^2$ is a gradient vector (column) that is independent of x. From the vector variable $x(t)$ in (17.11), we obtain the following theorem.

Theorem 2..1 *Let $y(t)$ be given by (17.2) with $y^0 = y(0)$ known, and $x(t)$ be defined by (17.11). Suppose that the control $u(t)$ is programmed (Definition 2..1). Then $x(t)$ is the unique solution of the differential equation given $u(t)$:*

$$\dot{x}(t) = \tilde{A}(u(t))x + \tilde{B}(u(t), v(t)), \quad x^0 = x(0), \tag{17.14}$$

where

$$\tilde{A}(u(t)) = \begin{pmatrix} A(u(t)) & 0 \\ 2M(u(t)v(t)) & \sigma^2 - 2u(t) \end{pmatrix}_{2\times 2}, \tag{17.15}$$

$$\tilde{B}(u(t), v(t)) = \begin{pmatrix} M(u(t), v(t)) \\ 0 \end{pmatrix}_{2\times 1}, \tag{17.16}$$

and $A(\cdot)$ and $M(\cdot)$ are given in (17.2).

Proof. See Appendix.

We note that Theorem 2..1 holds for the Lebesque measurable $A(u(t))$ given in (17.2), which is indeed the case with stepwise-programmed control $u(t)$. Using Theorem 2..1 and equation (17.12), the original DQR-stochastic problem can be converted to the following deterministic quality regulation problem (DQR-deterministic).

DQR-deterministic Model:

$$\min_{u(t)\in\mathcal{P}} \max_{v(t)\in\mathcal{Q}} J(x(t), u(t), v(t)) = \int_0^{\tau} L(x(t), u(t), v(t))dt, \tag{17.17}$$

s.t.

$$\dot{x}(t) = \tilde{A}(u(t))x + \tilde{B}(u(t), v(t)),$$
$$t \in [0, \tau], \quad x \in \Omega \subset R^2, \quad x^0 = x(0) \text{ (known)}$$

3. SINGULAR CHARACTERISTICS OF DQR

We define a value function for the DQR-deterministic problem, denoted by $\pi(x(t))$, as follows:

$$\pi(x(t)) = \min_{u\in\mathcal{P}} \max_{v\in\mathcal{Q}} J(x(t), u(t), v(t)). \tag{17.18}$$

Whenever convenient, we drop the time index t. Then according to control theory, the value function π is a solution to the following terminal value problem (TVP) of an HJBI equation:

$$H(x,p) = 0, x(0) = x^0, t \in [0,\tau], p = \frac{\partial \pi}{\partial x}, \pi(x(\tau)) = 0, \qquad (17.19)$$

where the adjoint state vector $p = \pi_x$ is the gradient of π with regard to x. The Hamiltonian $H(x,p)$ is defined as

$$H(x,p) = \min_{u \in \mathcal{P}} \max_{v \in \mathcal{Q}} \{\langle p, \dot{x} \rangle + L(x,u,v)\}, \qquad (17.20)$$

and $\langle \cdot, \cdot \rangle$ denotes an inner product. Note that TVP (17.19) contains a nonlinear first-order PDE, which is typically associated with the Cauchy problem: Find a twice-differentiable solution π to the TVP of (17.19). The classical solution to the Cauchy problem in (17.19) can then be obtained by solving the following system of *characteristic equations* (see more in Melikyan 1998):

$$\dot{x} = H_p, \;\; \dot{p} = -H_x - pH_\pi = -H_x, \;\; \dot{\pi} = \langle p, H_p \rangle. \qquad (17.21)$$

where H_x denotes the partial differential $\frac{\partial H}{\partial x}$. The existence and uniqueness of the classical solutions to the differential equations in (17.21) require that the Hamiltonian $H(x,p)$ be differentiable, or equivalently, that the value function π be twice-differentiable. In what follows, we see that the differentiability of the Hamiltonian is not preserved for the DQR problem. Thus, classical (twice-differentiable) solutions to (17.19) do not exist at those non-differentiable points. Such non-smoothness in the Hamiltonian causes singular characteristics which necessitate a search for generalized (viscosity) solutions (Fleming and Soner 1993). In differential pursuit games, the generalized solutions are also termed minimax solutions (Subbotin 1995). First, we introduce the general concepts of *minimax* solution and singularity in optimal control (OC) and game theory.

Definition 3..1 (minimax solution and singular point) *A general value function $\pi(x(t)) \in C$ is a minimax solution if for any $p \in R^n$ it solves*

$$\dot{\pi}(x(t)) = \langle p, \dot{x} \rangle - H(x(t), p), \qquad (17.22)$$

for almost all $t \in [0,\tau]$ with initial conditions $\pi(x(0)) = \pi(x^0)$ and any $x(0) = x^0 \in \Omega \subset R^n$, where C denotes the class of all continuous real functions of $x(t) : [0,\tau] \to \Omega$. A point $(x,t) \in \Omega \times [0,\tau]$ at which a minimax function $\pi(x(t))$ is non-differentiable is called a singular point. A singular curve, surface, or characteristic contains singular point(s).

Note that a minimax solution solves (17.22) for almost all t, not necessarily for every t. Let $x(t_j)$ be the system state at review point $t_j \in \partial S$. The characteristic equations of (17.21) are singular only at each review pint t_j, and are regular elsewhere.

Theorem 3..1 *Suppose that the control $u(t)$ and target $v(t)$ are both stepwise-programmed according to the same review schedule ∂S. Then, a minimax solution $\pi(x(t))$ is twice differentiable for $x(t)$ for all $t \in I$ and is continuous at $x(t_j)$ for every $t_j \in \partial S$.*

Proof. According to definition 2..2, stepwise-programmed regulation, the control and target are fixed at some constants, such as $\bar{u}^j$ and $\bar{v}^j$ respectively, within each regulation period S_j ($j = 1, \cdots, m-1$). It is easy to verify from (17.14) and (17.20) that the state and adjoint state equations ($\dot{x}$ and $\dot{p}$ in (17.21)) form a system of linear ODE for $t \in (t_j, t_{j+1})$, $t_j \in \partial S + \{0\}$. Then, according to Theorem 11.4 of Subbotin (1995), the value function $\pi(x(t))$ is a minimax solution to DQR and solves characteristic equations of (17.21) for $t \in I$. Because of the uniqueness and differentiability of the solution to the linear ODEs, $\pi(x(t))$ is at least twice differentiable for all $t \in I$. Since $\pi(x(t))$ is a minimax solution, thus we can conclude that $\pi(x(t))$ is continuous at $x(t_j)$ for every $t \in \partial S$. ■

According to Theorem 3..1, the DQR problem is *singular* only at a review point $t_j \in \partial S$ ($j = 1, \cdots, m$). For each feedback review point t_j, a feedback *review path*, denoted by Γ^j, can be constructed as

$$\Gamma^j = \{\hat{z}^j = (\hat{x}^j, \hat{p}^j) : H(\hat{x}^j, \hat{p}^j) = 0, t_j \in \partial S\}. \tag{17.23}$$

That is, a *minimax* solution is *singular* (non-differentiable) at every point $\hat{z}^j = (\hat{x}^j, \hat{p}^j) \in \Gamma^j$. Clearly, a *review path* Γ^j divides the two consecutive regulation periods $S_{j-1} = (t_{j-1}, t_j)$ and $S_j = (t_j, t_{j+1})$. According to Theorem 3..1, the classical (i.e., twice-differentiable) solutions exist on both side of the path Γ^j, but they fail to exist on the review path Γ^j. In differential pursuit theory, this type of solution is said to be a dispersal around a singular path (or surface) Γ^j, as depicted in Figure 17.2.

Figure 17.2 depicts an open domain of solutions for (17.21), which is then divided by Γ^j into two half-domains, D_{j-1} and D_j, corresponding to S_{j-1} and S_j respectively. A position (or a point) $z^{j-1} \in D_{j-1}$ is associated with a solution for a TVP of the left Hamiltonian $H^-(x,p)$ which moves in the direction of t, while a position $z^j \in D_j$ is associated with a solution for an IVP of the right Hamiltonian $H^+(x,p)$ which moves in the opposite direction of $t_{j+1} - t$ for $t \in [t_j, t_{j+1}]$ (Melikyan

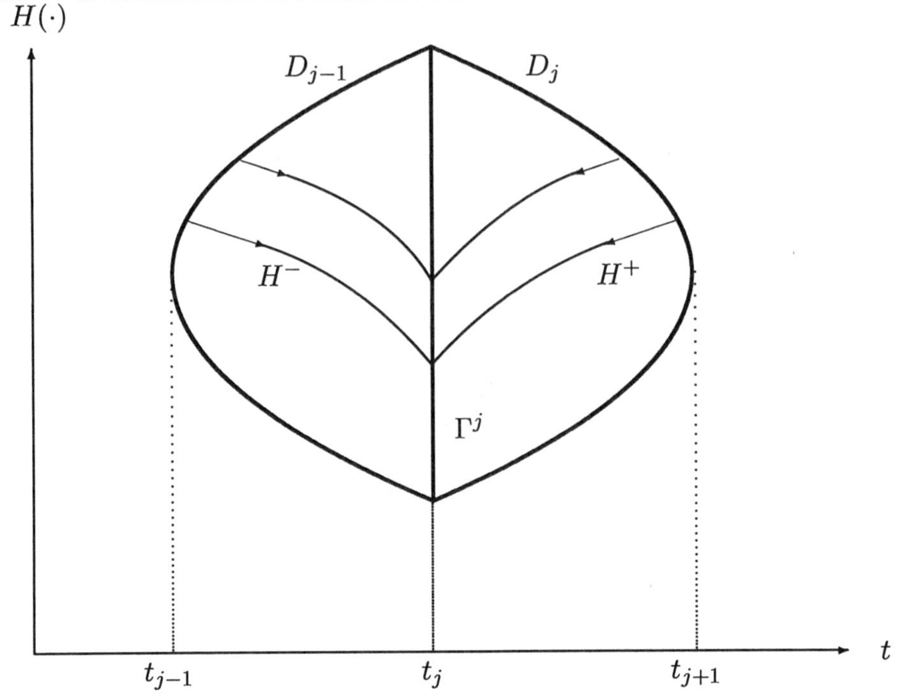

Figure 17.2 Dispersal singular paths and characteristics

1998). The motions of H^- and H^+ coincide at a point $\hat{z}^j$ on the review path Γ^j. Note that H^- takes the same time reference as the original Hamiltonian H (i.e., $H^-(x,p) = H(x,p)$). It is shown in Melikyan (1998) that the left and right Hamiltonians are connected as follows:

$$H^-(x,p) = H(x,p) = -H^+(x,-p),$$

where $H(x,p)$ is given by (17.20), and

$$H^+(x,p) = \max_{u \in \mathcal{P}} \min_{v \in \mathcal{Q}} \{\langle p, \dot{x} \rangle - L(x,u,v)\}. \tag{17.24}$$

Specifically, the value function $\pi^+(t) = \pi(x(\tau - t))$ is the IVP solution of

$$H^+(x,p^+) = 0, \;\; p^+ = \frac{\partial \pi^+}{\partial x}, \tag{17.25}$$

for $t \in (t_j, t_{j+1})$.

4. SINGULAR DISPERSAL SOLUTIONS OF DQR

In this section, we develop a forward-iterative solution algorithm using the method of *discrete approximation* (Subbotin 1995). Recall that the initial condition is given as $x(0) = x^0$.

We start with the case $j = 1$. That is, t_j is the first feedback review point within the open interval $(0, \tau)$ (i.e., $t_{j-1} = 0$), corresponding to which there is a feedback *review path* Γ^j. Suppose that there is an arbitrary point $\hat{z}^j = (\hat{x}^j, \hat{p}^j) \in \Gamma^j$ at which H^- and H^+ meet, as shown in Figure 17.2. Let $z^-(t) = (x^-(t), p^-(t))$ be characteristic solutions for $t \in (t_{j-1}, t_j)$, and $z^+(t) = (x^+(t), p^+(t))$ be characteristic solutions for the rest, $t \in (t_j, \tau)$. Thus, we are faced with the following two boundary-value problems over two adjacent time intervals partitioned from $[0, \tau]$: a TVP of H^- for $t \in [t_{j-1}, t_j]$ and an IVP of H^+ for $t \in [t_j, \tau]$.

$$TVP: \quad H^-(z^-(t)) = 0, \;\; t \in [t_{j-1}, t_j] \qquad (17.26)$$
$$\text{s.t. } x^-(t_{j-1}) = x^0, z^-(t_j) = \hat{z}^j$$
$$IVP: \quad H^+(z^+(t)) = 0, \;\; t \in [t_j, \tau] \qquad (17.27)$$
$$\text{s.t. } z^+(t_j) = \hat{z}^j, \pi(x^+(\tau)) = 0$$

According to Theorem 3..1, the characteristic equations (17.21) of H^- hold for $t \in (t_{j-1}, t_j)$, while equations (17.21) of H^+ hold for $t \in (t_j, \tau)$. Let u^- be an optimal strategy (Definition 2..1) for the TVP of (17.26). Thus, its optimal trajectory satisfies the following ODE system for $t \in [0, t_j]$:

$$\begin{cases} \dot{x}^-(t) = \tilde{A}(u^-)x^-(t) + \tilde{B}(u^-, v), & x^-(t_{j-1}) = x^0 \\ \dot{p}^-(t) = -p^-(t)\tilde{A}(u^-) - D(u^-, v), & p^-(t_j) = \hat{p}^j \end{cases} \qquad (17.28)$$

where $\tilde{A}$ is given in (17.15) and $D(\cdot)$ is given in (17.13). Similarly, using u^+ as an optimal strategy, the IVP of (17.27) reduces to the ODE system for $t \in [t_j, \tau]$:

$$\begin{cases} \dot{x}^+(t) = \tilde{A}(u^+)x^+(t) + \tilde{B}(u^+, v), & x^+(t_j) = \hat{x}^j \\ \dot{p}^+(t) = -p^+(t)\tilde{A}(u^+) + D(u^+, v), & p^+(\tau) = 0. \end{cases} \qquad (17.29)$$

According to control theory, the solutions to (17.28) can be expressed as follows:

$$x^-(t) = \Phi(t, t_{j-1}|u^-)x(t_{j-1}) + \int_{t_{j-1}}^{t} \Phi(t, s|u^-)\tilde{B}(u^-, v)ds \qquad (17.30)$$

$$p^-(t) = \hat{p}^j \Phi(t_j, t|u^-) - \int_t^{t_j} D(u^-, v)\Phi(s, t|u^-)ds \qquad (17.31)$$
$$t \in [t_{j-1}, t_j],$$

where $\Phi(t, t_0)|u^-)$ is the transition matrix of (17.28) under an optimal strategy u^-. It can be verified that the solutions $(x^-(t), p^-(t))$ are unique functions of boundary values of $x(t_{j-1})$ and $\hat{p}^j$. When u^- is fixed at a constant (i.e., $\tilde{A}$ becomes a constant matrix), it can be seen that

$$\Phi(t, t_0) = e^{\tilde{A}(t-t_0)} = I + \sum_{n=1}^{\infty} \frac{\tilde{A}^n (t - t_0)^n}{n!}, \quad t \geq t_0. \qquad (17.32)$$

When the analytical expressions of Φ in (17.32) are unattainable, they can be obtained easily by numerical methods. From (17.15), we can obtain the following.

Lemma 2 *For any fixed $\bar{u}$ in $[t_0, t]$, the transition matrix of (17.14) is derived as*

$$\Phi(t, t_0|\bar{u}) = \begin{pmatrix} \Phi_0 & 0 \\ 0 & \Phi_1 \end{pmatrix}_{2\times 2}, \qquad (17.33)$$

where

$$\Phi_0 = e^{-\bar{u}(t-t_0)} \qquad (17.34)$$

$$\Phi_1 = e^{(\sigma^2 - 2\bar{u})(t-t_0)}. \qquad (17.35)$$

Proof. The proof is immediate by verifying that the differential equation $\dot{\Phi} = \tilde{A}\Phi$ is satisfied. ■

Since the terminal adjoint state $\hat{p}^j$ is arbitrary on the path Γ^j, additional boundary conditions at the singular review point t_j must be properly constructed in order to secure a unique solution p^- for (17.31). In what follows, we propose a specific boundary construction. Note that the solutions $(x^-(t), p^-(t))$ in (17.30) and (17.31) are only applicable for $t \in [t_{j-1}, t_j]$. The purpose of the proposed boundary construction is to augment the TVP solution (17.26) with a characteristic solution of IVP (17.27). Denoting the augmented segment as (x^+, p^+) solves the following IVP (17.29) for $t \in [t_j, \tau]$ with $\hat{x}^j = x^-(t_j)$, where $x^-(t_j)$ is evaluated from (17.30). It is easy to verify that the equations (17.29) have the same transition matrix as (17.28). Thus, the augmented solutions for $t \in [t_j, \tau]$ can be expressed as

$$x^+(t) = \Phi(t, t_j|u^-)x^+(t_j) + \int_{t_j}^{t} \Phi(t, s|u^+)\tilde{B}(u^+, v)ds, \quad (17.36)$$

$$p^+(t) = \int_{t}^{\tau} D(u^-, v)\Phi(s, t|u^-)ds, \quad (17.37)$$

$t \in [t_j, \tau]$.

Therefore, we can conclude the following.

Lemma 3 *For any given u^- and v, the singular $\hat{p}^j$ is uniquely determined from the terminal condition $p^-(t_j) = p^+(t_j) = \hat{p}^j$ as follows:*

$$\hat{p}^j = \int_{t_j}^{\tau} D(u^-, v)\Phi(s, t|u^-)ds, \quad (17.38)$$

where D is given by (17.13) and Φ is given by (17.33). If the review point t_j is free, then an optimal review point t_j^* can be determined from

$$\frac{\partial H^-(x^-(t_j), p^-(t_j))}{\partial t_j} = 0. \quad (17.39)$$

Once a unique p^j is determined by (17.38), then a unique solution $(x^-(t), p^-(t))$ can be determined from (17.30) and (17.31). In summary, for any $(j-1)$th regulation period $\bar{S}_{j-1} = [t_{j-1}, t_j]$ $(j = 0, 1, \cdots, m-1)$, the motion of DQR for regulation period $\bar{S}_{j-1}$ is obtained by solving the TVP of (17.26), where the singular terminal value determined from (17.38) requires solving the IVP of (17.27) for the remaining time interval $[t_j, \tau]$.

Obviously, for any $j \geq 1$, the above solution $(x^-(t), p^-(t))$ for $t \in [t_{j-1}, t_j]$ can be obtained by solving the TVP of the H^- over $[t_{j-1}, t_j]$, where p^j is determined by solving the IVP of the H^+ over $[t_j, \tau]$, as long as a stepwise-programmed control u^- is specified for $t \in [t_{j-1}, t_j]$. Next, we consider the determination of an optimal control strategy in terms of solving the HJBI equation in the minimax sense.

5. OPTIMAL QUALITY REGULATION STRATEGY

First, we note that the quality regulation is confined to the programmed regulation policies. According to Definition 2..1, programmed regulation is solely dependent on the determination of a positional regulation strategy. Suppose that $\tilde{u}$ is an optimal strategy for $t = [t_{j-1}, t_j]$, with a known deterministic target $v(t)$. By optimal, we mean satisfying characteristic equations in the minimax sense. Denote the left Hamiltonian under such conditions by $H^-(x^-, p^-, \tilde{u})$. Thus, the associated left HJB equation can be written as

$$H^-(x^-, p^-, \tilde{u}) = \min_{u \in \mathcal{P}} \{\langle p^-, \dot{x}^- \rangle + L(x^-, u)\} = 0, \ \ t \in (t_{j-1}, t_j). \quad (17.40)$$

We consider a specific type of strategy $\tilde{u}$ such that the gradient vector of H^- satisfies the following for every $t \in (t_{j-1}, t_j)$:

$$\nabla_{\tilde{u}} H^- = \frac{\partial H^-}{\partial \tilde{u}} = F(x^-, p^-, \tilde{u}). \quad (17.41)$$

According to the verification theorem in optimal control (e.g., Theorem 5.1, Fleming and Soner 1993), if a strategy $\tilde{u}$ can be determined from (17.41), then $\tilde{u}$ is indeed an optimal strategy in the sense of minimax solutions. It is known that $\tilde{u}$ exists if the Jordanian determinant of (17.41) is non-zero, i.e.,

$$\frac{\partial \nabla_{\tilde{u}} H^-}{\partial \tilde{u}} = \frac{\partial F}{\partial \tilde{u}} \neq 0. \quad (17.42)$$

The condition in (17.42) is necessary and sufficient for an optimal regulation strategy $\tilde{u}(t)$ to exist.

6. CONCLUSIONS

We show in this paper that a stochastic quality diffusion system can be converted to a deterministic dispersal quality regulation (DQR) model, which is, in fact, a *minimax* problem of non-linear first order PDEs. The Hamiltonian of a DQR is non-smooth, and its characteristics are singular. Thus, generalized solutions, as opposed to classic solutions, must be considered. We develop a singular solution method, which can be used to obtain optimal minimax solutions to DQR in viscosity sense (i.e., for almost all t).

Appendix

Proof of Theorem 2..1 Proof. Since the control $u(t)$ is programmed by Definition 2..1 and mapping $A : \mathcal{P} \to R$ is continuous, it is easy to verify that the $A(u(t))$ in (17.2) is measurable in t. According to Theorem 8.5.2 in Arnold (1992), $K(t) = E\{y(t)\}$ is the unique nonnegative solution of the differential equation:

$$\dot{K}(t) = A(u(t))K(t) + M(u(t), v(t)), K(0) = y^0. \quad (17.43)$$

According to Theorem 8.5.5 of Arnold (1992), $P(t) = E\{y^2(t)\}$ is the unique non-negative-definite symmetric solution of the following equation:

$$\dot{P}(t) = 2A(\cdot)P(t) + 2M(\cdot)K + \sigma^2 P(t). \tag{17.44}$$

Therefore, by rearranging the terms in the above equations, we that the theorem follows. ■

References

[1] Arnold, Ludwig *Stochastic Differential Equations: Theory and Applications.* New York: Wiley & Sons, 1992.

[2] Fleming, W.H. and H.M. Soner, H.M. *Controlled Markov Process and Viscosity Solutions.* New York: Springer Verlag, 1993.

[3] Krasovskii N.N. and Subbotin, A.I. *Game-Theoretical Control Problems.* New York: Springer Verlag, 1988.

[4] Liu, J.J. (2000), Manufacturing Quality Regulation Under Discrete Audits. working paper series, School of Business Administration, University of Wisconsin-Milwaukee, Milwaukee.

[5] Liu, J.J. and Nam, S.H. Regulated Quality Diffusion Under Quadratic Costs. *Int'l Trans. in Operations Research*, **6** (1999) 393-409.

[6] Liu, J.J. and Nam, S.H. (1998), Regulation of Manufacturing Quality with Time-Variant Targets. working paper series, School of Business Administration, University of Wisconsin-Milwaukee, Milwaukee.

[7] Melikyan A. *Generalized Characteristics of First Order PDEs: Applications in Optimal Control & Differential Games.* Boston: Birkhauser, 1998.

[8] Subbotin, A.I. *Generalized Solutions of First Order PDEs: The Dynamical Optimization Perspective Games.* Boston: Birkhauser, 1995.

Chapter 18

SIMULTANEOUS MONITORING OF MEAN AND VARIANCE THROUGH OPTIMALLY DESIGNED SPRT CHARTS

Felipe Pachano-Azuaje and Tapas K. Das

Department of Industrial and Management Systems Engineering

University of South Florida

Tampa, Fl, 33620

(fpachano,das@eng.usf.edu)

Abstract Four essential properties that a control chart must have are applicability, high sensitivity for detecting shifts, cost-optimality, and the capability of jointly controlling all the important parameters that characterize a process. This paper introduces a quality monitoring system design for jointly monitoring the mean and the variance of a Gaussian process. The design of this monitoring system, the Economic Mean and Variance - Sequential Probability Ratio Tests (EMV-SPRT) charts, possesses all of the desired properties mentioned above. The applicability of the design comes from the utilization of assumptions that fit many actual processes. The sensitivity is a consequence of the use of SPRT. The determination of the design parameters through an economic model ensures cost optimality.

This study extends the original concept of SPRT in order to deal with two parameters. The resulting monitoring scheme benefits from the fact that SPRT charts indicate in-control or out-of-control states using the minimum number of sample units. In addition to having this inherent optimality, the EMV-SPRT sampling procedure uses neither a predetermined sample size, nor a set of predetermined bounds for Type I and Type II errors: instead, it uses an economic approach. The EMV-SPRT design is compared numerically with two other relevant designs that exist in the literature. The sensitivity of the EMV-SPRT design with respect to some of the key input parameters is also studied.

Keywords: Control charts, quality monitoring, sequential probability ratio test, economic model, control chart design.

1. INTRODUCTION

When should a production process be halted? When should major maintenance be performed? Major decisions like these must be based on numerous factors, the quality of the output being one of the most important. Naturally, a process that has deviated far from its quality requirements must be halted and repaired, whereas a process with minor variations of the target quality may proceed without interruption. Two questions arise: what is the threshold between the decisions, and how we determine that the threshold has been crossed? The role of Statistical Quality Control (SQC) is to provide decision tools that support production and maintenance activities of an organization. A quality monitoring system (QMS) must be sensitive to possible process changes (shifts) and at the same time should make a rational use of the information. Since data collection and analysis is costly, it is essential that a QMS provide as much information as possible (sensitivity) with relatively small samples (efficiency).

This paper presents the development of a model that optimizes the selection of the design parameters of a sensitive and efficient sampling scheme to jointly monitor the mean and the variance of a process working within a practical set of assumptions. The model developed here has the following characteristic features: optimality, achieved through an economic/statistical-economic design approach; sensitivity and efficiency, achieved through the use of control charts that use sequential probability ratio test (SPRT); statistical validity, achieved through a joint analysis of the mean and the variance; and finally plausibility, achieved through the use of a practical set of assumptions. We, henceforth, refer to the control chart resulting from our model as the "EMV-SPRT charts", where 'E' stands for the economic feature and 'M' and 'V' stand for the joint analysis of the mean and the variance. The main contributions of this paper are the mathematical development of the EMV-SPRT control charts and a numerical comparison of its performance with some of the other existing design approaches. In particular, the EMV-SPRT charts are compared with a variant of Shewhart's classical control charts, in which the design parameters of the mean and variance charts are selected optimally as proposed by Costa [1993], and with a non-economic SPRT control chart design for monitoring a process mean as proposed by Stoumbos and Reynolds [1997].

Ever since Shewhart developed the classical $\bar{x}$ and R charts, many different designs of control charts have been proposed in the literature in a search for higher flexibility and/or efficiency in the monitoring function. CUSUM [Page, 1954] and EWMA [Roberts, 1959] control charts are examples of monitoring schemes that are more sensitive than Shewhart's control charts, especially for smaller shifts. Several enhancements of the classical policy with variable sampling intervals and variable sample sizes, referred to as adaptive sampling, were also proposed in the literature; (see Costa [1998], Das and Jain [1997], Das *et al.* [1997], Runger and Montgomery [1993], Runger and Pigniatello [1991], Reynolds *et al.* [1988]).

The monitoring scheme underlying EMV-SPRT is the sequential probability ratio test (SPRT) control charts scheme [Wald, 1947]. An SPRT control chart takes consecutive single samples until the required information for making the decision to stop the sampling and declare the process to be either in control (IC) or out of control (OOC) is obtained. It is well known that SPRT represents a statistical test that requires the minimum expected number of sample units for testing a simple hypothesis with predetermined sensitivity. However, the superiority of the SPRT scheme over the adaptive extensions to the classical approach and even the pure classical approach is not very clear. It is conjectured that in many situations the performance of SPRT would be better, but it is not possible to ensure the uniform superiority of SPRT. Later, in this paper, a discussion on the conditions for which the use of SPRT is a preferred choice is presented.

The proposed EMV-SPRT is designed to work for processes that satisfy the following assumptions.

1. The process is monitored via a single control variable, which is distributed normally with no auto-correlation.

2. At the beginning of each process cycle, the process is in control ($\mu = \mu_0$, $\sigma = \sigma_0$) and in a steady state operating condition. A process cycle ends at the completion of a repair that is undertaken following the detection of an out-of-control condition of the process.

3. Separate independent assignable causes affect the mean and the variance of the process in a known way. Some assignable causes affect the process mean, and others affect the process variance. The occurrence of an assignable cause does not force or block the occurrence of an assignable cause of the other kind.

4. The change in the mean, to a value μ_1, occurs after an exponential random time T_1 with rate λ_1. In a similar way, the shift in the standard deviation, to a value σ_1, happens after an exponential random time T_2 with rate λ_2.

5. The time required to perform each SPRT is negligible compared to the cycle time.

6. The production process stops during a search for assignable causes triggered by an alarm and also during the repair time.

7. The repair process consists of removal of assignable causes and operating the system through the initial transient phase after restart. The time epoch when the process reaches steady state will henceforth be considered as the beginning of a new cycle.

There exist many variations of economic models for control charts. A survey and summary of the models developed between 1981 and 1991 can be found in Ho and Case [1994]. Most of those models follow common guidelines that were suggested by Duncan [1956]. Duncan first noticed the presence of cycles of quality. Just after a process is repaired, the process returns to the in-control state. After a random period of time, the process changes to an out-of-control state. The monitoring system detects the shift. After necessary repairs, the cycle begins again. Duncan used a common measure of performance of a quality-monitoring scheme as the expected cost per unit time, which he defined as the ratio of the cost per cycle to the expected time per cycle.

A process cycle time can be divided into the following phases.

1. *In-control phase*: The process is in a steady state operating condition and is producing at the best quality level. This phase ends when at least one of the parameters shifts to an unacceptable level.

2. *Searching phase*: It is the time spent in verifying if an alarm corresponds to an actual problem in the process. We assume for our model that during this phase the production process stops.

3. *Out-of-control phase*: It is the time between the moments of occurrence of the first parameter shift until the time when the shift is detected. During this period, the process usually incurs losses resulting from the fraction of products that could fall outside the specification limits.

4. *Repair phase*: During this phase, the assignable causes are eliminated, and the process is run through its transient operating phase.

5. *Sampling time*: This is related to the collection and analysis of the sample. (It is assumed to be negligible in our model).

The total cost per cycle can be divided into the following components.

1. *Cost of nonconformities*: During the in-control and especially during the out-of-control phases, the process is likely to produce some non-conforming output.

2. *Cost of false alarm*: Checking for alarms implies some costs. In addition, if the process stops during that period, there is a cost associated with the loss of production.

3. *Cost of repairing*: It includes the repair cost and the production loss cost.

4. *Cost of sampling*: There is usually a fixed cost, such as labor, and some variable costs, such as materials and testing equipment.

1.1 Sequential Probability Ratio Test (SPRT) .

Wald [1947] designed SPRT as a statistical tool for deciding between two simple hypotheses. If a random variable X is distributed $f(x, \theta)$, it is possible to contrast the simple hypothesis $H_0 : \theta = \theta_0$ with $H_1 : \theta = \theta_1$ using SPRT. This test is based on the Neyman-Pearson (N-P) Lemma [Hoel, 1984] that states that, for a fix sample of size n, the optimal design (most powerful test) for simple hypotheses can be obtained from the likelihood ratio (λ_n) as follows:

$$\begin{array}{lll} \text{Accept } H_0 & \text{if} & \lambda_n < k \\ \text{Accept } H_1 & \text{if} & \lambda_n \geq k, \end{array} \tag{18.1}$$

where

$$\lambda_n = \prod_{i=1}^{n} \frac{f(x_i, \theta_1)}{f(x_i, \theta_0)}, \tag{18.2}$$

k is the decision limit associated with significance α, and i denotes the observation index.

The SPRT based on the N-P Lemma can be given as follows. When the nth sample is being tested, SPRT makes one of the following three decisions.

$$\begin{array}{ll} \text{Accept } H_0 & \text{if } \lambda_n < B \\ \text{Accept } H_1 & \text{if } \lambda_n \geq A \\ \text{Continue Sampling} & \text{otherwise} \end{array} \tag{18.3}$$

The third decision implies that another observation must be added to the sample and that the test must be performed again. This sequential procedure stops whenever H_0 is accepted or rejected.

For different density function f and parameter θ, particular SPRTs can be designed and, with some work, it may be possible to obtain the magnitude of the α and β errors associated with the test for a given value of A and B. However, Wald showed that an approximation of those errors could be obtained using just the decision limits A and B as follows [Siegmund, 1985]:

$$\alpha \approx \frac{1-A}{B-A} \tag{18.4}$$

and

$$\beta \approx \frac{A(B-1)}{B-A}. \tag{18.5}$$

It can be shown that (18.4) and (18.5) overestimate α and β [Wald, 1947]. Some empirical results indicated that the actual errors are 30% less than the above approximations [Wetherill and Glazebrook, 1986]. This suggests that these approximations can be used as upper bounds of the errors. Hence, the economic performance of a QMS design based on (18.4) and (18.5) would always be better than what is predicted by the model developed here. Clearly, the performance of EMV-SPRT can be improved with better estimations of α and β. The numerical results, obtained from our model, although using overestimates of the errors, show the superiority of EMV-SPRT over other approaches (see Section 4).

Another important property of the SPRT is its optimality in reference to the average sampling number, ASN. It is shown [Wetherill and Glazebrook, 1996] that for all sequential tests of $f(x, \theta_0)$ versus $f(x, \theta_1)$ with errors α and β, SPRT has the least possible values of $E(n|\theta_0)$ and $E(n|\theta_1)$. That is, SPRT provides the minimum sample sizes under any value of θ within the hypotheses. However, there are two important considerations to make about the sample size. First, it is known [Wetherill and Glazebrook, 1996] that the sample size distribution is very skewed positively. Therefore, although SPRT provides a small ASN, problematic bottleneck samples may appear. One alternative discussed by Wald is to truncate the SPRT through a sample size limit that it is known to have negligible effects on the global SPRT performance. Secondly, SPRT is optimal when $\theta = \{\theta_0, \theta_1\}$, but it is far from optimal when θ is close to the mid-point between θ_0 and θ_1. In the latter case, it is possible to find some non-sequential tests that provide better power. This point is very important since the value of a parameter is never known and since the transition from one value to another may be sudden or drift continuously. Based on these facts, it is preferable to use composite hypotheses. Although SPRT is designed for simple hypotheses, Wald

[1947] developed an analysis of the consequences of applying SPRT to hypotheses such as $H_0 : \theta \leq \theta_0$ and $H_1 : \theta > \theta_0$.

The literature is rich in discussions about the scope, applications, and problems of SPRT, referring to simple hypotheses of one or more parameters and composite hypotheses of only one parameter. The focus of this paper is the study of a simple hypothesis versus composite hypothesis of two parameters. Specifically, it is intended to design and evaluate a sequential test for the simple null hypothesis that a Gaussian process is in control $(\mu, \sigma) = (\mu_0, \sigma_0)$ versus the composite hypothesis that the process is out of control $(\mu, \sigma) \in \{(\mu_0, \sigma_1), (\mu_1, \sigma_0), (\mu_1, \sigma_1)\}$. We have not found any published reference that deals with this problem.

2. JOINT SPRT CONTROL CHARTS FOR MEAN AND VARIANCE

The purpose of this section is to combine an SPRT control chart for the mean and an SPRT control chart for the variance to operate together to monitor a Gaussian process. Two different designs are proposed. The first, "reference design," consists of an SPRT for the mean, under the assumption that the variance is known and simultaneously, an SPRT for the variance, under the assumption that the mean is unknown. This model is not applicable to actual processes since the test for the mean requires knowledge of the variance when it is actually unknown. To avoid this problem, one possibility considered was to use a test for the mean with no knowledge of the variance. However, we discarded this possibility because of the complexity involved in the development of the associated decision errors. Hence, the approach that we use in this research is to modify in certain ways the test for the mean in the reference design. We call this new design an "operational design," details of which are given later in this section. Table (18.1) contains the glossary of the symbols introduced in this section.

2.1 SPRT Designs.

In this section, we describe our SPRT designs for the mean and the variance.

2.1.1 SPRT for the Mean.

We first summarize the development of the SPRT for the mean in the reference design and then describe the modifications necessary to obtain the operational design.

If we assume that the variance is known, it is possible to develop a SPRT design for testing the mean using the Neyman-Pearson Lemma. Set the hypothesis $H_0 : \mu = \mu_0, H_1 : \mu = \mu_1 (\mu_1 > \mu_0)$, for a population $X \sim N(\mu, \sigma^2)$ for which σ^2 is known. Suppose that after $n - 1$ obser-

Table 18.1 Glossary of Symbols.

Symbol	Description
μ, σ	Parameters of the Gaussian process
μ_0, μ_1	In-control and out-of-control values of the mean
σ_0, σ_1	In-control and out-of-control values of the standard deviation
T_1, T_2	Random interval of time during which the mean and the variance stay in control, respectively
λ_1, λ_2	Rate of shift (from in-control to out-of-control conditions) of the process mean and the process variance, respectively
D	Sampling interval
D_0, D_1	Time for searching alarms and time for repairs
H_0, H_1	Null and alternative hypotheses
L	Likelihood function
$a_\mu, b_\mu,$ a_σ, b_σ	Lower and upper limit of the likelihood ratio tests for mean and variance
$h_\mu, k_\mu,$ h_σ, k_σ	Lower and upper limit of the log-likelihood ratio tests for mean or variance
$\bar{X}_n$, S_n^2	Moving average and moving sampling variance
$\bar{\mu}$	Mid-point between μ_0 and μ_1
R_1, R_2	Constants that appear in the control limits of the variance chart

vations the test has indicated that there is no evidence for accepting or rejecting H_0. Define λ_n^μ, the nth likelihood ratio for testing the mean, as

$$\lambda_n^\mu = \frac{L(x_1, x_2, \cdots, x_n; \mu_1, \sigma^2)}{L(x_1, x_2, \cdots, x_n; \mu_0, \sigma^2)}. \tag{18.6}$$

Thus, the decision criterion is as follows.

Accept H_0 (Reject H_1)	if $\lambda_n^\mu < a_\mu$	
Reject H_0 (Accept H_1)	if $\lambda_n^\mu > b_\mu$	
Stay undecided and keep on sampling	otherwise	(18.7)

where a_μ and b_μ are design variables. Substituting the expression for the likelihood ratio, (18.6) can be written as

$$\lambda_n^\mu = \frac{\left(1/\sqrt{2\pi}\sigma\right)^n EXP\left(-1/2\sigma^2 \sum_{k=1}^{n} (x_k - \mu_1)^2\right)}{\left(1/\sqrt{2\pi}\sigma\right)^n EXP\left(-1/2\sigma^2 \sum_{k=1}^{n} (x_k - \mu_0)^2\right)}. \tag{18.8}$$

Applying a logarithm to both sides, the log-likelihood ratio $l_n^\mu = ln(\lambda_n^\mu)$ can be written as

$$\begin{aligned} l_n^\mu &= -\frac{1}{2\sigma^2}\left[\sum_{k=1}^{n} (x_k - \mu_1)^2 - \sum_{k=1}^{n} (x_k - \mu_0)^2\right] \\ &= -\frac{1}{2\sigma^2}\left[-2(\mu_1 - \mu_0)\sum_{k=1}^{n} x_k + n\left(\mu_1^2 - \mu_0^2\right)\right] \end{aligned} \tag{18.9}$$

Defining $\bar{x}_n = \frac{\sum_{k=1}^{n} x_k}{n}$ as the moving sample mean and $\bar{\mu} = \frac{\mu_1 + \mu_0}{2}$, l_n^μ can be written as

$$l_n^\mu = n\left(\frac{\mu_1 - \mu_0}{\sigma^2}\right)(\bar{x}_n - \bar{\mu}). \tag{18.10}$$

Hence, it is possible to have a test in terms of $\bar{x}_n$.

Accept H_0	if $\bar{x}_n < \bar{\mu} + \dfrac{h_\mu \sigma^2}{n(\mu_1 - \mu_0)}$	
Reject H_0	if $\bar{x}_n > \bar{\mu} + \dfrac{k_\mu \sigma^2}{n(\mu_1 - \mu_0)}$	
Remain undecided	otherwise.	(18.11)

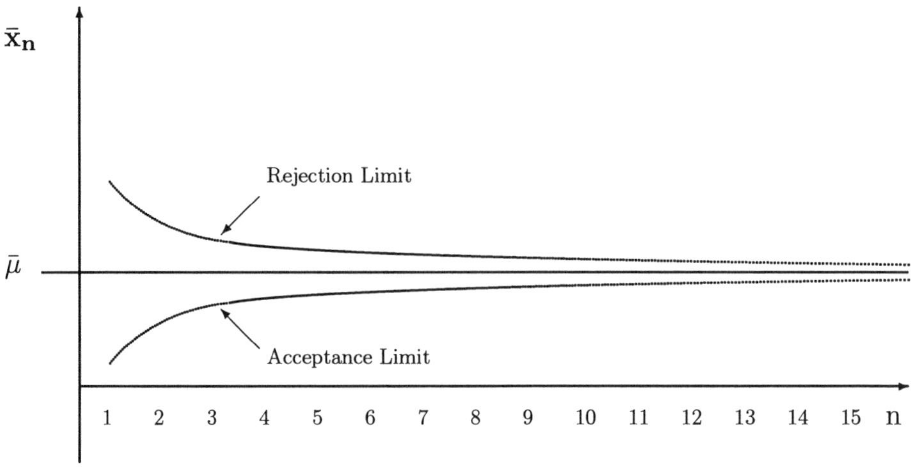

Figure 18.1 A sequential probability ratio test chart for the mean (as in (18.11)).

Here, h_μ and k_μ, the logarithms of a_μ and b_μ (defined in (18.7)) respectively, are design variables. This SPRT design is valid for $n \geq 1$ and has the advantage of providing a very special control chart since the rejection and acceptance limits for the test converge to $\bar{\mu}$ as n converges to infinity. Figure 18.1 shows a typical control chart for this design.

In the operational design, the test for the mean is also constructed with known variance. The difference at this stage, is that the variance is assumed to be constant and always equal to the in-control variance. That is, instead of using the test given in (18.11), the test used is obtained by replacing the actual variance of the process σ^2 by σ_0^2.

$$\begin{array}{lll} \text{Accept } H_0 & \text{if } \bar{X}_n < \bar{\mu} + \dfrac{h_\mu \sigma_0^2}{n\,(\mu_1 - \mu_0)} & \\ \text{Reject } H_0 & \text{if } \bar{X}_n > \bar{\mu} + \dfrac{k_\mu \sigma_0^2}{n\,(\mu_1 - \mu_0)} & \\ \text{Remain undecided} & \text{otherwise.} & (18.12) \end{array}$$

It might seem that the modeling assumption to use σ_0^2 instead of σ^2 represents an additional source of error that may affect the performance of the global test in the operational design. However, it will be shown later in this paper that in most practical situations the operational design is more cost-efficient.

2.1.2 SPRT for the Variance. The test for the variance, with no knowledge of the mean (as in both the reference and the operational design), is developed here by modifying the test for the variance when the mean is known. Set the hypothesis as $H_0 : \sigma^2 = \sigma_0^2$, $H_1 : \sigma^2 = \sigma_1^2$ ($\sigma_0^2 < \sigma_1^2$), for a population $X \sim N(\mu, \sigma^2)$ (μ is known). Suppose that, after $n-1$ observations, the test has indicated that there is no evidence for accepting or rejecting H_0. Define λ_n^σ, the nth likelihood ratio for testing the variance as follows:

$$\lambda_n^\sigma = \frac{L(x_1, x_2, \cdots, x_n; \mu, \sigma_1^2)}{L(x_1, x_2, \cdots, x_n; \mu, \sigma_0^2)}$$

$$= \frac{\left(1/\sqrt{2\pi}\sigma_1\right)^n EXP\left(-1/2\sigma_1^2 \sum_{k=1}^{n} (x_k - \mu)^2\right)}{\left(1/\sqrt{2\pi}\sigma_0\right)^n EXP\left(-1/2\sigma_0^2 \sum_{k=1}^{n} (x_k - \mu)^2\right)}$$

$$= \left(\frac{\sigma_0}{\sigma_1}\right)^n EXP\left[\frac{1}{2}\sum_{k=1}^{n}(x_k - \mu)^2 \left(\frac{1}{\sigma_0^2} - \frac{1}{\sigma_1^2}\right)\right]. \tag{18.13}$$

Converting the above to a log likelihood ratio, it can be seen that

$$l_n^\sigma = \ln(\lambda_n^\sigma) = -n \ln\left(\sigma_1/\sigma_0\right) + \frac{1}{2}\sum_{k=1}^{n}(x_k - \mu)^2 \left(\frac{1}{\sigma_0^2} - \frac{1}{\sigma_1^2}\right). \tag{18.14}$$

Define

$$R_1 = \ln\left(\sigma_1/\sigma_0\right), \tag{18.15}$$

$$R_2 = \frac{1}{2}\left(\frac{1}{\sigma_0^2} - \frac{1}{\sigma_1^2}\right), \tag{18.16}$$

and

$$\hat{\sigma}_n^2 = \frac{\sum_{k=1}^{n}(x_k - \mu)^2}{n}. \tag{18.17}$$

Then (18.14) can be given as

$$l_n^\sigma = -nR_1 + n\hat{\sigma}_n^2 R_2. \tag{18.18}$$

Thus, the SPRT for the variance with known mean is as follows.

$$\begin{array}{lll} \text{Accept} \quad H_0 & \text{if} \quad \hat{\sigma}_n^2 < \dfrac{R_1}{R_2} + \dfrac{h_\sigma^{'}}{nR_2}, & \\ \text{Reject} \quad H_0 & \text{if} \quad \hat{\sigma}_n^2 > \dfrac{R_1}{R_2} + \dfrac{k_\sigma^{'}}{nR_2}, & \\ \text{Remain undecided} & \text{otherwise.} & (18.19) \end{array}$$

Here $k_n^{'}$ and $h_n^{'}$ are decision variables. This test is valid for $n \geq 1$.

For obtaining the test of the variance when the mean is unknown, we replace μ in (18.19) with its maximum likelihood estimator ($\bar{x}_n$). Also, instead of $\hat{\sigma}_n^2$, $s_n^2 = \frac{\sum_{k=1}^{n} (x_k - \bar{x}_n)^2}{n-1}$ may be used. Thus, the SPRT design for the variance with unknown mean becomes as follows.

$$\begin{array}{lll} \text{Accept} \quad H_0 & \text{if} \quad S_n^2 < \dfrac{R_1}{R_2} + \dfrac{h_\sigma}{nR_2}, & \\ \text{Reject} \quad H_0 & \text{if} \quad S_n^2 > \dfrac{R_1}{R_2} + \dfrac{k_\sigma}{nR_2}, & \\ \text{Remain undecided} & \text{otherwise.} & (18.20) \end{array}$$

Here, k_σ and h_σ are decision variables and the test needs $n > 1$. Figure 18.2 shows a typical SPRT control chart for variance.

3. DEVELOPMENT OF ECONOMIC MODELS

In what follows, we present two economic models for joint monitoring of the mean and the variance using the reference and the operational SPRT designs described in the previous section. Our economic model, which has a structure similar to the model developed by Costa [1993], is essentially a non-linear programming model. This is designed with the objective of minimizing the overall cost per unit time that results from the application of the quality monitoring system. The decision variables of this programming model are the design parameters. In the particular cases of the reference and operational designs, the design parameters are the sampling interval (d), the acceptance and rejection limit for the log-likelihood ratio of the mean chart (h_μ and k_μ), and the acceptance and rejection limit for the log-likelihood ratio of the variance chart (h_σ and k_σ). An alternative approximate economic modeling approach that eliminates the need for non-linear optimization by using some simplifying assumptions can be found in Collani [1997].

There are some errors associated with the four control limits, h_μ, k_μ, h_σ and k_σ. The computation of these errors, commonly referred to as

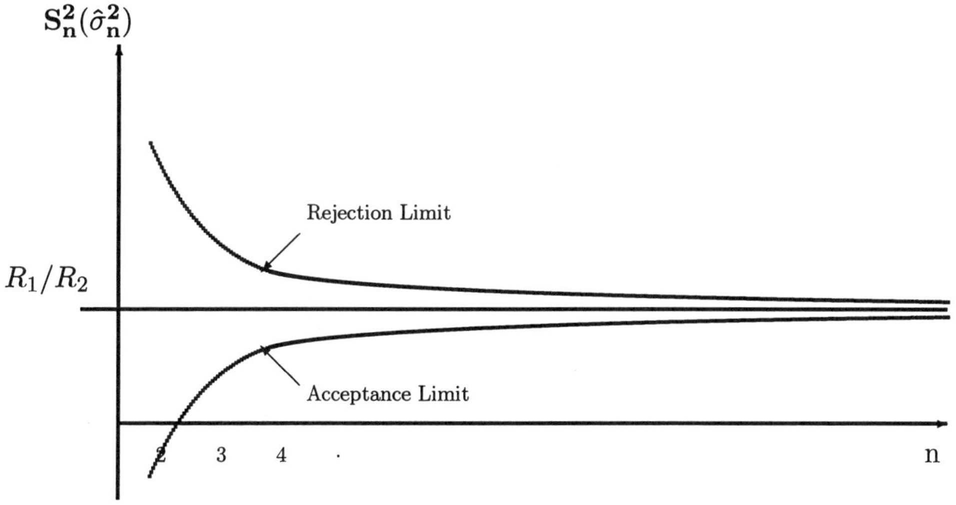

Figure 18.2 A sequential probability ratio test chart for the variance.

Type I and the Type II decision errors, along with the sampling interval d are required to compute the expected length of each phase within the process cycle and also to compute the different costs incurred during each cycle. Then, it is possible to compute the expected cost per unit time that represents the objective to be optimized.

3.1 Notation. Table 18.2 contains a glossary of the symbols introduced in this chapter and used throughout the remaining part of this paper. The table contains the symbols and a brief description of their meanings.

3.2 Error Probabilities. This subsection discusses the computation of the Type I and Type II errors (α and β). The exact values of α and β are very difficult to obtain. However, Wald [1947] showed that it is possible to obtain approximations of those values that provide upper bounds. This implies that every computation that depends on this approximation will be affected. For instance, the cost per unit time of the control system will be an upper bound, and the objective function value will present a lower bound. Therefore, the results of implementing this method in an actual process will not be worse than the values presented here.

3.2.1 Errors for Reference Design. In accordance with Wald's theory, the computation of the decision errors for the reference design is given as follows:

$$\alpha \approx \frac{1 - e^{h}}{e^{k} - e^{h}}, \tag{18.21}$$

and

$$\beta \approx \frac{\left(e^{k} - 1\right) e^{h}}{e^{k} - e^{h}}, \tag{18.22}$$

where h and k represent the acceptance and rejection limits for the log-likelihood ratio, respectively. These general expressions (18.21) and (18.22) can be applied to the separate and independent tests for mean and variance. Thus, it is possible to obtain the probability of errors in the partial tests, α_μ, β_μ, α_σ, and β_σ, as a function of h_μ, k_μ, h_σ, and k_σ. For instance, α_μ can be shown to be

$$\alpha_\mu \approx \frac{1 - e^{h_\mu}}{e^{k_\mu} - e^{h_\mu}}. \tag{18.23}$$

The global errors can be computed as follows:

$$\alpha = \alpha_\mu + \alpha_\sigma - \alpha_\mu \alpha_\sigma, \tag{18.24}$$

$$\beta\left(\mu_1, \sigma_0\right) = \beta_\mu \left(1 - \alpha_\sigma\right), \tag{18.25}$$

$$\beta\left(\mu_0, \sigma_1\right) = \left(1 - \alpha_\mu\right) \beta_\sigma, \tag{18.26}$$

and

$$\beta\left(\mu_1, \sigma_1\right) = \beta_\mu \beta_\sigma. \tag{18.27}$$

3.2.2 Errors for Operational Design. The approximation formulae given by Wald were applied in the reference model without any problem. In the case of the operational design, there is a problem when the actual variance is out of control since the control limits are given by $\bar{\mu} + \frac{h_\mu \sigma_0^2}{n(\mu_1 - \mu_0)}$ and $\bar{\mu} + \frac{k_\mu \sigma_0^2}{n(\mu_1 - \mu_0)}$, when they should be given by $\bar{\mu} + \frac{h_\mu^* \sigma_1^2}{n(\mu_1 - \mu_0)}$ and $\bar{\mu} + \frac{k_\mu^* \sigma_1^2}{n(\mu_1 - \mu_0)}$. Note that h_μ^* and k_μ^* represent the actual control limits of the log-likelihood ratio tests when the variance is out of control and the they can be written so that $h_\mu^* = h_\mu \frac{\sigma_0^2}{\sigma_1^2}$ and $k_\mu^* = k_\mu \frac{\sigma_0^2}{\sigma_1^2}$. When the variance is in control, those actual control limits are simply $h_\mu^* = h_\mu$ and $k_\mu^* = k_\mu$. The computation of the errors is then modified

Table 18.2 Glossary of Symbols.

Symbol	**Description**
$\alpha_\mu, \alpha_\sigma, \beta_\mu, \beta_\sigma$	Type 1 and Type 2 errors for partial tests
$\alpha, \beta(\mu, \sigma)$	Global Type 1 error and global Type 2 error for each condition
T_{IC}	Expected time that the process stays in control
T_F	Expected time used searching for alarms
$T_\mu, T_\sigma,$ $T_{\mu\sigma}$	Expected time that the process operates with the specified parameter out of control
$a_1, a_2,$ a_3, a_4	Costs of each SPRT, each sample unit, each false alarm and each reparation of the system
D_0, D_1	Time required to search for alarms and time for repairs
P_1, P_2, P_3	Power of the test for the different shift types
q_1, q_2, q_3	Probability of a shift of each type occurring during a sampling interval, given that the process goes out of control in that period
τ_1, τ_2	Expected time of occurrence of a shift (of Type 1 or 2) within a particular sampling interval
S_μ, S_σ	Expected time of a shift of the specified parameter, given that both parameters did not shift in the same sampling interval
$C_\mu, C_\sigma,$ $C_{\mu\sigma}$	Expected time of a shift of the specified parameter, given that both parameters shifted in the same sampling interval
$SL_1, SL_2,$ SL_3	Expected time that both parameters stay out-of control for the different types of shifts
ψ	Global average sample size per SPRT
$ASN_{IC},$ ASN_μ, ASN_σ $ASN_{\mu\sigma}$	Average sample number when the process is in control or when the specified parameter is out of control
Y_i, Z_i	The random processes related with the log-likelihood of the tests for the mean and the variance
$V(\mu, \sigma)$, V	Income per unit time for any process condition (μ, σ) and income per unit time for perfect production

Table 18.3 Process Conditions.

Condition	Shift Type	Parameter	Notation
In-Control	No Shift	μ_0, σ_0	T_{IC}
Searching caused by False Alarms	No Shift	μ_0, σ_0	T_F
Mean Only OOC	Type 1	μ_1, σ_0	T_μ
Variance Only OOC	Type 2	μ_0, σ_1	T_σ
Both Parameters OOC	Type 3	μ_1, σ_1	$T_{\mu\sigma}$
Search and Repair of Assignable Causes	Type 1, 2 or 3	$\mu = \mu_1$ and/or $\sigma = \sigma_1$	D_1

by replacing h_μ by h_μ^* and k_μ by k_μ^* in the equations (18.21) through (18.27). The remaining developments of the cost model are not specific to either the reference or the operational design with the exception of the computation of the average sample number (ASN).

3.3 Estimation of Average Cycle Time. The possible operating conditions of the system during a process cycle are listed in Table 18.3. The average cycle time is obtained by summing the expected length of time the process stays in each of the conditions given in Table 18.3. The last column of the table gives the notation for the mean values of the random lengths of stay at all process conditions.

The expected length of the in-control condition (T_{IC}) is just the expected value of the minimum of two exponential variables T_1 and T_2, given as $1/\lambda$ (where $\lambda = \lambda_1 + \lambda_2$). The expected time spent searching for false alarms (T_F) can be given as

$$T_F = D_0 \alpha \frac{e^{-\lambda d}}{1 - e^{-\lambda d}}, \tag{18.28}$$

where D_0 is the average searching time for a false alarm, and $\frac{e^{-\lambda d}}{1-e^{-\lambda d}}$ gives the expected number of SPRTs during the IC phase. The expected length of a search and repair period is considered to be D_1 (see Table 18.3).

3.3.1 Expected Time the Process is Out of Control. Let the power of the test for different shift conditions be denoted as

$$P_1 = 1 - \beta(\mu_1, \sigma_0), \tag{18.29}$$

$$P_2 = 1 - \beta(\mu_0, \sigma_1), \tag{18.30}$$

and

$$P_3 = 1 - \beta(\mu_1, \sigma_1). \tag{18.31}$$

Let q_i denote the probability of a shift of type i (See Table 18.3) occurring during a sampling interval, given that the process goes OOC in that period. Then, for any sampling interval m, it is possible to obtain

$$q_1 = \frac{\left(1 - e^{-\lambda_1 d}\right)\left(e^{-\lambda_2 d}\right)}{\left(1 - e^{-\lambda d}\right)}, \tag{18.32}$$

$$q_2 = \frac{\left(1 - e^{-\lambda_2 d}\right)\left(e^{-\lambda_1 d}\right)}{\left(1 - e^{-\lambda d}\right)}, \tag{18.33}$$

and

$$q_3 = \frac{\left(1 - e^{-\lambda_1 d}\right)\left(1 - e^{-\lambda_2 d}\right)}{\left(1 - e^{-\lambda d}\right)}. \tag{18.34}$$

Also, for the shift types i $(i = 1, 2)$, define τ_i as the expected time of occurrence of the shift within the mth sampling interval. Then τ_i can be expressed as

$$\tau_i = \frac{1 - (1 + \lambda_i d)\left(e^{-\lambda_i d}\right)}{\lambda_i\left(1 - e^{-\lambda_i d}\right)}. \tag{18.35}$$

In the case of a shift of Type 1, Costa [1993] showed that the expected time when only the mean is out of control is

$$T_\mu = q_1 S_\mu + q_3 C_\mu, \tag{18.36}$$

where

$$S_\mu = \frac{d}{1 - (1 - P_1)\, e^{-\lambda_2 d}} - \tau_1 + \tau_2 \left(1 - \frac{P_1}{1 - (1 - P_1)\, e^{-\lambda_2 d}}\right), \tag{18.37}$$

and

$$C_\mu = \frac{\left(\frac{\lambda_2 - \lambda_1}{\lambda_1 \lambda_2}\right) e^{-\lambda_2 d} - d\, e^{-\lambda_2 d} + \frac{1}{\lambda}\left(\frac{\lambda_1}{\lambda_2} - \frac{\lambda_2}{\lambda_1} e^{-\lambda_2 d}\right)}{\left(1 - e^{-\lambda_1 d}\right)\left(1 - e^{-\lambda_2 d}\right)}. \tag{18.38}$$

Similarly, for a shift of Type 2, the expected time when only the variance is out of control is

$$T_\sigma = q_2 S_\sigma + q_3 C_\sigma, \tag{18.39}$$

where

$$S_\sigma = \frac{d}{1-(1-P_2)\,e^{-\lambda_1 d}} - \tau_2 + \tau_1\left(1 - \frac{P_2}{1-(1-P_2)\,e^{-\lambda_1 d}}\right), \tag{18.40}$$

and

$$C_\sigma = \frac{\left(\frac{\lambda_1-\lambda_2}{\lambda_1\lambda_2}\right)e^{-\lambda_1 d} - de^{-\lambda_1 d} + \frac{1}{\lambda}\left(\frac{\lambda_2}{\lambda_1} - \frac{\lambda_1}{\lambda_2}\,e^{-\lambda_1 d}\right)}{(1-e^{-\lambda_1 d})\,(1-e^{-\lambda_2 d})}. \tag{18.41}$$

The expected time when both mean and variance are out of control is

$$T_{\mu\sigma} = q_1 SL_1 + q_2 SL_2 + q_3 SL_3, \tag{18.42}$$

where

$$SL_1 = \left({}^{d}\!/\!P_3 - \tau_2\right)\left(1 - \frac{P_1}{1-(1-P_1)\,e^{-\lambda_2 d}}\right), \tag{18.43}$$

$$SL_2 = \left({}^{d}\!/\!P_3 - \tau_1\right)\left(1 - \frac{P_2}{1-(1-P_2)\,e^{-\lambda_1 d}}\right), \tag{18.44}$$

$$SL_3 = d\left({}^{1}\!/\!P_3 - 1\right) + C_{\mu\sigma}, \tag{18.45}$$

and

$$C_{\mu\sigma} = \frac{d + \frac{e^{-\lambda_1 d}}{\lambda_1} + \frac{e^{-\lambda_2 d}}{\lambda_2} + \frac{1-e^{-\lambda d}}{\lambda} - \frac{\lambda}{\lambda_1\lambda_2}}{(1-e^{-\lambda_1 d})\,(1-e^{-\lambda_2 d})}. \tag{18.46}$$

This completes the formulation of the components of the cycle time. To summarize, the expected cycle time $E(T)$ can be given as

$$E(T) = 1/\lambda + T_F + T_\mu + T_\sigma + T_{\mu\sigma} + D_1. \tag{18.47}$$

3.4 Estimation of Net Income per Cycle. The net income per cycle is the gross income made during a quality cycle minus the costs incurred in the quality process. The quality cost is comprised of the sampling cost, the quality loss cost, the cost of searching for assignable causes, and the repair cost. Let $V(\mu, \sigma)$ denote the income per unit when the process mean is $\mu \in \{\mu_0, \mu_1\}$ and the process variance is $\sigma \in \{\sigma_0, \sigma_1\}$. This income may be expressed as (Lorenzen and Vance, 1986)

$$V(\mu, \sigma) = V - 1000FD(\mu, \sigma), \tag{18.48}$$

where $FD(\mu, \sigma)$ represents the fraction of non-conforming products (out of the specification limits, set at $\mu_0 \pm 3.5\sigma_0$) for any combination of μ and σ. $FD(\mu, \sigma)$ can be given as

$$FD(\mu, \sigma) = \Phi\left(\frac{\mu_0 - 3.5\sigma_0 - \mu}{\sigma}\right) + 1 - \Phi\left(\frac{\mu_0 + 3.5\sigma_0 - \mu}{\sigma}\right). \tag{18.49}$$

V is the income per unit time when no products fall out of the specification limits. The factor of 1000, chosen arbitrarily, may be interpreted as a cost of \$1.00 for each 0.1% increase in the percentage defective. Thus, the expected value of the net income per cycle (I) is given as

$$E(I) = V(\mu_0, \sigma_0)\left({}^{1}\!/_{\lambda}\right) + V(\mu_1, \sigma_0) T_\mu + V(\mu_0, \sigma_1) T_\sigma$$

$$+V(\mu_1, \sigma_1) T_{\mu\sigma} - \left(SC + a_3 \alpha \frac{e^{-\lambda d}}{1 - e^{-\lambda d}} + a_4\right). \tag{18.50}$$

The sampling cost (SC) can be formulated as

$$SC = (a_1 + a_2 \psi) \frac{{}^{1}\!/_{\lambda} + T_\mu + T_\sigma + T_{\mu\sigma}}{d}, \tag{18.51}$$

where a_1 and a_2 represents the fixed and variable costs of sampling, and ψ represents the global average sample size per SPRT.

3.4.1 Computation of the Global Average Sample Size (ψ).

The global average sample size per SPRT, ψ, can be expressed as

$$\psi = \frac{ASN_{IC}/\lambda + ASN_\mu * T_\mu + ASN_\sigma * T_\sigma + ASN_{\mu\sigma} * T_{\mu\sigma}}{1/\lambda + T_\mu + T_\sigma + T_{\mu\sigma}}, \tag{18.52}$$

where ASN_{IC}, ASN_μ, ASN_σ, and $ASN_{\mu\sigma}$ represent the average sample number for each one of the conditions of the process, namely the in-control condition, and out the three types of the out-of-control condition. The next section explains how the average sample number is computed.

3.4.2 Computation of the ASN. Like the computation of the error probabilities, the ASN has been obtained using Wald's approach. It is important to mention that the following development corresponds to the SPRT of the mean with known variance and the SPRT of the variance with known mean. However, the design for monitoring the variance that is modeled in this paper requires no knowledge of the mean. The ASN for the variance test differs for cases where the mean is known and unknown. With reference to Girshick, Wald [1946] argued that the ASN for the test of the variance is larger by one unit when the mean is unknown. This information has been used in obtaining the expression for the ASN.

Define the two processes Y_i and Z_i as

$$Y_i = \left\{ \ln \frac{f\left(x_i; \mu_1, \sigma^2\right)}{f\left(x_i; \mu_0, \sigma^2\right)} \right\}, i = 1, 2, ... \tag{18.53}$$

and

$$Z_i = \left\{ \ln \frac{f\left(x_i; \mu, \sigma_1^2\right)}{f\left(x_i; \mu, \sigma_0^2\right)} \right\}, i = 1, 2, ... \tag{18.54}$$

where, $Y_i(Z_i)$ represents the log-likelihood ratio process for testing H_0^{μ} : $\mu = \mu_0$ and $H_1^{\mu} : \mu = \mu_1$ ($H_0^{\sigma} : \sigma = \sigma_0$ and $H_1^{\sigma} : \sigma = \sigma_1$) when σ^2 is known (μ is known). Let N denote the random number of observations required by the test (stopping time). The expected value of N (ASN) is computed depending on the decisions based on the individual tests. Let N^* be an integer sufficiently large to ensure that $P(N \geq N^*) \approx 0$. Hence, it is assumed that $N < N^*$.

Thus,

$$Y_1 + Y_2 + .. + Y_{N^*} = Y_1 + Y_2 + ... + Y_N + Y_{N+1} + ... + Y_{N^*}, \tag{18.55}$$

and

$$Z_1 + Z_2 + ... + Z_{N^*} = Z_1 + Z_2 + ... + Z_N + Z_{N+1} + ... + Z_{N^*}. \tag{18.56}$$

Applying expectation on both sides of (18.55) and (18.56), we obtain

$$N^* E(Z) = E(Z_1 + Z_2 + ... + Z_N) + E(Z_{N+1} + ... + Z_{N^*}), \tag{18.57}$$

and a similar expression for $N^* E(Y)$. Since N and $Z_i (i > N)$ are independent random variables,

$$E(Z_{N+1} + ... + Z_{N^*}) = E(N^* - N)E(Z). \tag{18.58}$$

Thus, it follows from (18.57) that

$$E(Z_1 + Z_2 + ... + Z_N) = N^*E(Z) - E(N^* - N)E(Z), \tag{18.59}$$

which gives

$$E(N) = \frac{E(Z_1 + Z_2 + ... + Z_N)}{E(Z)}. \tag{18.60}$$

Similarly,

$$E(N) = \frac{E(Y_1 + Y_2 + ... + Y_N)}{E(Y)}. \tag{18.61}$$

Hence, there are two ways of computing $E(N)$. Wald [1947; pp. 123, 132] computed the values of $E(Y)$ and $E(Z)$ as

$$E(Y) = \frac{\mu_1 - \mu_0}{\sigma^2}(\mu - \bar{\mu}), \tag{18.62}$$

and

$$E(Z) = R_2\sigma^2 - R_1, \tag{18.63}$$

where $\bar{\mu}$ is the mid point between μ_0 and μ_1, and R_1 and R_2 are given in (18.15) and (18.16). Note that when $\mu = \mu_0$, $E(Y) < 0$, and when $\mu = \mu_1$, $E(Y) > 0$. Similarly, it can be shown that when $\sigma = \sigma_0$, $E(Z) < 0$, and when $\sigma = \sigma_1$, $E(Z) > 0$.

Define A_μ and A_σ (R_μ and R_σ) as the events associated with the acceptance (rejection) of the partial null hypotheses of mean and variance, respectively. Note that N indicates the stopping time, which is determined by the sample number for which either one of the in-control hypotheses is rejected, or both in-control hypotheses are accepted. In the remaining part of this subsection, for analytical reasons only, we consider allowing the sample process to continue until both charts have made their indications. Thus, the sampling process, if allowed to continue until both charts have made their indications, would end with any of the combinations of decisions given by $A_\mu A_\sigma$, $A_\mu R_\sigma$, $R_\mu A_\sigma$, and $R_\mu R_\sigma$, where the orders of the pairs are irrelevant. Thus, the value of the summation $Y_1 + Y_2 + ... + Y_N$ is approximated by

$$Y_1 + ... + Y_N \approx \begin{cases} k_\mu & \text{if } \mathrm{R}_\mu A_\sigma, \\ k_\mu & \text{if } \mathrm{R}_\mu R_\sigma \text{ and the mean test finishes first,} \\ h_\mu & \text{if } \mathrm{A}_\mu A_\sigma \text{ and the mean test finishes last,} \\ J_\sigma < k_\mu & \text{Otherwise.} \end{cases} \tag{18.64}$$

Similarly, we obtain

$$Z_1 + ... + Z_N \approx \begin{cases} k_\sigma & \text{if } \mathrm{A}_\mu R_\sigma, \\ k_\sigma & \text{if } \mathrm{R}_\mu R_\sigma \text{ and the mean test finishes last,} \\ h_\sigma & \text{if } \mathrm{A}_\mu A_\sigma \text{ and the mean test finishes first,} \\ J_\mu < k_\sigma & \text{Otherwise,} \end{cases} \tag{18.65}$$

Figure 18.3 provides two examples of the cases included in (18.64) and (18.65). Figure 18.3(a) shows the case where neither the mean nor the variance control charts support the in-control hypothesis. The mean chart gives a signal before the variance chart; consequently, at the stopping time, $Y_1 + Y_2 + ... + Y_N$ is at or above k_μ, and $Z_1 + Z_2 + ... + Z_N$ is at certain value $J_\mu < k_\mu$. Figure 18.3(b) depicts the case where both the mean and the variance chart support the hypothesis that the process is in control and that the mean chart reaches the boundary line first. However, this does not satisfy the stopping rule, and the sampling continues until the variance test ends by indicating acceptance of the in-control hypothesis. Hence, at the stopping time in this case, $Z_1 + Z_2 + ... + Z_N$ is at or below h_σ, and $Y_1 + Y_2 + ... + Y_N$ is at certain value J_σ. Note that it is not necessary in the development of the model to know the exact values of J_μ and J_σ in (18.64) and (18.65).

Up to this point, it has been considered that both partial tests operate until a decision is reached independent of the result of the other test. However, the actual global test finishes when either a rejection happens in any of the partial tests, or acceptances happen in both tests. The following analysis contributes to the identification of the value of the ASN for the global test. We consider four (4) possible situations ($A_\mu R_\sigma$, $R_\mu A_\sigma$, $R_\mu R_\sigma$, and $A_\mu A_\sigma$). Using (18.60), (18.61), (18.64) and (18.65) we can write the expression for the ASN as follows:

$$E[N] = E(N\,|A_\mu R_\sigma)\,P(A_\mu R_\sigma) + E(N\,|A_\sigma R_\mu)\,P(A_\sigma R_\mu) + ...$$

$$... + E(N\,|R_\mu R_\sigma)\,P(R_\mu R_\sigma) + E(N\,|A_\mu A_\sigma)\,P(A_\mu A_\sigma). \tag{18.66}$$

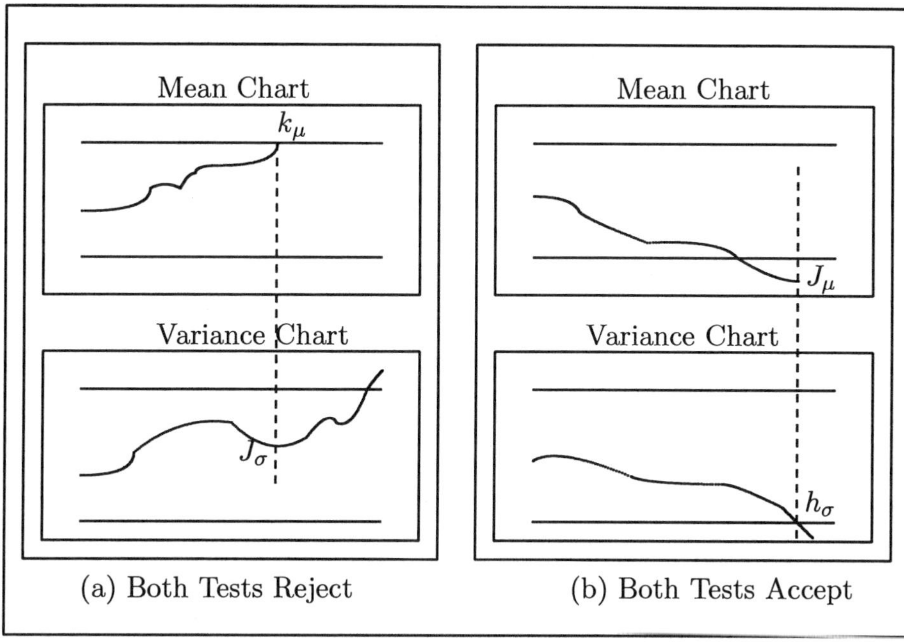

Figure 18.3 Two sample cases of joint SPRT from (18.64) and (18.65).

The probabilities in (18.66) can be computed easily once it is known that the partial tests work independently. Note that it is required to evaluate those probabilities for all combinations of $\mu = \{\mu_0, \mu_1\}$ and $\sigma = \{\sigma_0, \sigma_1\}$. For instance, the joint probability $P(A_\mu A_\sigma)$, when $\mu = \mu_0$ and $\sigma = \sigma_0$, can be computed as

$$P\left(A_\mu A_\sigma \left|\mu_0\sigma_0\right.\right) = P\left(A_\mu \left|\mu_0\sigma_0\right.\right) P\left(A_\sigma \left|\mu_0\sigma_0\right.\right) = \left(1-\alpha_\mu\right)\left(1-\alpha_\sigma\right). \tag{18.67}$$

In a similar way, it is possible to obtain the rest of the probabilities for all combinations of the parameters.

The expected values in (18.66) are computed as follows:

$$E\left(N\left|A_\mu R_\sigma\right.\right) = \frac{K_\sigma}{E\left(Z\right)}, \tag{18.68}$$

$$E\left(N\left|R_\mu A_\sigma\right.\right) = \frac{K_\mu}{E\left(Y\right)}, \tag{18.69}$$

$$E\left(N\left|R_{\mu}\,R_{\sigma}\right.\right)=\begin{cases}\frac{k_{\mu}}{E(Y)} & \text{if}\left|\frac{k_{\mu}}{E(Y)}\right|<\left|\frac{k_{\sigma}}{E(Z)}\right| \\ \frac{k_{\sigma}}{E(Z)} & \text{otherwise,}\end{cases} \tag{18.70}$$

and

$$E\left(N\left|A_{\mu}A_{\sigma}\right.\right)=\begin{cases}\frac{h_{\sigma}}{E(Y)} & \text{if}\left|\frac{h_{\mu}}{E(Y)}\right|<\left|\frac{h_{\sigma}}{E(Z)}\right| \\ \frac{h_{\mu}}{E(Z)} & \text{otherwise.}\end{cases} \tag{18.71}$$

Note that in equation (18.70) the value of the expected sample number is indicated by the minimum of $|\frac{k_{\mu}}{E(Y)}|$ and $|\frac{k_{\sigma}}{E(Z)}|$, since the minimum of those absolute values indicates which one of the partial tests ends first. A similar analysis is made for the case when both partial tests indicate acceptance, as shown in (18.71). In this case, the maximum of $|\frac{h_{\mu}}{E(Y)}|$ and $|\frac{h_{\sigma}}{E(Z)}|$ indicates the partial test that ends last.

The expression of the conditioned expected values in (18.68), (18.70) and (18.71) must be modified to operate with unknown mean in the test of the variance. In this case, those expected values should be adjusted [Wald, 1947] as follows:

$$E\left(N\left|A_{\mu}R_{\sigma}\right.\right)=\frac{K_{\sigma}}{E\left(Z\right)}+1, \tag{18.72}$$

$$E\left(N\left|R_{\mu}\,R_{\sigma}\right.\right)=\begin{cases}\frac{k_{\mu}}{E(Y)} & \text{if}\left|\frac{k_{\mu}}{E(Y)}\right|<\left|\frac{k_{\sigma}}{E(Z)}+1\right| \\ \frac{k_{\sigma}}{E(Z)}+1 & \text{otherwise,}\end{cases} \tag{18.73}$$

and

$$E\left(N\left|A_{\mu}A_{\sigma}\right.\right)=\begin{cases}\frac{h_{\sigma}}{E(Y)}+1 & \text{if}\left|\frac{h_{\mu}}{E(Y)}\right|<\left|\frac{h_{\sigma}}{E(Z)}+1\right| \\ \frac{h_{\mu}}{E(Z)} & \text{otherwise.}\end{cases} \tag{18.74}$$

This completes the development of a general expression for computing the ASN, from which ASN_{IC}, ASN_{μ}, ASN_{σ}, and $ASN_{\mu\sigma}$ can be found. For example, $ASN_{\mu\sigma}$ corresponding to the case where both mean and variance are out of control can be obtained by evaluating (18.66) for $\mu=\mu_1$ and $\sigma=\sigma_1$.

3.5 The Optimization Model. In this section, we present the optimization model from which the optimal design parameters d, h_{μ}, k_{μ}, h_{σ}, and k_{σ} are to be derived. Let denote the average net income of a process cycle given by $\bar{V} = E(I)/E(T)$, where I denotes the net income per cycle and T denotes the cycle time. Let F denote lost-income per unit time, where

$$F = V(\mu_0, \sigma_0) - \bar{V}. \tag{18.75}$$

The optimization model can be given as

$$\text{Minimize} : F \tag{18.76}$$

$$\text{Subject to} : d > 0;\ h_\mu, h_\sigma < 0;\ k_\mu, k_\sigma > 0. \tag{18.77}$$

The negativity of h_μ and h_σ, as well as the non-negativity of k_μ, and k_σ, was assumed in the work of Wald from which much of our modeling steps are derived. The above pure economic model can be easily modified to be an economic statistical model by the addition of the following constraints:

$$\alpha < \alpha_c, \beta_\mu < \beta_{\mu c}, \beta_\sigma < \beta_{\sigma c}, \tag{18.78}$$

where α_c, $\beta_{\mu c}$, $\beta_{\sigma c}$ are the critical upper bounds of the errors. It may be noted from (18.75) that since $V(\mu_0, \sigma_0)$ is constant for a given process, minimizing F is equivalent to maximizing I.

4. NUMERICAL RESULTS

The objective of this section is twofold: 1) to provide empirical evidence of the performance of the EMV-SPRT control charts and 2) to provide an analysis of the sensitivity of the performance measures to the process parameters. The issue of the EMV-SPRT control chart performance is first addressed through the comparison of the pure economic performance of the operational design and the reference design. If the cost of the operational design is not significantly higher than the cost of the reference design, it implies that every subsequent comparison can be based on the operational model. The economic performance of EMV-SPRT control charts is also compared with other available designs, namely an economic design of Shewhart's models proposed by Costa [1993] and a non-economic SPRT design proposed by Stoumbos and Reynolds [1997]. In addition, EMV-SPRT was also evaluated under an economic statistical scenario where constraints on the decision errors were added to the optimization model. Such an economic statistical model (for a different control chart) was first introduced by Saniga [1995].

The inclusion of economic statistical and pure economic models and the presence of operational and reference designs makes it possible to consider several combination of designs in the numerical study. In particular, the six combinations that were investigated were formed using three levels of the constraints for the decision errors and two design types (reference and operational). The three levels of constraints are referred to as "free", "loose", and "tight." A pure economic model with no statistical constraints is referred to as free level. The tight level corresponds to bounds of 5% for the Type I error and 10% for the Type II error. The corresponding numbers for the loose level are 10% and 20% respectively.

All of the numerical results for the different EMV-SPRT models come from a MATLAB code that was developed utilizing the multivariate constrained gradient search algorithm available in the MATLAB optimization toolbox [The MathWorks, 1992]. This kind of algorithm provides an optimal solution for non-linear problems with convex properties. Since the convexity of the cost function derived in this study was not determined, the solutions presented here are not guaranteed to be optimal. However, the primary focus of this study is to develop new designs that perform better than the existing ones. The optimal configurations of the EMV-SPRT design, if different from what is given here, would provide more improvements.

4.1 Scenarios for Analysis. The numerical study conducted here includes a performance comparison of the reference and the operational designs, comparison of the pure economic model with the economic statistical model, comparisons of the performance of EMV-SPRT control charts with the designs proposed by Costa[1993] and Stoumbos and Reynolds[1997], and the sensitivity of the input parameters. The sample problems used for the numerical study are adopted from Costa [1993] and Stoumbos and Reynolds [1997], and are summarized in Table 18.4.

4.2 Comparison of the Reference and the Operational Designs. Table 18.5 shows the comparative results for all the three levels of decision error constraints (free, loose, and tight). As evident from the table, for all of the studied cases the operational design gives better economic performance. A careful analysis of the impact of the decision errors, as given below, helps to explain the superiority of the operational design. First, we compare the limits for the sequential test for both the reference and the operational designs given in (18.11) and (18.12), respectively. If $\sigma_0^2 < \sigma^2$, then the limits in the mean chart for the operational design are narrower than those of the reference design

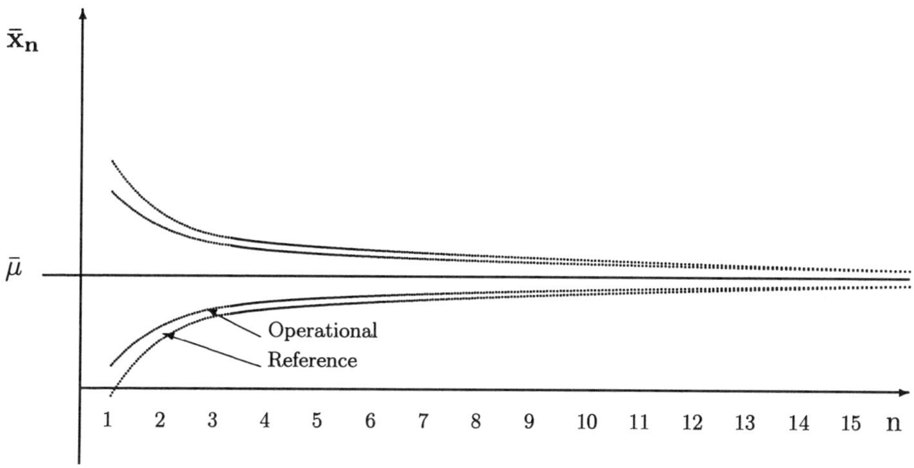

Figure 18.4 Decision Limits in Mean Charts for Operational and Reference Designs.

(See Figure 18.4). In the numerical examples considered in this study, the actual variance is never smaller than the in-control variance. Consequently, it is expected to have higher Type I and Type II errors in the operational design. In general, narrow (wide) limits are associated with high (low) errors of both types and small (large) sample sizes.

Despite the fact that the errors in the partial test for the mean are higher in the operational design, these errors affect the global test only when the variance is out of control since there is no difference between the reference and the operational design when the variance is in control. Consequently, α and $\beta(\mu_1, \sigma_0)$ values in the reference design are the same as the operational designs. Note that the errors, when the variance is out of control, are given by $\beta(\mu_0, \sigma_1) = (1 - \alpha_\mu)\beta_\sigma$ and $\beta(\mu_1, \sigma_1) = \beta_\mu \beta_\sigma$. Since α_μ is greater in the operational design than in the reference design, and β_σ is equal in both designs, it is implied that $\beta(\mu_0, \sigma_1)$ is smaller in the operational design. Conversely, since β_μ is greater in the operational design, $\beta(\mu_1, \sigma_1)$ is larger in the operational design. In conclusion, the global test in the operational design has relatively smaller errors when only the variance is out of control, and larger errors when both the mean and the variance are out of control. Since the process encounters the "variance only out of control" state more often than the "both mean and variance out of control" state, the gain due to smaller β error in the former overshadows the loss due to β error in the latter.

Table 18.4 Input parameters used for numerical study.

Input parameters used in evaluating performances of EMV-SPRT comapred to Costa's design					
Fixed parameters			Variable parameters		
Symbol	Meaning	Value	Symbol	Meaning	Values
a_1	Costs of each SPRT	0.5	μ_1	OOC Mean	1.0,1.5, 2.0
a_2	Cost of each sample unit	0.1	σ_1	OOC St. Deviation	1.5, 2.0
a_3	Cost of each false alarm	2.0	λ_2	Rate of Change of the Variance	0.05-λ_1
a_4	Cost of each reparation of the system	1.0	λ_1	Rate of Change of the Mean	0.00001, 0.001, 0.005, 0.025, 0.045, 0.049, 0.04999
D_0	Time for searching for alarms	0.1			
D_1	Time for repairs	0.3			
Input parameters used in comparison with S&R's design					
Fixed parameters			Variable parameters		
μ_1	OOC Mean	0.3	IMPACT	Cost per each 0.1% increase in fraction of non-conf. items	1, 20, 40
σ_1	OOC St. Deviation	1.1			
λ_2	Rate of Change of the Variance	0.05-λ_1			
λ_1	Rate of Change of the Mean	.04999			

4.3 Comparison of the Pure Economic and the Economic Statistical Designs. It has been argued in the literature that a pure economic design often ignores important practical issues such as reduction of worker confidence in the control chart due to frequent false alarms [Woodall, 1986]. Saniga [1989, 1995] presents models for the economic statistical design of $\bar{X}$, R, and attribute charts.

Table 18.6 shows the impact of adding statistical constraints to the optimization model for all three designs. The columns that contain the penalty cost show the increases in lost income due to statistical constraints relative to unconstrained (free) design. As evident from the table, these penalty costs range from 0.0% to 41.7%. This indicates that, for some processes, statistical constraints are economically viable, while for others they are prohibitive.

Note that in an unconstrained design, although the decision error statistics are higher, other statistics such as the average sample number, the expected time spent in searching for alarms, and the global out-of-control time are found to be lower. For the remaining numerical analysis presented in this chapter, only the unconstrained (free) design (i.e. the pure economic model) is considered.

4.4 Comparison of EMV-SPRT with the Economic Design of Shewhart's Charts by Costa. In this section, the performance of the EMV-SPRT control charts is compared with the economic design of $\bar{X}$ and R charts developed by Costa [1993]. Table 18.7 presents a comparison of the costs per unit time of applying Costa's economic design using Shewhart's $\bar{X}$ and R charts and the cost of using EMV-SPRT control charts. The numbers that appear in the column "Costa" were obtained from Costa's paper. From the table, it is clear that, for the given framework of analysis, EMV-SPRT is a superior tool for jointly monitoring the mean and variance of a process that satisfy the assumptions of this study. It is found that a 95% confidence interval for the savings obtained with EMV-SPRT is (12.54%, 14.10%) with an average of 13.32%.

4.5 Comparison of EMV-SPRT Design with a Non-Economic SPRT Design. Stoumbos and Reynolds [1997] present a control chart for monitoring (only) the mean using a sequential probability ratio test at fixed intervals. Their experiments were associated with a scenario characterized by a jump in the mean of 0.3σ. Hence, if the Gaussian parameters are standardized, then the in-control mean is zero, the out-of-control mean is 0.3, and the standard deviation is 1. Stoumbos and Reynolds [1997] suggest the use of some rational guide-

Table 18.5 Comparison of the Reference and the Operational Designs.

Parameters			Free		
σ_1	μ_1	λ_1	Reference	Operational	Savings
1.5	1	0.00001	4.0351	4.0054	0.7%
		0.001	4.0186	3.9889	0.7%
		0.005	3.9493	3.9203	0.7%
		0.025	3.5355	3.5110	0.7%
		0.045	2.8691	2.8612	0.3%
		0.049	2.6234	2.6238	0.0%
		0.04999	2.5386	2.5382	0.0%
	1.5	0.00001	4.0357	4.0051	0.8%
		0.001	4.0347	4.0050	0.7%
		0.005	4.0323	3.9996	0.8%
		0.025	3.8996	3.8586	1.1%
		0.045	3.7379	3.7077	0.8%
		0.049	3.2377	3.2346	0.1%
		0.04999	3.1610	3.1610	0.0%
	2	0.00001	4.0358	4.0055	0.8%
		0.001	4.0772	4.0454	0.8%
		0.005	4.2139	4.1760	0.9%
		0.025	4.5070	4.4412	1.5%
		0.045	4.2771	4.2300	1.1%
		0.049	4.0688	4.0579	0.3%
		0.04999	3.9832	3.9821	0.0%
2	1	0.00001	5.3495	5.2969	1.0%
		0.001	5.3230	5.2650	1.1%
		0.005	5.1876	5.1325	1.1%
		0.025	4.3971	4.3587	0.9%
		0.045	3.1675	3.1537	0.4%
		0.049	2.7104	2.7054	0.2%
		0.04999	2.5368	2.5367	0.0%
	1.5	0.00001	5.3710	5.2759	1.8%
		0.001	5.3449	5.2493	1.8%
		0.005	5.2297	5.1409	1.7%
		0.025	4.5903	4.5192	1.5%
		0.045	3.6428	3.6075	1.0%
		0.049	3.2955	3.2843	0.3%
		0.04999	3.1617	3.1617	0.0%
	2	0.00001	5.3737	5.2725	1.9%
		0.001	5.3640	5.2616	1.9%
		0.005	5.3185	5.2152	1.9%
		0.025	5.0266	4.9169	2.2%
		0.045	4.4283	4.3539	1.7%
		0.049	4.1230	4.0949	0.7%
		0.04999	3.9811	3.9807	0.0%

Parameters			Loose		
σ_1	μ_1	λ_1	Reference	Operational	Savings
1.5	1	0.00001	4.0350	4.0053	0.7%
		0.001	4.0186	3.9889	0.7%
		0.005	3.9493	3.9203	0.7%
		0.025	3.5366	3.5144	0.6%
		0.045	2.9142	2.9075	0.2%
		0.049	2.7195	2.7176	0.1%
		0.04999	2.6603	2.6601	0.0%
	1.5	0.00001	4.0356	4.0051	0.8%
		0.001	4.0360	4.0050	0.8%
		0.005	4.0307	4.0002	0.8%
		0.025	3.9457	3.9283	0.4%
		0.045	3.7796	3.7757	0.1%
		0.049	3.7330	3.7322	0.0%
		0.04999	3.7202	3.7202	0.0%
	2	0.00001	4.0357	4.0055	0.7%
		0.001	4.0771	4.0454	0.8%
		0.005	4.2162	4.1887	0.7%
		0.025	4.7427	4.7301	0.3%
		0.045	5.1175	5.1152	0.0%
		0.049	5.1799	5.1795	0.0%
		0.04999	5.1944	5.1944	0.0%
2	1	0.00001	5.3542	5.2969	1.1%
		0.001	5.3230	5.2675	1.0%
		0.005	5.1875	5.1332	1.0%
		0.025	4.3971	4.3587	0.9%
		0.045	3.1674	3.1537	0.4%
		0.049	2.7105	2.7054	0.2%
		0.04999	2.5409	2.5604	-0.8%
	1.5	0.00001	5.3710	5.2762	1.8%
		0.001	5.3410	5.2493	1.7%
		0.005	5.2298	5.1409	1.7%
		0.025	4.5927	4.5192	1.6%
		0.045	3.6529	3.6275	0.7%
		0.049	3.3813	3.3754	0.2%
		0.04999	3.3053	3.3052	0.0%
	2	0.00001	5.3740	5.2725	1.9%
		0.001	5.3640	5.2616	1.9%
		0.005	5.3185	5.2152	1.9%
		0.025	5.0153	4.9189	1.9%
		0.045	4.5514	4.5295	0.5%
		0.049	4.4447	4.4402	0.1%
		0.04999	4.4173	4.4172	0.0%

Parameters			Tight		
σ_1	μ_1	λ_1	Reference	Operational	Savings
1.5	1	0.00001	4.0596	4.0422	0.7%
		0.001	4.0441	4.0267	0.7%
		0.005	3.9784	3.9622	0.7%
		0.025	3.5906	3.5781	0.7%
		0.045	2.9960	2.9920	0.7%
		0.049	2.8078	2.8069	0.7%
		0.04999	2.7511	2.7511	0.7%
	1.5	0.00001	4.0594	4.0425	0.7%
		0.001	4.0637	4.0472	0.7%
		0.005	4.0779	4.0630	0.7%
		0.025	4.0871	4.0790	0.7%
		0.045	4.0014	3.9996	0.7%
		0.049	3.9707	3.9704	0.7%
		0.04999	3.9620	3.9620	0.7%
	2	0.00001	4.0594	4.0430	0.7%
		0.001	4.1129	4.0969	0.7%
		0.005	4.3077	4.2944	0.7%
		0.025	5.0257	5.0200	0.7%
		0.045	5.5344	5.5334	0.7%
		0.049	5.6204	5.6202	0.7%
		0.04999	5.6408	5.6408	0.7%
2	1	0.00001	5.3536	5.3026	0.7%
		0.001	5.3189	5.2693	0.7%
		0.005	5.1839	5.1367	0.7%
		0.025	4.3967	4.3633	0.7%
		0.045	3.1729	3.1617	0.7%
		0.049	2.7255	2.7228	0.7%
		0.04999	2.5668	2.5677	0.7%
	1.5	0.00001	5.3597	5.2812	0.7%
		0.001	5.3329	5.2559	0.7%
		0.005	5.2290	5.1483	0.7%
		0.025	4.5907	4.5413	0.7%
		0.045	3.7300	3.7159	0.7%
		0.049	3.4958	3.4926	0.7%
		0.04999	3.4314	3.4315	0.7%
	2	0.00001	5.3497	5.2803	0.7%
		0.001	5.3500	5.2707	0.7%
		0.005	5.3043	5.2307	0.7%
		0.025	5.0519	5.0086	0.7%
		0.045	4.7477	4.7383	0.7%
		0.049	4.6782	4.6763	0.7%
		0.04999	4.6604	4.6604	0.7%

Table 18.6 Comparison of Pure Economic and Economic Statistical Models

Parameters			Cost per unit time			Penalty cost(%)	
σ_1	μ_1	λ_1	Free	Loose	Tight	Loose	Tight
1.5	1	0.00001	4.0054	4.0053	4.0422	0.0%	0.9%
		0.001	3.9889	3.9889	4.0267	0.0%	0.9%
		0.005	3.9203	3.9203	3.9622	0.0%	1.1%
		0.025	3.5110	3.5144	3.5781	0.1%	1.9%
		0.045	2.8612	2.9075	2.9920	1.6%	4.6%
		0.049	2.6238	2.7176	2.8069	3.6%	7.0%
		0.04999	2.5382	2.6601	2.7511	4.8%	8.4%
	1.5	0.00001	4.0051	4.0051	4.0425	0.0%	0.9%
		0.001	4.0050	4.0050	4.0472	0.0%	1.1%
		0.005	3.9996	4.0002	4.0630	0.0%	1.6%
		0.025	3.8586	3.9283	4.0790	1.8%	5.7%
		0.045	3.7379	3.7757	3.9996	1.0%	7.0%
		0.049	3.2346	3.7322	3.9704	15.4%	22.7%
		0.04999	3.1610	3.7202	3.9620	17.7%	25.3%
	2	0.00001	4.0055	4.0055	4.0430	0.0%	0.9%
		0.001	4.0454	4.0454	4.0969	0.0%	1.3%
		0.005	4.1760	4.1887	4.2944	0.3%	2.8%
		0.025	4.4412	4.7301	5.0200	6.5%	13.0%
		0.045	4.2300	5.1152	5.5334	20.9%	30.8%
		0.049	4.0579	5.1795	5.6202	27.6%	38.5%
		0.04999	3.9821	5.1944	5.6408	30.4%	41.7%
2	1	0.00001	5.2969	5.2969	5.3026	0.0%	0.1%
		0.001	5.2650	5.2675	5.2693	0.0%	0.1%
		0.005	5.1325	5.1332	5.1367	0.0%	0.1%
		0.025	4.3587	4.3587	4.3633	0.0%	0.1%
		0.045	3.1537	3.1537	3.1617	0.0%	0.3%
		0.049	2.7054	2.7054	2.7228	0.0%	0.6%
		0.04999	2.5367	2.5604	2.5677	0.9%	1.2%
	1.5	0.00001	5.2759	5.2762	5.2812	0.0%	0.1%
		0.001	5.2493	5.2493	5.2559	0.0%	0.1%
		0.005	5.1409	5.1409	5.1483	0.0%	0.1%
		0.025	4.5192	4.5192	4.5413	0.0%	0.5%
		0.045	3.2843	3.3754	3.4926	2.8%	6.3%
		0.04999	3.1617	3.3052	3.4315	4.5%	8.5%
	2	0.00001	5.2725	5.2725	5.2803	0.0%	0.1%
		0.001	5.2616	5.2616	5.2707	0.0%	0.2%
		0.005	5.2152	5.2152	5.2307	0.0%	0.3%
		0.025	4.9169	4.9189	5.0086	0.0%	1.9%
		0.045	4.3539	4.5295	4.7383	4.0%	8.8%
		0.049	4.0949	4.4402	4.6763	8.4%	14.2%
		0.04999	3.9807	4.4172	4.6604	11.0%	17.1%

lines to ensure short average time to signal (ATS) when the process is out of control, after establishing some acceptable levels of the ASN and ATS when the process is in control. In particular, they fixed the ASN to 3 observations and the ATS to 740.80 units.

To ensure a fair comparison between the design proposed in Stoumbos and Reynolds [1997] and the EMV-SPRT design, the following assumptions must be made.

1. When the process is in control, the mean is zero, and the standard deviation is one.

2. The rate of change of the mean, 0.0499999, and the rate of change of the variance, 0.0000001, equal a total rate of change in the process of 0.05. Note that since Stoumbos and Reynolds [1997] do not consider a change in the process variance, the variance change has been kept close to zero. (Setting the variance shift rate equal to zero will render the EMV-SPRT computationally infeasible.)

3. The process may get out of control due to a jump in the mean to a value of 0.3 and/or a jump in the standard deviation to a value of 1.1. Note that Stoumbos and Reynolds [1997] do not consider a shift in the variance. Though a jump in the standard deviation of 1.1 could be considered large, the effective change here is negligible considering the very low rate of change in the variance.

4. The different cost parameters are the same as those (in Table 18.4) used in the comparison with Costa's result.

Equation (18.48) was modified to include the factor "*IMPACT*": $V(\mu,\sigma) = V - 1000(IMPACT)(FD(\mu,\sigma))$, where IMPACT represents how severely the income per unit time of a process deviates from its in-control income rate due to an out-of-control condition. The use of different levels of IMPACT factor allows an understanding of how the economic model can adapt to different cost configurations.

To complete the comparison, we need to know the ARL of the process when the mean is equal to the out-of-control reference value (in this case 0.3). Unfortunately, Stoumbos and Reynolds [1997] did not provide the required ARL but instead provided the ARL for some other magnitudes of shift of the mean. In particular, they computed an ARL of 12.08 when the mean is 0.25 and an ARL of 6.18 for a mean of 0.35. Although, the value of the ARL for a mean of 0.3 could perhaps be interpolated, we assumed in our study that the ARL, for a shift of 0.3, is also 6.18, giving an optimistic estimation of the actual ARL when the guidelines

Table 18.7 Comparison of EMV-SPRT and Costa's Designs

Parameters			Cost		**Savings**
σ_1	μ_1	λ_1	EMV-SPRT	COSTA	
1.5	1	0.00001	4.0054	4.74	15.5%
		0.001	3.9889	4.72	15.5%
		0.005	3.9203	4.64	15.5%
		0.025	3.5110	4.11	14.6%
		0.045	2.8612	3.24	11.7%
		0.049	2.6238	2.92	10.1%
		0.04999	2.5382	2.78	8.7%
	1.5	0.00001	4.0051	4.74	15.5%
		0.001	4.0050	4.74	15.5%
		0.005	3.9996	4.72	15.3%
		0.025	3.8586	4.48	13.9%
		0.045	3.7379	3.91	4.4%
		0.049	3.2346	3.69	12.3%
		0.04999	3.1610	3.59	11.9%
	2	0.00001	4.0055	4.74	15.5%
		0.001	4.0454	4.78	15.4%
		0.005	4.1760	4.91	14.9%
		0.025	4.4412	5.05	12.1%
		0.045	4.2300	4.77	11.3%
		0.049	4.0579	4.62	12.2%
		0.04999	3.9821	4.55	12.5%
2	1	0.00001	5.2969	6.26	15.4%
		0.001	5.2650	6.22	15.4%
		0.005	5.1325	6.06	15.3%
		0.025	4.3587	5.10	14.5%
		0.045	3.1537	3.59	12.2%
		0.049	2.7054	3.01	10.1%
		0.04999	2.5367	2.78	8.8%
	1.5	0.00001	5.2759	6.26	15.7%
		0.001	5.2493	6.22	15.6%
		0.005	5.1409	6.08	15.4%
		0.025	4.5192	5.28	14.4%
		0.045	3.6075	4.11	12.2%
		0.049	3.2843	3.73	11.9%
		0.04999	3.1617	3.59	11.9%
	2	0.00001	5.2725	6.26	15.8%
		0.001	5.2616	6.24	15.7%
		0.005	5.2152	6.17	15.5%
		0.025	4.9169	5.69	13.6%
		0.045	4.3539	4.91	11.3%
		0.049	4.0949	4.65	11.9%
		0.04999	3.9807	4.55	12.5%

Table 18.8 Comparison of EMV-SPRT and Non-Economic SPRT Designs

IMPACT	1	20	40
In-Control Profit per time	119.5347	110.6948	101.3897
Out-of-control Profit per Time	119.2405	104.8103	89.6206
EMV-SPRT			
Sampling Interval	Infinite	29.7797	6.5571
Average Sample	0	2.5884	4.7571
Net Income	Infinite	4,156.70	2,340.20
Sample Cost	0	0.9827	3.5042
Time Spent for False Alarms	0	0.024	0.166
Cycle Time	Infinite	38.889	24.015
Cost per Time	0.294	3.808	3.941
S&R Design			
Sampling Interval	1	1	1
Average Sample	3	3	3
Net Income	3105.60	2839.63	2559.65
Sample Cost	20.944	20.944	20.944
Time Spent for False Alarms	0.003	0.003	0.003
Cycle Time	26.483	26.483	26.483
Cost per Time	2.731	12.774	23.346

of Stoumbos and Reynolds are followed. It is also assumed that the ASN is three in both cases, when the mean is in control or out of control.

Table 18.8 presents the results of the comparison for three levels of the *IMPACT* parameter. In the first case, there is not much difference between the income per time when the process is in control or out of control. Since the cost of producing non-conformities is relatively small compared to the cost of sampling, the optimal policy, as indicated by EMV-SPRT, is to leave the process without monitoring, unlike, the Stoumbos and Reynolds' design, which requires some sampling.

In the second case, where $IMPACT = 20$ (which means a cost of \$20 for each 0.1% increase in the fraction of defectives), the reduction of income between in control and out of control is more significant. EMV-SPRT indicates the use of a very sporadic sampling scheme, causing few false alarms and small sampling costs, but resulting in a relatively large out-of-control time, which is evident from the long cycle time. Although the Stoumbos and Reynolds' design has a lower cycle time, it incurs a much higher cost primarily due to higher sampling.

In the case where $IMPACT = 40$, there is a significant loss of income when the system runs out of control, suggesting that the process deserves a closer examination. EMV-SPRT sets a sampling interval larger than that suggested by Stoumbos and Reynolds' design. We also note that EMV-SPRT assumes a higher α error (evident from the time spent for false alarms) than the S&R design. Larger sample size combined with higher α error implies less β error. Consequently, the cycle time and, hence, the out-of-control time indicated by EMV-SPRT is shorter than that of Stoumbos and Reynolds' design, resulting in lower cost.

This section contains insights about the value of using economic designs. The three cases discussed in this section show the serious consequences of not using an economic model. The formulation of an economic model is not an easy task. However, it is clear that in many cases it may be worth the price.

4.6 Sensitivity Analysis. The purpose of the sensitivity analysis is to see how the EMV-SPRT design-parameters (d, h_μ, k_μ, h_σ, and k_σ) and also some of the performance measures (ASN, α, β_1, β_2, and β_3) vary with the input parameters (μ_1, σ_1, and λ_1). The results of our sensitivity analysis is given in the fairly self-explanatory Table 18.9, i.e., for several combinations of μ_1, σ_1, and λ_1 (and hence λ_2), optimal EMV-SPRT design parameters and performance measures are presented. An analysis of the results in Table 18.9 reveals the following.

1. When the rate of change of the variance, λ_2, tends to zero, the decision variable h_σ tends to zero. Hence, the variance test indicates a fast acceptance and leaves the mean test operating alone.

2. When the rate of change of the mean (λ_1) tends to zero, the mean chart does not become unnecessary, in contrast to the case in point 1 above when the rate of change of the variance tends to zero. This is because the mean chart has a lot of power for detecting shifts in the variance.

3. The sampling interval, d, decreases when μ_1 or σ_1 increases. The effect of λ_1 interacts with the value of μ_1 in the following way. When μ_1 is low, the sampling interval increases as λ_1 increases, and the reverse occurs when μ_1 is high.

4. As expected, the average sample number decreases when μ_1 or σ_1 increases. The average sample number also decreases when λ_1 increases.

5. The Type I error is very small, less than 0.1%, in all the cases studied. For smaller shift parameters (μ_1 and σ_1), the design allows for higher a error and a corresponding higher power of detection.

6. The Type II error for any one of the out-of-control conditions increases, naturally, with the increase of λ_1.

5. CONCLUSIONS

This highly competitive world demands a high levels of understanding of production processes in order to obtain optimal productivity and continuously improving quality. As a result, researchers are continuously looking for approaches to enhance the level of assistance that the quality monitoring systems (QMS) can provide to the production systems.

The enhancements in the literature have followed two lines. Some enhancements are directed to allow better selection of the decision variables through either optimization of the global cost performance of the QMS or optimizing some statistical performances of the process. The other line of enhancement is oriented toward increasing the sensitivity of the QMS to the actual conditions of the process without worsening its efficiency, or conversely, to reduce the costs of implementing the QMS without affecting the sensitivity.

EMV-SPRT control charts, as proposed in this paper, are eclectic tools that combine the main virtues of Shewhart's approach with an economic decision criterion for selecting the chart parameters and a statistical tool for hypotheses testing that allows an excellent trade-off of efficiency and sensitivity. In other words, the EMV-SPRT control chart combines features such as feasible assumptions, joint analysis of mean and variance, economic design, and SPRT. The numerical results confirm the expectation that EMV-SPRT should provide superior performance compared to an economically designed Shewhart chart and also when compared to a non-economic SPRT design.

The economic design approach is essentially adapted from Costa's [1993] design with some important modifications specific to the use of SPRT. In addition, the optimization model was modified in order to include some statistical constraints to the model, as suggested by Saniga [1995]. The effect of adding statistical constraints is portrayed via numerical results, indicating that such constraints should be evaluated in light of their economic consequences which tend to vary widely.

The proposed approach considers optimization of the control charts, parameters on the basis of long-term performance. However, it is possible for some SPRTs to require very large sample sizes. To avoid this problem, a truncated test can be used. The determination of a proper

Table 18.9 Sensitivity Analysis

σ_1	μ_1	λ_1	0.00001	0.001	0.005	0.025	0.045	0.049	0.04999
1.5	1	h_μ	-6.81	-6.77	-6.71	-5.80	-3.89	-2.78	-2.42
		k_μ	5.74	5.73	5.70	5.58	5.58	5.91	5.99
		h_σ	-1.62	-1.62	-1.59	-1.35	-0.87	-0.58	0.00
		k_σ	5.08	5.09	5.14	5.39	6.18	5.46	3.90
		d	1.86	1.86	1.89	2.10	2.68	2.98	3.16
		ASN	14.07	14.00	13.77	11.91	8.16	6.01	5.43
		α	0.008	0.008	0.008	0.007	0.005	0.004	0.002
		β_1	0.001	0.001	0.001	0.003	0.020	0.062	0.089
		β_2	0.182	0.183	0.188	0.237	0.390	0.528	0.952
		β_3	0.009	0.009	0.010	0.018	0.069	0.153	0.325
1.5	2	h_μ	-9.20	-9.20	-9.20	-9.21	-5.11	-3.84	-3.30
		k_μ	5.84	5.85	5.89	5.96	6.56	7.73	9.14
		h_σ	-1.62	-1.54	-1.27	-0.57	-0.20	-0.12	0.00
		k_σ	5.04	5.07	5.17	5.27	5.47	6.15	4.63
		d	1.86	1.75	1.45	0.86	0.66	0.64	0.63
		ASN	14.07	13.42	11.37	5.98	2.80	2.04	1.70
		α	0.008	0.008	0.007	0.005	0.002	0.001	0.000
		β_1	0.000	0.000	0.000	0.000	0.006	0.021	0.037
		β_2	0.182	0.198	0.261	0.523	0.779	0.865	0.986
		β_3	0.003	0.003	0.004	0.009	0.080	0.157	0.227
2	1	h_μ	-3.66	-3.61	-3.59	-3.58	-3.31	-2.86	-2.43
		k_μ	6.68	6.67	6.63	6.46	6.11	5.99	6.06
		h_σ	-2.01	-1.98	-1.97	-1.96	-1.79	-1.50	-1.23
		k_σ	7.83	7.86	7.90	8.18	9.04	10.03	14.40
		d	0.72	0.72	0.75	0.98	1.83	2.66	3.16
		ASN	7.36	7.26	7.23	7.26	6.85	6.13	5.47
		α	0.002	0.002	0.002	0.002	0.002	0.002	0.002
		β_1	0.026	0.027	0.028	0.028	0.036	0.057	0.088
		β_2	0.118	0.122	0.123	0.123	0.145	0.194	0.259
		β_3	0.047	0.049	0.050	0.050	0.063	0.095	0.141
2	2	h_μ	-9.21	-9.21	-9.21	-9.21	-5.70	-4.06	-3.29
		k_μ	6.32	6.32	6.31	6.31	6.74	7.59	9.18
		h_σ	-1.87	-1.85	-1.77	-1.30	-0.59	-0.33	-0.21
		k_σ	7.83	7.84	7.90	8.24	8.84	9.52	12.26
		d	0.71	0.71	0.70	0.67	0.64	0.63	0.62
		ASN	6.92	6.86	6.62	5.13	2.89	2.07	1.69
		α	0.002	0.002	0.002	0.002	0.001	0.001	0.000
		β_1	0.000	0.000	0.000	0.000	0.003	0.017	0.037
		β_2	0.125	0.128	0.138	0.221	0.473	0.648	0.766
		β_3	0.013	0.013	0.014	0.022	0.114	0.235	0.337

level of truncation that avoids too large samples while maintaining the decision errors is a matter for future research.

Another reasonable concern related with the applicability of SPRT is the fact that data must be collected individually. If this is considered impractical, one option could be to configure rational subgroups and averages of the subgroups which can be plotted. In this way, the number of points plotted on the charts is reduced. Subgrouping also makes the assumption of normality more reliable.

All the results in this study were obtained through the utilization of a multivariate constrained gradient search algorithm that provides, perhaps, pseudo-optimal (but highly satisfactory) solutions. It was found that the time required for optimizing the EMV-SPRT cost model is, in general, less than one minute when using the ULTRA5 SUN workstation. Hence, EMV-SPRT control charts can be implemented on line with actual processes to accommodate any change in the process parameters and to modify the control chart accordingly. If online computer are not available, the monitoring process can be performed by providing the operators with control charts similar to those given in Figure 18.2 and Figure 18.3 and by the use of common calculators for computing the sample mean and variance that are plotted on these charts.

Based on the results of the economic comparisons and the insights provided by the joint analysis, the use of EMV-SPRT control charts under the necessary assumptions seems very worthwhile. Of the assumptions, perhaps the most restrictive is that each parameter can have only two values, one value when in control and another value when out of control. For many processes, this assumption does not hold strictly since smaller or larger shifts may be observed. Also, Wald showed that, while SPRT is more powerful than traditional techniques, the power of SPRT is smaller in the presence of small shifts. Finally, we assume that the time for performing each SPRT is negligible when compared to the cycle time. If it is not possible to make this assumption, it will be necessary to modify the development of the economic model, but the application of SPRT will remain unaltered.

References

[1] Collani, E.v. Determination of the Economic Design of Control Charts Simplified. In: *Optimization in Quality Control*, eds, K.S. Al-Sultan and M. A. Rahim, Kluwer Academic, Boston, 1997.

[2] Costa, A. Joint Economic Design of $\bar{X}$ and R Charts for Process subject to Two Independent Assignable Causes. *IIE Transitions*, **25**(6) (1993) 27-33.

[3] Costa, A. Joint $\bar{X}$ and R Charts with variable parameters. *IIE Transitions, Quality and Reliability*, **30**(6) (1998) 505-514.

[4] Das, T. K., Jain, V., and Gosavi, A. Economic design of dual-sampling-interval policies for $\bar{X}$ charts with and without run rules. *IIE Transactions*, **29**(6) (1997) 497-506.

[5] Das, T. K., and Jain, V. An economic design model for X-bar charts with random sampling policies. *IIE Transactions*, **29** (1997) 507-518.

[6] Duncan, A. J. The Economic Design of X-bar Charts used to Maintain Current Control of a Process. *Journal of the American Statistical Association*, **51** (1956) 228-242.

[7] Girshick, M. A., 1946.Contribution to the Theory of Sequential Analysis. *Annals of Math Statistics*, June 1996.

[8] Ho, C. and Case, K. E. Economic design of control charts: a literature review for 1981-1991. *Journal of Quality Technology*, **26**(1) (1994) 39-53.

[9] Hoel , P. G. *Introduction to Mathematical Statistics*. Wiley, New York, 1984.

[10] Lorenzen, T. and Vance, L. The Economic Design of Control Charts: A Unified Approach. *Technometrics*, **28**(1)(1986) 1-13.

[11] Page E. S. Cumulative sum control charts. *Technometrics*, **3** (1954).

[12] Reynolds, M., Armin, R, Arnold, J., and Nachlas, J. X-bar charts with variable sampling intervals. *Technometrics*, **30** (1988) 181-92.

[13] Roberts S. W. Control charts based on geometric averages. *Technometrics*, **1**, (1959) 239-250.

[14] Runger, G. and Montgomery D. Adaptive sampling enhancements for Shewhart control charts. *IIE Transactions*, **25** (1993) 41-51.

[15] Runger, G. and Pignatiello, J. Adaptive sampling for process control. *Journal of Quality Technology*, **23**(2) (1991) 133-155.

[16] Saniga, E. M. Economic Statistical Control Chart Designs with Application to X-bar and R Charts. *Journal of Quality Technology*, **31**(3) (1989) 313-320.

[17] Saniga, E. M. and McWilliams, T. Economic, statistical, and economic-statistical design of attribute charts. *Journal of Quality Technology*, **27**(1) (1995) 56-73.

[18] Shewhart, W.A. *Economic Control of Quality of Manufactured Products*. D. Van Nostrand, New York, 1931.

[19] Siegmund D. *Sequential Analysis: Tests and Confidence Intervals*. Springer, New York, 1985.

[20] Stoumbos, Z. and Reynolds, M. Control Charts Applying a Sequential Test at Fixed Sampling Intervals. *Journal of Quality Technology*, **29**(1)(1997) 21-40.

[21] The MathWorks, Inc. (1992). *Optimization Toolbox User's Guide.*

[22] Wald, A. *Sequential Analysis.* Dover Publications, New York, 1947.

[23] Wetherill, G. B. and Glazebrook, K. *Sequential Methods in Statistics.* Chapman and Hall, New York, 1996.

[24] Wheeler, D., 1991. Myths, Foundations and Competitors for Shewhart's Control Charts. *Statistical Process Controls, Inc., Knoxville.*

[25] Woodall, W. H. Weakness of the Economic Design of Control Charts. *Technometrics*, **28**(4) (1986) 408-409.

(s S) inventory —036 (s, S) inventory —036

Index

List of Referees

F. Al-Sunni, King Fahd University of Petroleum & Minerals, Saudi Arabia
F. J. Arcelus, University of New Brunswick, Fredericton, N.B., Canada.
E. K. Boukas, Ecole Polytechnique de Montreal, Montreal, Canada.
Y. H. Chun, Louisiana State University, U.S.A.
E. v. Collani, University of Würzburg, Germany.
A. F. B. Costa, UNESP-Campus of Guaratiguet, S. P., Brazil.
E. Del Castillo, The Pennsylvania State University, U.S.A.
E. A. Elsayed, Rutgers University, New Jersey, U.S.A.
A. Gunasekaran, University of Massachusetts-Dartmouth, U.S.A.
M. Hariga, King Saud University, Saudi Arabia.
Y. Hong, Pohang Univ. of Science & Technology, Pohang, South Korea.
L. F. Hsu, The City University of New York, New York, U.S.A.
A. K. Jain, Lucent Technologies, Bell Labs Innovations, N. J., U.S.A.
M. E. Johnson, University of Central Florida, Orlando, U.S.A.
M. J. Khouja, University of North Carolina at Charlotte, U.S.A.
V. Makis, University of Toronto, Toronto, Canada.
T. McWilliams, Drexel University, Philadelphia, U.S.A.
J. E. Middle, Loughborough University, Leicestershire, U.K.
D. N. P. Murthy, The University of Queensland, Brisbane, Australia.
P. R. Nelson, Clemson University, Clemson, U.S.A.
H. Ohta, Osaka Prefacture University, Osaka, Japan.
M. Parlar, McMaster University, Hamilton, Ontario, Canada.
A. Rolstadås, Norwegian U. of Science & Technology, Trondheim, Norway.
E. M. Saniga, University of Delaware, Newark, Delaware, U.S.A.
S. P. Sethi, The University of Texas at Dallas, U.S.A.
E. A. Silver, University of Calgary, Calgary, Alberta, Canada.
G. Srinivasan, University of New Brunswick, Fredericton, N.B., Canada.
M. M. Srinivasan, The University of Tennessee, Knoxville, U.S.A.
G. Tagaras, Aristoteles, University of Thessaloniki, Thessaloniki, Greece.
M. Taksar, State University of New York at Stonybrook, U.S.A.
K. Tang, Purdue University, U.S.A.
S. T. Tseng, National Tsing-Hua University, Hsinchu, Taiwan.
Q. Zang, The University of Georgia, Athens, U.S.A.